Springer Series on
ATOMIC, OPTICAL, AND PLASMA PHYSICS 49

Springer Series on
ATOMIC, OPTICAL, AND PLASMA PHYSICS

The Springer Series on Atomic, Optical, and Plasma Physics covers in a comprehensive manner theory and experiment in the entire field of atoms and molecules and their interaction with electromagnetic radiation. Books in the series provide a rich source of new ideas and techniques with wide applications in fields such as chemistry, materials science, astrophysics, surface science, plasma technology, advanced optics, aeronomy, and engineering. Laser physics is a particular connecting theme that has provided much of the continuing impetus for new developments in the field. The purpose of the series is to cover the gap between standard undergraduate textbooks and the research literature with emphasis on the fundamental ideas, methods, techniques, and results in the field.

A. Voitkiv   J. Ullrich

Relativistic Collisions of Structured Atomic Particles

With 63 Figures

 Springer

Dr. Alexander Voitkiv
Professor Dr. Joachim Ullrich
Max-Planck-Institut für Kernphysik
Saupfercheckweg 1, 69117 Heidelberg, Germany
E-mail: alexander.voitkiv@mpi-hd.mpg.de, joachim.ullrichv@mpi-hd.mpg.de

Springer Series on Atomic, Optical, and Plasma Physics   ISSN 1615-5653

ISBN 978-3-540-78420-3   e-ISBN 978-3-540-78421-0

Library of Congress Control Number: 2008927184

Typesetting and production: SPi Publisher Services
Cover concept: eStudio Calmar Steinen
Cover design: WMX Design GmbH, Heidelberg

SPIN 12099078   57/3180/SPi
Printed on acid-free paper

9 8 7 6 5 4 3 2 1

springer.com

Preface

During the last two decades the explorations of different processes accompanying ion–atom collisions at high-impact energies have been a subject of much interest. This interest was generated not only by the advent of accelerators of relativistic heavy ions which enabled one to investigate these collisions in an experiment and possible applications of obtained results in other fields of physics, but also by the variety of physical mechanisms underlying the atomic collisional phenomena at high impact energies.

Often highly charged projectiles produced at accelerators of heavy ions are not fully stripped ions but carry one or more very tightly bound electrons. In collisions with atomic targets, these electrons can be excited or lost and this may occur simultaneously with electronic transitions in the target. The present book concentrates on, and may serve as an introduction to, theoretical methods which are used to describe the projectile–electron transitions occurring in high-energy collisions between ions and neutral atoms. Special attention is given to relativistic impact energies and highly charged projectiles. Experimental results are used merely as illustrations and tests for theory.

This book will be useful to graduate students and professional scientists who are interested in studying atomic collisions occurring at high-impact energies. It assumes that the reader possesses the basic knowledge in classical electrodynamics and nonrelativistic and relativistic quantum mechanics.

This book could not have been written without the cooperation and support of our professional colleagues B. Najjari, N. Grün, E. Montenegro, R. Moshammer, C. Müller, and W. Scheid. We are especially indebted to B. Najjari and N. Grün for the close and long-term collaboration on atomic collision theory, numerous discussions and the careful readings of the draft version of this book.

Heidelberg,   *Alexander Voitkiv*
May 2008   *Joachim Ullrich*

Contents

Part II Relativistic Collisions

1

Introduction

During the last several decades the ion–atom collisions, occurring at impact velocities substantially exceeding the typical orbiting velocities of outer-shell atomic electrons, have been a subject of extensive research, both experimental and theoretical. The studies of different processes, accompanying such collisions, are of great interest not only for the basic atomic physics research but also have many applications in other fields of physics such as plasma physics, astrophysics and radiation physics.

With the advent of accelerators of relativistic heavy ions much higher impact energies and projectile charge states had become accessible for the explorations in experiments on ion–atom collisions. This, as well as the variety of physical processes governing atomic collisions at high impact energies, have triggered both great interest from and, simultaneously, become a source of substantial challenges for atomic physics theorists.

Three basic atomic physics processes can occur in collisions between a bare projectile-nucleus and a target-atom. (i) The atom can be excited or ionized by the interaction with the projectile. (ii) One or more atomic electrons can be picked up by the projectile-nucleus and form bound or low-lying continuum states of the corresponding projectile-ion. The pick-up process can proceed with or without emission of radiation and is called radiative or nonradiative electron capture, respectively. A combination of (i) and (ii) can also occur. Besides, in relativistic collisions the pair production becomes possible with cross sections reaching quite substantial values in the case of extreme relativistic impact energies when the collision velocity approaches very closely the speed of light in vacuum.

Highly charged projectiles produced at accelerators of heavy ions often are not fully stripped ions but carry one or more very tightly bound electrons. If such projectiles collide with atomic targets these electrons can be excited and/or lost. In the rest frame of the ion this can be viewed as excitation or 'ionization' of the ion by the impact of the incident atom. In addition to the nucleus the atom has electrons which may influence the motion of the electrons of the ion in different ways. As a result, the physics of the ion excitation and 'ionization' by the neutral atom impact will in general strongly differ from that for excitation and ionization in collisions with a bare atomic nucleus. Thus, in collisions of partially stripped ions with neutral atoms, a qualitatively new process – the projectile-electron excitation and/or loss – becomes possible.

Recently there have been published several books on energetic ion–atom collisions, [1–5]. Two of them, [3,4], are almost entirely devoted to the field of relativistic atomic collisions. Besides, some aspects of such collisions were reviewed in an earlier book [6].

In the present book we concentrate our consideration on the projectile-electron transitions occurring in high-energy collisions between ions and atoms whose electrons also actively or passively participate in the collision process. The especial attention is given to relativistic impact energies and highly charged projectiles. Except for a very short discussion in [3], in the previous books on the relativistic atomic collisions this subject was practically untouched.

For nonrelativistic ion–atom collisions the projectile-electron transitions were considered in [1,2]. However, the theoretical considerations in these books were mostly restricted just to the simplest theoretical model, the first-order (first Born) approximation. Besides, since the time when these books were published, there have been new interesting developments in the field. Therefore we feel that a more extensive coverage in a book-format manuscript of certain theoretical models, applied to study the nonrelativistic collisions of structured atomic particles, may be quite appropriate.

The present book describes various theoretical methods which can be applied to consider collisions between an ion and an atom, both of which have initially electrons actively or passively participating in the collision process. Experimental results are used mainly as illustrations and tests for theory.

We hope that the book can be useful to graduate students and professional scientists who are interested in studying atomic collisions occurring at high impact energies.

In ion–atom collisions nuclear reactions may also take place. However, compared to the atomic processes these reactions are normally characterized by much smaller cross sections and will not be considered. Throughout the book the nuclei of the colliding particles will be regarded just as point-like charges which cannot be excited or broken in the collision.

The book consists of two parts which are organized as follows. In Part I we discuss several theoretical approaches which were developed to study the projectile-electron transitions occurring in the nonrelativistic ion–atom collisions. In Chap. 2 we consider descriptions of these collisions within the framework of the first order perturbation theory in the projectile–target interaction. In Chap. 3 a number of theoretical approaches, which go beyond the first order theory, is introduced. Part I by no means pretends to be an exhaustive review of all up-to-date developments in this vast field with long history. Instead, in this part we focus our consideration on introducing and discussing some basic ideas and theoretical methods. The literature cited in this part is extremely far from being complete, but can furnish a starting point for a further search.

In Part II we consider descriptions of projectile-electron transitions in relativistic collisions. This part represents the main subject of the book.

It begins with Chap. 4, where we very briefly discuss the three 'corner-stones' – the special theory of relativity, the Maxwell and Dirac equations – which form the basis of the theory of relativistic atomic collisions. This chapter cannot serve as a substitute for textbooks. Its intention is rather modest: merely to remind the reader about some basic ideas and facts which he or she is already well aware of from the textbook literature.

In Chap. 5 we give a detailed consideration of relativistic collisions between a projectile-ion and a target-atom, which both initially have electrons actively or passively participating in the collision. In this chapter the ion–atom interaction is treated within the first order perturbation approach, i.e. by assuming that the interaction occurs via just a single virtual photon exchange. In Chap. 5, amongst several topics, we discuss the quantum and semi-classical versions of the first-order theory and show their equivalence, consider its non-relativistic limit, touch upon questions concerning the choice of appropriate gauges and the gauge (in)dependence of obtained results, possible simplifications in the full expression for the first order relativistic transition amplitude, etc.

Chapter 6 describes several methods for treating the projectile-electron excitation and loss in relativistic collisions which extend beyond the framework of the first order theory. In the case of a strong ion–atom interaction these methods, compared to the first order theory, enable one to treat this interaction in a better way. Therefore, they can be applied to collisions in which the fields of the colliding particles become effectively too strong, leading to the failure of the first order considerations. In this chapter we discuss (a) a recently developed symmetric eikonal model in which the four-transition current of the atom includes distortions of the initial and final atomic states caused by the field of a highly charged ion, (b) the so called light-cone approximation, which becomes 'exact' when the collision velocity is equal to the speed of light, and (c) three-body distorted-wave models which are based on the reduction of the projectile-electron excitation and loss processes to a three-body problem. Besides, Chap. 6 also contains a very brief discussion of relativistic nonperturbative methods, represented by coupled channel approaches and numerical solutions of the Dirac equation on a lattice, and a comparison between results of these methods and the distorted-wave models for the projectile-electron excitation and loss cross sections.

Chapters 7 and 8 contain mainly applications of the theoretical approaches, considered in the previous chapters of Part II, to concrete collision processes. In Chap. 7 we discuss impact parameter dependencies for probabilities of the projectile-electron excitation and loss in relativistic collisions. Cross sections are considered in Chap. 8, where results of calculations are also compared with available experimental data.

Chapter 8 also includes sections in which we consider bound-free and free-free electron-positron pair production in collisions between a bare nucleus and a neutral atom and the excitation and break-up of pionium colliding with neutral atoms at relativistic velocities. All these processes have much in common with the projectile-electron excitation and loss and can be treated using rather similar methods. For example, in the Dirac sea picture the bound-free pair production is very closely related to the projectile-electron loss process. On the other hand, although the strong interaction is of crucial importance for the physics of the π^+ and π^- pions constituting pionium, the interaction between the pionium and atoms, which is responsible for the pionium excitation and break-up, occurs predominantly via the electromagnetic interaction and in this sense the pionium–atom collisions just represent some exotic case of relativistic atomic collisions.

Chapter A is an Appendix. Besides two sections, in which rather technical questions are considered, it also contains in-depth discussions of two topics which both had been a subject of much controversy in the atomic collision physics community for a very long time. The first is a proper form of the wave equation for a nonrelativistic electron, which is initially bound by an attracting center and is subjected in the collision to the field of an extreme relativistic projectile. The second topic concerns a very delicate and intimate interrelation between Galilean and gauge transformations in the case when radiative atomic processes, like e.g. the radiative electron capture, are considered. Although this topic is not directly related to the rest of this book, it touches upon some of the most fundamental questions in physics – reference frame and gauge transformations – and we found it appropriate to include the discussion of this topic into the present book.

This book presupposes that the reader possesses the university level knowledge in Classical Electrodynamics and in Nonrelativistic and Relativistic Quantum Mechanics, including Quantum Theory of Scattering. The familiarity with the basic principles of Quantum Electrodynamics is also very desirable.

There exist a number of good textbooks in which the above fields of physics are nicely discussed. Based on our own experience we may recommend the following books: [7–17]. Besides, such topics like the correspondence between the field-theoretical approach of Quantum Field Theory and less general methods used in the theory of atomic collisions, the Dirac equation and its solutions in the case of a relativistic electron moving in an external Coulomb field can be also found in [3, 4].

Our last remark in this introductory part concerns the system of units. Unless otherwise is stated, throughout the book we shall use the atomic system of units in which $\hbar = m_e = e = 1$, where $\hbar$ is the Planck's constant, m_e is the electron rest mass and e is the absolute value of the electron charge. In this system the typical velocity of an electron, which is bound in the ground state of an hydrogen-like ion with a nuclear charge Z, is equal to Z and the speed of light in vacuum is approximately equal to 137.

Part I

Nonrelativistic Collisions

2

First Order Considerations

2.1 Quantum Plane-Wave Born Approximation

In quantum considerations of atomic collisions all atomic particles, electrons and nuclei, are treated quantum mechanically. The simplest quantum mechanical approach for considering nonrelativistic collisions of two structured atomic particles – first order plane-wave (plane-wave Born) approximation – was formulated long ago by Bates and Griffing [18–20]. Later on this approximation was used in many papers which were devoted to the different aspects of the projectile-electron excitation and loss in collisions with neutral atoms and simplest molecules. Earlier reviews, discussing the applications of the first order approximations to the projectile–target collisions, are presented in [1, 2, 21, 22], where also references to very many original papers, published before the middle of the nineties, can be found.

Although already first order calculations can be quite formidable for practical implementations, the formulation of the plane-wave Born approximation for nonrelativistic collisions is *per se* elementary. Here we sketch very briefly how one can derive first order cross sections using an approach which permits a natural generalization for the case of relativistic collisions.

Let us consider a collision between a projectile-ion and a target-atom. The charges of the nuclei of the colliding particles are Z_I and Z_A, respectively, and v is the collision velocity. For simplicity we will assume for the moment that each of the colliding atomic particles has initially only one electron. The S-matrix element, describing transitions in the colliding system, can be quite generally written as

$$S_{fi} = -\mathrm{i} \int_{-\infty}^{+\infty} \mathrm{d}t \int \mathrm{d}^3\mathbf{x} \varrho_\mathrm{I}(\mathbf{x}, t)\, \varphi_\mathrm{A}(\mathbf{x}, t). \qquad (2.1)$$

Here $\varrho_\mathrm{I}(\mathbf{x}, t)$ is the transition charge density, created by the projectile at time t and space point $\mathbf{x}$, and $\varphi_\mathrm{A}(\mathbf{x}, t)$ is the transition scalar potential, generated by the target atom at the same t and $\mathbf{x}$.[1] Throughout the book the indices A and I stand for the atom and ion, respectively. The scalar potential, created by the target in the collision, is a solution of Poisson's equation

$$\Delta \varphi_A(\mathbf{x}, t) = -4\pi \varrho_A(\mathbf{x}, t), \qquad (2.2)$$

where $\varrho_A(\mathbf{x}, t)$ is the transition charge density of the target.

Assuming that the collision velocity is sufficiently high, such that the electrons belonging to the ion and atom can be treated as distinguishable particles, the charge densities are written according to

$$\varrho_I(\mathbf{x}, t) = \int d^3 \mathbf{R}_I \, d^3 \mathbf{r} \Psi^*_{I,f}(\mathbf{R}_I, \mathbf{r}, t) \left[Z_I \delta(\mathbf{x} - \mathbf{R}_I) - \delta(\mathbf{x} - \mathbf{r}) \right] \Psi_{I,i}(\mathbf{R}_I, \mathbf{r}, t),$$

$$\varrho_A(\mathbf{x}, t) = \int d^3 \mathbf{R}_A \, d^3 \boldsymbol{\lambda} \Psi^*_{A,f}(\mathbf{R}_A, \boldsymbol{\rho}, t) \left[Z_A \delta(\mathbf{x} - \mathbf{R}_A) - \delta(\mathbf{x} - \boldsymbol{\rho}) \right] \Psi_{A,i}(\mathbf{R}_A, \boldsymbol{\lambda}, t).$$

$$(2.3)$$

Within the first-order treatment $\Psi_{I,i}$, $\Psi_{A,i}$ and $\Psi_{I,f}$, $\Psi_{A,f}$ are approximated by unperturbed initial and final states, respectively, of the colliding particles. The form of these states is well known: they are a product of a plane-wave, representing the motion of the center of mass of the atomic particle, and a function describing the internal motion of the electron in the particle. Further, in (2.3) $\mathbf{R}_I$ is the coordinate of the projectile nucleus, $\mathbf{r}$ is the coordinate of the projectile electron with respect to the projectile nucleus, $\mathbf{R}_A$ the coordinate of the target nucleus and $\boldsymbol{\rho}$ the coordinate of the target electron with respect to the target nucleus.

The target scalar potential and the integrals in (2.1) are conveniently evaluated by using Fourier transforms for the charge densities ϱ_I, ϱ_A and the scalar potential φ_A, e.g.

$$\varrho_A(\mathbf{x}, t) = \frac{1}{4\pi^2} \int d\omega d^3\mathbf{k} \exp(i\mathbf{k} \cdot \mathbf{x} - i\omega t)\xi_A(\mathbf{k}, \omega),$$

where ξ_A is the Fourier transform of ϱ_A. Using the standard procedure of obtaining a cross section from a known S-matrix transition element, one can show that the cross section for a collision, in which the electron of the projectile makes a transition from an initial internal state ψ_0 into a final internal state ψ_n and the electron of the target makes a transition from its internal initial state u_0 to a final state u_m, is given by

$$\sigma^{0 \to m}_{0 \to n} = \frac{4}{v^2} \int d^2\mathbf{q}_\perp \frac{|F^I_{0n}(\mathbf{q}) \, F^A_{0m}(-\mathbf{q})|^2}{q^4}. \qquad (2.4)$$

Here $\mathbf{q} = (\mathbf{q}_\perp, q_{min})$ is the momentum transfer to the projectile where, $\mathbf{q}_\perp$ is the two-dimensional part of the momentum, which is perpendicular to the

---

[1] Of course, one can take the S-matrix element in a fully equivalent form where the target charge density is coupled with the projectile scalar potential.

collision velocity $\mathbf{v}$, and q_{min} is the minimum momentum transfer to the projectile given by

$$q_{min} = \frac{\varepsilon_n - \varepsilon_0 + \epsilon_m - \epsilon_0}{v}. \tag{2.5}$$

In (2.5) $\varepsilon_{0(n)}$ and $\epsilon_{0(m)}$ are the initial (final) electron energies in the internal states of the projectile and target, respectively. Further, in (2.4)

$$F_{0n}^{I}(\mathbf{q}) = Z_I \delta_{n0} - \langle \psi_n | \exp(i\mathbf{q} \cdot \mathbf{r}) | \psi_0 \rangle,$$
$$F_{0m}^{A}(\mathbf{q}) = Z_A \delta_{m0} - \langle u_m | \exp(i\mathbf{q} \cdot \boldsymbol{\rho}) | u_0 \rangle \tag{2.6}$$

are the form-factors of the ion and atom.

When we consider collisions of a projectile carrying initially an electron with a target, we will be interested in the study of those collisions, where the projectile electron makes a transition, i.e. when it gets excited or lost and $n \neq 0$. In what follows we will not consider collisions where $n = 0$, i.e. collisions which are elastic for the projectile. While final states of the projectile are observed in experiment, there is often no experimental information about the final state of the target. Therefore, in order to describe theoretically such a situation, one has to calculate the cross section

$$\sigma_{0 \to n} = \sum_m \sigma_{0 \to n}^{0 \to m}, \tag{2.7}$$

where the summation has to be performed over all possible final states of the target including the continuum. It is convenient to split the first order cross section (2.7) into two parts and discuss them separately.

2.1.1 Elastic Target Mode

One part represents the contribution to the cross section (2.7) from collisions in which the target electron remains in the initial state, i.e. from collisions, where this electron can be considered as 'passive'. This part reads

$$\sigma_{0 \to n}^{s} = \frac{4}{v^2} \int d^2 q_\perp Z_{A,eff}^2(\mathbf{q}_0) \frac{|\langle \psi_n | \exp(i\mathbf{q}_0 \cdot \mathbf{r}) | \psi_0 \rangle|^2}{q^4}. \tag{2.8}$$

Here $\mathbf{q}_0 = (\mathbf{q}_\perp, \frac{\varepsilon_n - \varepsilon_0}{v})$ and $Z_{A,eff} = Z_A - \langle u_0 | \exp(-i\mathbf{q}_0 \cdot \boldsymbol{\rho}) | u_0 \rangle$ is the effective charge of the target which is 'seen' by the electron of the projectile in collisions where the target does not change its internal state. Considering this effective charge as a function of the momentum transfer, one can note the following important points (see also [2, 21]). The value of the effective charge $Z_{A,eff}$ varies in the limits $Z_A - 1 < Z_{A,eff} < Z_A$.[2] The charge $Z_{A,eff}$ approaches its lower and upper limits in collisions where the momentum transfer q_0 is

---

[2] If the target contains N_A electrons then $Z_A - N_A < Z_{A,eff} < Z_A$.

much lower and much larger, respectively, than a typical electron momentum in the initial target state. It is seen that the effect of the target electron(s) in collisions, where the target remains in its initial internal state, is to weaken the field of the target nucleus acting on the projectile electron, i.e. to partially or completely screen the nucleus.

The projectile–target collision mode, in which the target does not change its internal state (while the projectile does), is often called the *elastic mode*, implying that it is elastic only for the target. This mode is also referred to as *screening* because of the screening (or shielding) role of the atomic electrons, which in this mode counteract to the field of the atomic nucleus and reduce the total atomic field acting on the electron of the ion. Below we will use both these expressions.

2.1.2 Inelastic Target Mode

The second part of the cross section (2.7) describes collisions in which the target electron makes transitions. It reads

$$
\sigma^{\mathrm{a}}_{0 \to n} = \sum_{m \neq 0} \sigma^{0 \to m}_{0 \to n}
$$

$$
= \frac{4}{v^2} \sum_{m \neq 0} \int d^2 \mathbf{q}_\perp \frac{|\langle u_m \mid \exp(-i\mathbf{q} \cdot \boldsymbol{\rho}) \mid u_0 \rangle \langle \psi_n \mid \exp(i\mathbf{q} \cdot \mathbf{r}) \mid \psi_0 \rangle|^2}{q^4}.
$$

$$(2.9)$$

Equation (2.9) deals with the collision mode where not only the electron of the projectile but also that of the target are 'active' in the collision. This collision mode is called *doubly inelastic* or simply *inelastic*.

According to the first order approximation, the inelastic mode is not influenced by the interaction between the electron of the projectile and the nucleus of the target and the projectile electron undergoes a transition solely due to the interaction with the electron of the target. The latter is sometimes referred to as the two-center dielectronic interaction (TCDI) [1].

Contributions from collisions, in which the target changes its initial internal state, increase the total cross section (2.7). This action of the target electron is just opposite to that in the elastic mode, where the electron by screening the target nucleus decreases the cross section value compared to that in collisions with the bare atomic nucleus. Therefore, the inelastic collision mode is also often termed as *antiscreening*.

2.1.3 Collisions with Large Momentum Transfer. Free Collision Model

Let us consider collisions in which the minimum momentum transfer $q_{\min}$, given by (2.5), and, thus, the total momentum transfer q are much larger than a typical momentum of the target electron in the initial target state. Such a situation can occur if the atomic number Z_I of the projectile substantially exceeds that of the target Z_A and the collision velocity is not too high.

Elastic Mode. For the elastic mode the effective charge $Z_{A,\text{eff}} = Z_A - \langle u_0 | \exp(-i\mathbf{q}_0 \cdot \boldsymbol{\rho}) | u_0 \rangle$, because of the rapid oscillations of the integrand due to the factor $\exp(-i\mathbf{q}_0 \cdot \boldsymbol{\rho})$, becomes approximately equal to the charge Z_A of the bare target nucleus. Therefore, in collisions with a large momentum transfer the shielding effect of the target electron is very weak and the transition of the electron of the projectile is almost solely caused by its interaction with the nucleus of the atomic target.

Inelastic Mode. In collisions with large momentum transfers the rapid oscillations of the term $\exp(-i\mathbf{q} \cdot \boldsymbol{\rho})$ in the integrands of the transition matrix elements $\langle u_m | \exp(-i\mathbf{q} \cdot \boldsymbol{\rho}) | u_0 \rangle$ can make them negligible. These oscillations, however, can be compensated in the case when final states of the target electron are continuum states, where the electron momentum $\mathbf{k}$ with respect to the target nucleus is close to $-\mathbf{q}$, i.e. where $\mathbf{k} \approx -\mathbf{q}$ or, by separating the transverse and longitudinal parts, $\mathbf{k}_\perp \approx -\mathbf{q}_\perp$ and $k_z \approx -q_{\text{min}}$.

The condition $\mathbf{k}_\perp \approx -\mathbf{q}_\perp$ simply implies that nearly the whole transverse momentum transfer to the target has to be taken by the target electron alone.

More insight into the collision physics can be obtained by considering the condition $k_z \approx -q_{\text{min}}$. Taking into account the explicit form of q_{min}, this condition can be rewritten as a quadratic equation for k_z with the solutions

$$k_z^\pm \approx -v \pm \sqrt{v^2 - k_\perp^2 - 2(\varepsilon_n - \varepsilon_0 - \epsilon_0)}. \tag{2.10}$$

If $\frac{v^2}{2} < (\varepsilon_n - \varepsilon_0 - \epsilon_0) \approx (\varepsilon_n - \varepsilon_0)$, then both roots in (2.10) are complex. Physically it means that in such a case, due to the restrictions imposed by the energy-momentum conservation in the collision, there are no target states where the rapidly oscillating factor $\exp(-i\mathbf{q} \cdot \boldsymbol{\rho})$ can be compensated by a similar term arising from the final motion of the target electron. As a result, the inelastic contribution (2.9) to the cross section (2.7) is negligible in this case.

The roots $k_z^\pm$, given by (2.10), become real if $\frac{v^2}{2} > (\varepsilon_n - \varepsilon_0 + k_\perp^2/2 - \epsilon_0)$. If, in addition, we assume that $\frac{v^2}{2} \gg (\varepsilon_n - \varepsilon_0 + k_\perp^2/2 - \epsilon_0)$, then these roots are given by $k_z^+ \approx -(0.5k_\perp^2 + \varepsilon_n - \varepsilon_0 - \epsilon_0)/v$ and $k_z^- \approx -2v$. In the rest frame of the projectile these roots correspond to an electron having the z-component of the momentum approximately equal to v and $-v$, respectively, where $v > 0$ is the velocity of the incident target. Analysis shows that the contribution of the electrons with $k_z \approx k_z^-$ to the inelastic cross section is much smaller than that of the electrons with $k_z \approx k_z^+$ and can be neglected. A rough estimate for the contribution to the inelastic cross section (2.9) from collisions in which $k_z \approx -(0.5k_\perp^2 + \varepsilon_n - \varepsilon_0)/v$ can be easily obtained if one neglects the dependence of q_{min} on the final energy of the target electron. In such a case the integration over the final continuum states of the target electron in (2.9) is elementary performed by assuming that, because of large k, these states can be approximated by plane waves. The result is

$$\sigma_{0\to n}^{\mathrm{a}} \simeq \frac{4}{v^2} \int \mathrm{d}^2\mathbf{q}_\perp \frac{|\langle \psi_n \mid \exp(\mathrm{i}\mathbf{q}_0 \cdot \mathbf{r}) \mid \psi_0 \rangle|^2}{q_0^4}, \qquad (2.11)$$

where $\mathbf{q}_0 = (\mathbf{q}_\perp, \varepsilon_n - \varepsilon_0/v)$. This cross section can be interpreted as describing transitions of the electron in the projectile under the action of a fast free electron which has initially velocity v with respect to the projectile.

Combining (2.11) and (2.8) and taking into account that $Z_{\mathrm{A,eff}} \approx Z_{\mathrm{A}}$, we see that the cross section (2.7) in collisions with large momentum transfers can be approximated by

$$\sigma_{0\to n} \approx (Z_{\mathrm{A}}^2 + 1)\sigma_{0\to n}^{\mathrm{pr}}. \qquad (2.12)$$

In the above expression $\sigma_{0\to n}^{\mathrm{pr}}$ is the cross section for collisions in which the projectile electron makes a transition $0 \to n$ due to the interaction with a point-like unit charge moving with velocity v in the projectile frame. According to (2.12) the target nucleus and the target electron act incoherently in the collision. If the atom has initially Z_{A} electrons the factor $Z_{\mathrm{A}}^2 + 1$ in (2.12) should be replaced by $Z_{\mathrm{A}}^2 + Z_{\mathrm{A}}$. Equation (2.12) is the essence of the free collision model introduced long ago by Bohr [23]. This model, in particular, suggests that the relative importance of the elastic mode in the projectile–target collisions should rapidly increase with increasing atomic number of the neutral target.

The free collision model is quite simple and physically appealing but not very accurate. Better results for the cross sections can be obtained by applying the so called impulse approximation which is closely related to the free collision model. The application of the impulse approximation to the projectile electron excitation and loss was discussed in a review article [22] where also references to original articles were given. The impulse approximation takes into account the inner motion of the electrons in the target atom by averaging the projectile cross sections over the momentum distribution of these electrons in their initial bound state. An insightful discussion of the relationship between the plane-wave Born and impulse approximations was presented in [24].

2.2 Semi-Classical Approach

In the theory of fast ion–atom collisions quite often only electrons are treated quantum mechanically whereas the nuclei of the colliding partners are regarded as classical particles and their relative motion is described in terms of a classical trajectory. Such an approach is called *semi-classical*. Although the impact parameter, according to quantum mechanics, in general does not represent a measurable quantity, the semi-classical approach has important merits. First, by considering transition probabilities as a function of the impact parameter, one can get an additional insight into the collision physics. Second, the impact parameter consideration is usually more convenient for developing treatments which go beyond the first order approximation in the projectile–target interaction. Third, formulating a theory in terms of impact parameter allows one to apply the independent electron approximation for evaluating cross sections of multielectron transitions.

Let us now consider the projectile–target collision using the semi-classical approach[3]. We shall assume that the target nucleus, having a charge Z_A, is at rest and taken as the origin of our reference frame. In this frame the nucleus of the projectile-ion with a charge Z_I ($Z_I \gg 1$) moves along a straight-line classical trajectory $\mathbf{R}(t) = \mathbf{b} + \mathbf{v}t$, where $\mathbf{b}$ is the impact parameter and $\mathbf{v}$ the projectile velocity. A straight-line trajectory becomes a good approximation starting with collision energies of a few thousand electron volts and is certainly an excellent approximation for the collision energies of interest for this book. The projectile initially carries an electron bound in the ground state.

For simplicity we shall consider that the target also has only one electron. We denote the coordinates of the electron of the target and that of the projectile, given with respect to the target nucleus, by $\boldsymbol{\rho}$ and $\boldsymbol{\xi}$, respectively (see Fig. 2.1). Further, $\mathbf{s}$ and $\mathbf{r}$ are coordinates of the target and projectile electrons with respect to the projectile nucleus.

The electronic system of the colliding particles is described by the time-dependent Schrödinger equation

$$\left(i\frac{\partial}{\partial t} - H_I^{el} - H_A^{el} - V \right) \Psi(\mathbf{r}, \boldsymbol{\rho}, t) = 0, \tag{2.13}$$

where $\Psi(\mathbf{r}, \boldsymbol{\rho}, t)$ is the time-dependent wave function describing the electronic degrees of freedom. In (2.13) H_I^{el} and H_A^{el} are the electronic Hamiltonians of the ion and atom, respectively, and

$$V = \frac{Z_I Z_A}{R(t)} - \frac{Z_I}{s} - \frac{Z_A}{\xi} + \frac{1}{|\boldsymbol{\xi} - \boldsymbol{\rho}|} \tag{2.14}$$

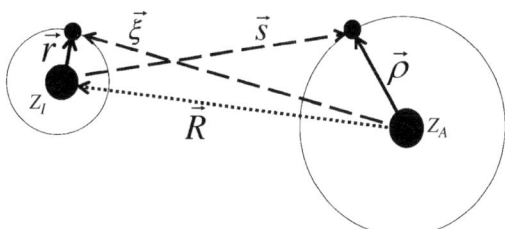

Fig. 2.1. Schematic representation of space coordinates characterizing the projectile–target collision.

---

[3] To our knowledge, for the projectile-electron excitation and loss the semi-classical approach was for the first time applied in [25].

is the interaction between the ion and the atom. The term $Z_I Z_A/R(t)$, representing the inter-nuclear interaction, is independent of the electron coordinates. In fast collisions this interaction does not influence cross sections for the electron transitions (integrated over the impact parameter) and below will be ignored.

Within the first order approximation, and assuming again that the electrons of the ion and atom are distinguishable, the initial and final states of the electrons of the colliding particles are approximated by the product of the undistorted initial and final states of the colliding particles

$$\chi_i(t) = u_0(\boldsymbol{\rho}) \exp(-i\epsilon_0 t)\, \psi_0(\boldsymbol{\xi} - \mathbf{R}(t)) \exp(-i\varepsilon_0 t) \exp(i\mathbf{v} \cdot \boldsymbol{\xi} - iv^2 t/2),$$
$$\chi_f(t) = u_m(\boldsymbol{\rho}) \exp(-i\epsilon_m t)\, \psi_n(\boldsymbol{\xi} - \mathbf{R}(t)) \exp(-i\varepsilon_n t) \exp(i\mathbf{v} \cdot \boldsymbol{\xi} - iv^2 t/2).$$
$$(2.15)$$

In (2.15) u_0 and u_m are the initial and final internal states of the target, respectively, given in the target frame. Further, ψ_0 and ψ_n have similar meanings but are for the projectile and given in the projectile rest frame. The term $\exp(i\mathbf{v} \cdot \boldsymbol{\xi}) \exp(-iv^2 t/2)$ is the so called translational factor[4] (see e.g. [6, 16]). The semi-classical first order amplitude reads

$$a_{fi}^{(1)}(\mathbf{b}) = -i \int_{-\infty}^{+\infty} dt \langle \chi_f(t) \mid V(t) \mid \chi_i(t) \rangle. \qquad (2.16)$$

Inserting the states (2.15) into expression (2.16) and keeping in mind that we consider only collisions, in which the internal state of the projectile changes ($n \neq 0$), it is not difficult to obtain that

$$a_{fi}^{(1)}(\mathbf{b}) = i \int_{-\infty}^{+\infty} dt \exp(i(\varepsilon_n + \epsilon_m - \varepsilon_0 - \epsilon_0)t)$$
$$\times \left\langle \psi_n u_m \left| \frac{Z_A}{|\mathbf{R}(t) + \mathbf{r}|} - \frac{1}{|\mathbf{R}(t) + \mathbf{r} - \boldsymbol{\rho}|} \right| \psi_0 u_0 \right\rangle. \qquad (2.17)$$

The straightforward generalization of expression (2.17) to the case, when the atom has Z_A electrons, yields

$$a_{fi}^{(1)}(\mathbf{b}) = i \int_{-\infty}^{+\infty} dt \exp(i(\varepsilon_n + \epsilon_m - \varepsilon_0 - \epsilon_0)t)$$
$$\times \left\langle \psi_n u_m \left| \frac{Z_A}{|\mathbf{R}(t) + \mathbf{r}|} - \sum_{j=1}^{Z_A} \frac{1}{|\mathbf{R}(t) + \mathbf{r} - \boldsymbol{\rho}_j|} \right| \psi_0 u_0 \right\rangle. \qquad (2.18)$$

---

[4] This factor appears because it is the wave function $\psi_n(\boldsymbol{\xi} - \mathbf{R}(t)) \exp(i\mathbf{v} \cdot \boldsymbol{\xi})$ $\exp(-iv^2 t/2 - i\varepsilon_n t)$ (and not merely $\psi_n(\boldsymbol{\xi} - \mathbf{R}(t)) \exp(-i\varepsilon_n t)$), which represents an exact solution of the Schrödinger equation for an undistorted atomic system moving with a constant velocity $\mathbf{v}$ in an inertial reference frame.

By applying the integral representation

$$\frac{1}{|\mathbf{x}|} = \frac{1}{2\pi^2} \int d^3\mathbf{k} \frac{\exp(i\mathbf{k} \cdot \mathbf{x})}{k^2} \tag{2.19}$$

to the Coulomb potentials $1/|\mathbf{R}(t) + \mathbf{r}|$ and $1/|\mathbf{R}(t) + \mathbf{r} - \boldsymbol{\rho}_j|$, the amplitude (2.18) is transformed into

$$a_{fi}^{(1)}(\mathbf{b}) = \frac{i}{\pi v} \int d^2\mathbf{q}_\perp \exp(-i\mathbf{q}_\perp \cdot \mathbf{b}) \left\langle u_m \left| Z_\mathrm{A} - \sum_j^{Z_\mathrm{A}} \exp(-i\mathbf{q} \cdot \boldsymbol{\rho}_j) \right| u_0 \right\rangle$$

$$\times \frac{\langle \psi_n |\exp(i\mathbf{q} \cdot \mathbf{r})| \psi_0 \rangle}{q^2}, \tag{2.20}$$

where $\mathbf{q} = (\mathbf{q}_\perp, q_{min})$ with q_{min} given by (2.5). It is easy to show [25] that the semi-classical first-order cross section

$$\sigma_{0 \to n}^{0 \to m} = \int d^2\mathbf{b} \left| a_{fi}^{(1)}(\mathbf{b}) \right|^2 \tag{2.21}$$

coincides with that following from the plane-wave Born approximation.

3

Considerations Beyond First Order Perturbation Theory

Theoretical considerations, based on the first-order perturbation theory in the projectile–target interaction, are expected to represent a good approximation to treat the different aspects of the projectile–target collisions only provided the conditions $Z_I/v \ll 1$ and $Z_A/v \ll 1$ are fulfilled. If at least one of them is violated, then one has to look for better approaches which go beyond the first order perturbation theory. It is the goal of this chapter to briefly discuss some of such approaches.

3.1 Second Order Approximation

The exact transition amplitude (in the prior form) may be written as

$$a_{fi} = -\mathrm{i} \int_{-\infty}^{+\infty} \mathrm{d}t \left\langle \Psi_{\mathrm{f}}^{(-)} \left| V \right| \chi_{\mathrm{i}} \right\rangle, \tag{3.1}$$

where $\Psi_{\mathrm{f}}^{(-)}$ is an exact solution of the full Schrödinger equation with the total Hamiltonian $\hat{H}$, which satisfies appropriate boundary conditions, and χ_{i} is the initial state of the system. The latter is a solution of the Schrödinger equation with the 'free' Hamiltonian $\hat{H}_0$ which does not contain the interaction term V.

The formal solution for the state Ψ_{f} is given by

$$|\Psi_{\mathrm{f}}\rangle = (1 + \hat{G}V)|\psi_{\mathrm{f}}\rangle, \tag{3.2}$$

where $\hat{G}$ is the Green operator of the full Schrödinger equation and ψ_{f} is an appropriate solution of the Schrödinger equation with the Hamiltonian $\hat{H}_0$. Taking into account that the full Green operator can be expressed as

$$\hat{G} = \hat{G}_0 + \hat{G}_0 V \hat{G}, \tag{3.3}$$

where $\hat{G}_0$ is the Green operator for the Schrödinger equation with the Hamiltonian $\hat{H}_0$, (3.2) can be expanded in the perturbation series in the interaction V. Terminating this expansion after the first order term, we obtain the second order transition amplitude:

$$a_{fi} = -\mathrm{i} \int_{-\infty}^{+\infty} dt \left\langle \chi_\mathrm{f} \left| \left(V + V\hat{G}_0 V \right) \right| \chi_\mathrm{i} \right\rangle = a_{fi}^{(1)} + a_{fi}^{(2)}, \qquad (3.4)$$

where $a_{fi}^{(1)}$ and $a_{fi}^{(2)}$ denote the parts of the transition amplitude which are of first and second order, respectively, in the interaction V.

The explicit expression for the second order transition amplitude is quite cumbersome even in simplest cases. For instance, within the semi-classical approach applied to collisions between an ion and an atom, each of which has initially just one electron, the second order part of this amplitude is given by

$$a_{fi}^{(2)} = (-\mathrm{i})^2 \sum_{\alpha}^{\mathrm{atom}} \sum_{\beta}^{\mathrm{ion}} \int_{-\infty}^{+\infty} dt \int_{-\infty}^{t} dt' \exp(\mathrm{i}(\varepsilon_n + \epsilon_m - \varepsilon_\beta - \epsilon_\alpha)t)$$

$$\times \exp(\mathrm{i}(\varepsilon_\beta + \epsilon_\alpha - \varepsilon_0 - \epsilon_0)t')$$

$$\times \left(\left\langle \psi_n u_m \left| \frac{1}{|\,\mathbf{R}(t)+\mathbf{r}-\boldsymbol{\rho}\,|} \right| \psi_\beta u_\alpha \right\rangle \left\langle \psi_\beta u_\alpha \left| \frac{1}{|\,\mathbf{R}(t')+\mathbf{r}-\boldsymbol{\rho}\,|} \right| \psi_0 u_0 \right\rangle \right.$$

$$+ \left\langle \psi_n u_m \left| \frac{1}{|\,\mathbf{R}(t)+\mathbf{r}-\boldsymbol{\rho}\,|} \right| \psi_\beta u_\alpha \right\rangle \left\langle \psi_\beta \left| \frac{-Z_\mathrm{A}}{|\,\mathbf{R}(t')+\mathbf{r}\,|} \right| \psi_0 \right\rangle \delta_{\alpha 0}$$

$$+ \left\langle \psi_n u_m \left| \frac{1}{|\,\mathbf{R}(t)+\mathbf{r}-\boldsymbol{\rho}\,|} \right| \psi_\beta u_\alpha \right\rangle \left\langle u_\alpha \left| \frac{-Z_\mathrm{I}}{|\,\mathbf{R}(t')-\boldsymbol{\rho}\,|} \right| u_0 \right\rangle \delta_{\beta 0}$$

$$+ \left\langle \psi_n \left| \frac{-Z_\mathrm{A}}{|\,\mathbf{R}(t)+\mathbf{r}\,|} \right| \psi_0 \right\rangle \delta_{ma} \left\langle \psi_\beta u_\alpha \left| \frac{1}{|\,\mathbf{R}(t')+\mathbf{r}-\boldsymbol{\rho}\,|} \right| \psi_0 u_0 \right\rangle$$

$$+ \left\langle u_m \left| \frac{-Z_\mathrm{I}}{|\,\mathbf{R}(t)-\boldsymbol{\rho}\,|} \right| u_\alpha \right\rangle \delta_{n\beta} \left\langle \psi_\beta u_\alpha \left| \frac{1}{|\,\mathbf{R}(t')+\mathbf{r}-\boldsymbol{\rho}\,|} \right| \psi_0 u_0 \right\rangle$$

$$+ \left\langle \psi_n \left| \frac{-Z_\mathrm{A}}{|\,\mathbf{R}(t)+\mathbf{r}\,|} \right| \psi_\beta \right\rangle \delta_{ma} \left\langle \psi_\beta \left| \frac{-Z_\mathrm{A}}{|\,\mathbf{R}(t')+\mathbf{r}\,|} \right| \psi_0 \right\rangle \delta_{\alpha 0}$$

$$+ \left\langle \psi_n \left| \frac{-Z_\mathrm{A}}{|\,\mathbf{R}(t)+\mathbf{r}\,|} \right| \psi_\beta \right\rangle \delta_{ma} \left\langle u_\alpha \left| \frac{-Z_\mathrm{I}}{|\,\mathbf{R}(t')-\boldsymbol{\rho}\,|} \right| u_0 \right\rangle \delta_{\beta 0}$$

$$+ \left\langle u_m \left| \frac{-Z_\mathrm{I}}{|\,\mathbf{R}(t)-\boldsymbol{\rho}\,|} \right| u_\alpha \right\rangle \delta_{n\beta} \left\langle \psi_\beta \left| \frac{-Z_\mathrm{A}}{|\,\mathbf{R}(t')+\mathbf{r}\,|} \right| \psi_0 \right\rangle \delta_{\alpha 0}$$

$$+ \left\langle u_m \left| \frac{-Z_\mathrm{I}}{|\,\mathbf{R}(t)-\boldsymbol{\rho}\,|} \right| u_\alpha \right\rangle \delta_{n\beta} \left\langle u_\alpha \left| \frac{-Z_\mathrm{I}}{|\,\mathbf{R}(t)-\boldsymbol{\rho}\,|} \right| u_0 \right\rangle \delta_{\beta 0} \right). \qquad (3.5)$$

Expression (3.5) describes the second order contribution to the transition amplitude from a collision in which the electron of the projectile-ion makes a transition $\psi_0 \rightarrow \psi_n$ while the electron of the target-atom undergoes a transition $u_0 \rightarrow u_m$. The transitions proceed via the double pair-wise interactions between the constituents of the ion and atom and the physical meaning of the each term in the amplitude (3.5) is quite transparent.

In (3.5) $\mathbf{R}(t) = \mathbf{b} + \mathbf{v}t$ denotes the coordinate of the nucleus of the ion with respect to the nucleus of the atom (the origin), $\boldsymbol{\rho}$ is the coordinate of the electron of the atom with respect to the origin and $\mathbf{r}$ is the coordinate of the electron of the ion with respect to the nucleus of the ion (see Fig. 2.1). The summations run over the complete sets of the intermediate states of the ion ($\{\psi_\beta\}$) and the atom ($\{u_\alpha\}$).

Note that the double summation in (3.5) is actually present only for the part of (3.5) whose integrand is proportional to the first term in the parentheses. This part yields the contribution to the transition amplitude caused by two consequent interactions between the electrons. In the part of the amplitude (3.5), which describes transitions due to the interaction between the electron of the ion and the nucleus of the atom accompanied by the interaction between the electron of the atom with the nucleus of the ion, the intermediate states simply coincide with either the initial or final states. In the other parts the summation goes over the complete set of either atomic or ionic intermediate states. Note also that (since it is assumed that $n \neq 0$) the last term in the parentheses in fact vanishes. Besides, the term which describes two interactions between the electron of the ion and the nucleus of the atom is nonzero only in the elastic atomic mode. Nevertheless, although expression (3.5) is not as 'scaring' as it might seem at the first glance, the practical evaluation of the second order transition amplitude becomes feasible provided some additional assumptions, like e.g. the closure approximation, are made in (3.5).

According to the first order approximation, the simultaneous electron transitions in the projectile and target may occur only due to the interaction between the ionic and atomic electrons. The inspection of the amplitude (3.5) shows that for such reactions there exist also other paths. In particular, the simultaneous electron transitions can be caused by the two-center electron–nucleus interactions: the electron of the target undergoes a transition due to its interaction with the projectile nucleus and, simultaneously, the electron of the projectile makes a transition induced by the interaction with the target nucleus. The importance of this channel increases when the atomic numbers of the projectile and/or target increase.[1] As we see, in a Born approximation formalism such a reaction channel becomes possible only starting with the second order terms in the projectile–target interaction.

Equation (3.4) represents the standard form of the second order Born amplitude.[2] One of the familiar features of this amplitude is that, according to it, the projectile electron in its intermediate states is governed by the field of the projectile nucleus while the interaction with the field of the atom acts merely as a perturbation. In the case of light ions impinging on atomic targets with relatively large atomic numbers, the screened field of the atomic

---

[1] In particular, in the next section we shall see that this channel becomes of especial importance in collisions involving highly charged ions when $Z_I \sim v$.

[2] This amplitude, for instance, was used in [26, 27] to evaluate the spectrum of the electron emitted from the projectile in collisions with neutral atoms.

nucleus, acting on the electron of the projectile during the collision, may become effectively stronger than the field of the ionic nucleus. Then, a possible alternative to the amplitude (3.4) would be to return to expression (3.3) and to replace the expansion for the full Green operator given by (3.3) by a new expansion

$$\hat{G} = \hat{G}'_0 + \hat{G}'_0 V' \hat{G}, \tag{3.6}$$

where the interaction between the electron of the projectile and the atom is included into the definition of the new 'free' Green operator $\hat{G}'_0$, corresponding to the new 'free' Hamiltonian $\hat{H}'_0$. The interaction with the nucleus of the ion is now a part of the new perturbation V'. In such a way one arrives at the so called second-order strong-field Born approximation which was used in [28] to consider the projectile-electron loss in collisions with relatively heavy atoms. The consideration of the second-order strong-field Born approximation also enables to establish a link (see [28]) with the impulse approximation of [29].

3.2 Distorted-Wave Approach

We shall now consider collisions between a hydrogen-like projectile-ion and a light atom. We shall suppose that, while the condition $Z_A/v \ll 1$ is fulfilled, the ion has a sufficiently high nuclear charge so that the condition $Z_I/v \ll 1$ is no longer met and instead we have a much softer condition $Z_I/v \stackrel{\sim}{<} 1$.

In the standard Born series the expansion parameter is essentially given by the ration $\nu = Z_I/v$. Since this ratio can be now close to unity, one has to develop an approach which does not rely on the standard Born expansion. At the same time, it is very desirable to keep such an approach from being too complicated. As we shall see below, suitable candidates for an approach satisfying the above two requirements can be found by using the ideology of distorted wave models.

Note that such models (see e.g. [30–36]) have been proved to be quite successful in considering atomic ionization, excitation and electron capture in energetic collisions with highly charged bare nuclei. The success of these models owes to the fact that in these models the initial and final states of an electron, which moves initially in the atomic field and is subjected in the collision to the field of the incident nucleus, already account for the important parts of the interaction between the electron of the atom and the incident nucleus. As a result, the residual interaction, treated in these models perturbatively, is effectively much weaker compared to that appearing in the standard Born expansion. This makes the distorted wave models much faster convergent and very often already the first term of the distorted-wave expansions does a very good job (for a review on the distorted-wave models in nonrelativistic base ion-atom collisions see [37–39]).

We again adopt the semi-classical treatment and assume that the target nucleus, having a charge Z_A, is at rest and taken as the origin. In the frame of the nucleus of the target the projectile nucleus with a charge Z_I ($Z_I \gg 1$) moves along a straight-line classical trajectory $\mathbf{R}(t) = \mathbf{b} + \mathbf{v}t$, where $\mathbf{b}$ is the impact parameter and $\mathbf{v}$ is the projectile velocity.

For simplicity we shall consider that the target has only one (active) electron. As before, we denote the coordinates of the electron of the target and that of the projectile, given with respect to the target nucleus, by $\boldsymbol{\rho}$ and $\boldsymbol{\xi}$, respectively, and $\mathbf{s}$ and $\mathbf{r}$ are the coordinates of the target and projectile electrons with respect to the projectile nucleus (see Fig. 2.1).

The prior form of the semi-classical transition amplitude is given by

$$a_{fi}(\mathbf{b}) = -\mathrm{i} \int_{-\infty}^{+\infty} \mathrm{d}t \langle \Psi^{(-)}(t) \mid \left(\hat{H} - \mathrm{i}\partial/\partial t \right) \phi_i(t) \rangle. \qquad (3.7)$$

In (3.7) $\Psi^{(-)}(t)$ is the solution of the Schrödinger equation (2.13) and $\phi_i(t)$ is the solution of

$$\mathrm{i}\frac{\partial \phi_i}{\partial t} = \left(\hat{H}_A + \hat{H}_I + W(t) \right) \phi_i, \qquad (3.8)$$

where $W(t)$ is a distortion potential.

In the simplest approach the states $\phi_i(t)$ and $\Psi^{(-)}(t)$ would be replaced by the undistorted initial and final states (2.15), which would lead to the first order amplitude (2.17). In order to treat collisions, in which the ratio Z_I/v may be not small, we take the initial and final states as [40].

$$\chi_i(t) = L_i\, u_0(\boldsymbol{\rho}) \exp(-\mathrm{i}\epsilon_0 t)\, \psi_0(\boldsymbol{\xi} - \mathbf{R}(t)) \exp(\mathrm{i}\mathbf{v} \cdot \boldsymbol{\xi}) \exp(-\mathrm{i}v^2 t/2 - \mathrm{i}\varepsilon_0 t),$$
$$\chi_f(t) = L_f\, u_m(\boldsymbol{\rho}) \exp(-\mathrm{i}\epsilon_m t)\, \psi_n(\boldsymbol{\xi} - \mathbf{R}(t)) \exp(\mathrm{i}\mathbf{v} \cdot \boldsymbol{\xi}) \exp(-\mathrm{i}v^2 t/2 - \mathrm{i}\varepsilon_n t).$$
$$(3.9)$$

These states differ from those given by expression (2.15) just in one, however, important point. Namely, they include the distortions of the initial and final states of the target electron caused by its interaction with the strong field of the projectile nucleus. The distortion factors L_i and L_f depend on the coordinates $\mathbf{s}$ of the target electron with respect to the projectile nucleus, $L_i = L_i(\mathbf{s})$ and $L_f = L_f(\mathbf{s})$, but at this point the explicit form of these factors is not yet specified.

Inserting the states (3.9) into (3.7) and remembering our assumption that $n \neq 0$ we obtain for the transition amplitude

$$a_{fi} = a_{fi}^{ee} + a_{fi}^{eN}, \qquad (3.10)$$

where

$$a_{fi}^{ee}(\mathbf{b}) = -\mathrm{i} \int_{-\infty}^{+\infty} \mathrm{d}t \exp(\mathrm{i}(\varepsilon_n + \epsilon_m - \varepsilon_0 - \epsilon_0)t)$$
$$\times \int \mathrm{d}^3\mathbf{r} \int \mathrm{d}^3\boldsymbol{\rho}\, \psi_n^*(\mathbf{r}) u_m^*(\boldsymbol{\rho}) L_f^*(\mathbf{s}) \frac{1}{|\mathbf{R} + \mathbf{r} - \boldsymbol{\rho}|} L_i(\mathbf{s}) u_0(\boldsymbol{\rho}) \psi_0(\mathbf{r})$$
$$(3.11)$$

and

$$a_{fi}^{\mathrm{eN}}(\mathbf{b}) = \mathrm{i} \int_{-\infty}^{+\infty} \mathrm{d}t \exp(\mathrm{i}(\varepsilon_n + \epsilon_m - \varepsilon_0 - \epsilon_0)t)$$

$$\times \int \mathrm{d}^3\mathbf{r} \int \mathrm{d}^3\boldsymbol{\rho}\, \psi_n^*(\mathbf{r}) u_m^*(\boldsymbol{\rho}) L_{\mathrm{f}}^*(\mathbf{s}) \frac{Z_A}{|\mathbf{R} + \mathbf{r}|} L_{\mathrm{i}}(\mathbf{s}) u_0(\boldsymbol{\rho}) \psi_0(\mathbf{r}). \quad (3.12)$$

The part a_{fi}^{ee} of the amplitude (3.10) describes transitions caused by the interaction between the electrons of the ion and atom. Such transitions may occur already within the first order consideration. However, in contrast to the latter, now the action of the electron–electron interaction is 'modulated' by the multiple interactions between the target electron and the projectile nucleus which are accounted for by the distortion factors. The remaining part, a_{fi}^{eN}, contains the contribution to the transition which is due to the interaction between the projectile electron and the target nucleus.

It is obvious that for collisions, which are inelastic also for the target ($m \neq 0$), the part a_{fi}^{eN} would simply vanish if the distortion factors are replaced by 1. In such a case we would recover the first order result which predicts that the simultaneous transitions in the projectile and target may be caused by the electron–electron interaction only. However, if we keep the distortion factors $\mathbf{s}$-dependent, then a_{fi}^{eN} becomes nonzero suggesting that the distorted-wave amplitude (3.10) contains yet another mechanism for the doubly inelastic collisions to proceed. Indeed, in general, the part a_{fi}^{eN} of the amplitude describes the simultaneous transitions in the projectile and target as occurring due to the joint effect of the single interaction between the projectile electron and the target nucleus and the multiple interactions of the target electron and the projectile nucleus. Thus, we see that merely by the introduction of the distortion factors for the target electron one is directly led to the transition amplitude, in which not only the two-center electron–electron interaction but also the two-center electron–nucleus interactions are automatically taken into account in a relatively simple way.

Another point of interest to be mentioned here is that the integrands in (3.11) and (3.12) do not contain derivatives of the distortion factors. Such derivatives are well known to contribute to the transition amplitude in the case when distorted-wave models are applied to collisions with bare projectile-nuclei ([31–36]). In the case under consideration, however, the terms containing the derivatives vanish because of orthogonality of the initial and final internal states of the structured projectile-ion.

The transition amplitude (3.10) can be converted to the momentum space by performing the Fourier transformation

$$S_{fi}(\mathbf{q}_\perp) = \frac{1}{2\pi} \int \mathrm{d}^2\mathbf{b}\, a_{fi}(\mathbf{b}) \exp(\mathrm{i}\mathbf{q}_\perp \cdot \mathbf{b}). \quad (3.13)$$

The quantity $\mathbf{q}_\perp$ can be thought of as the two-dimensional transverse ($\mathbf{q}_\perp \cdot \mathbf{v} = 0$) momentum transfer to the target. Using (3.10)–(3.13) one can show that the transition amplitude in the momentum space reads

$$S_{fi}(\mathbf{q}_\perp) = -\frac{i}{4\pi^3 v} \int d^3\boldsymbol{\kappa}\, \frac{1}{\kappa^2} I_d(\mathbf{q}+\boldsymbol{\kappa})\, I_p(\boldsymbol{\kappa})\, I_t(\mathbf{q},\boldsymbol{\kappa}), \qquad (3.14)$$

where

$$I_d(\mathbf{p}) = \int d^3\mathbf{s}\, \exp(-i\mathbf{p}\cdot\mathbf{s}) L_f^*(\mathbf{s}) L_i(\mathbf{s}),$$

$$I_p(\mathbf{p}) = \int d^3\mathbf{r}\, \psi_n^*(\mathbf{r})\, \exp(i\mathbf{p}\cdot\mathbf{r})\psi_0(\mathbf{r}),$$

$$I_t(\mathbf{p_1},\mathbf{p_2}) = \int d^3\rho\, u_m^*(\boldsymbol{\rho})\, \exp(i\mathbf{p_1}\cdot\boldsymbol{\rho})\, (1 - Z_A \exp(i\mathbf{p_2}\cdot\boldsymbol{\rho}))\, u_0(\boldsymbol{\rho}) \quad (3.15)$$

and $\mathbf{q} = (\mathbf{q}_\perp; q_{min})$ with q_{min} given by (2.5).

In order to move further one has to specify the distortion factors L_i and L_f. The choice of these factors is not unique. In theory of atomic ionization and excitation by collisions with bare ions the following distorted-wave approximations have been very extensively and successfully used: the continuum-distorted-wave model [31], the continuum-distorted-wave-eikonal-initial-state model [33] and the symmetric eikonal model [35,36]. Below, following the consideration of projectile–target collisions given in [40], we shall take the distortion factors L_i and L_f as in the symmetric eikonal model.

3.2.1 Symmetric Eikonal Model

In the spirit of the symmetric eikonal model we set

$$L_i(\mathbf{s}) = \exp(-i\nu \ln(vs + \mathbf{v}\cdot\mathbf{s})),$$
$$L_f(\mathbf{s}) = \exp(+i\nu \ln(vs - \mathbf{v}\cdot\mathbf{s})), \qquad (3.16)$$

where $\nu = Z_I/v$. From the point of view of computation such a choice of the distortion factors is the simplest one. This choice is also quite a natural one for considering collisions in which the target is finally in a bound state. Besides, (3.16) may also be applied to treat collisions leading to ionization of the target, provided the velocity of the target electron in the final state is much less than the projectile velocity. Note also that, since the net charge of a hydrogen-like projectile with $Z_I \gg 1$ is approximately equal to the charge of its nucleus, the distortion factors (3.16) can be viewed as imposing the Coulomb boundary conditions on the initial and final states of the target electron.

It is worth mentioning that the symmetric eikonal model represents the first order term of the corresponding distorted wave series with the expansion parameter $\sim Z_I/v^2$ and, therefore, seems to be well suited to consider the ion–atom collisions in the intermediate-to-high velocity regime $Z_I/v \lesssim 1$ where one has $Z_I/v^2 \ll 1$ for $v \gg 1$.

One can show [40] that with the distortion factors defined by (3.16) the transition amplitude (3.14) is given by

$$S_{fi}(\mathbf{q}_\perp) = -\frac{2\mathrm{i}}{v^{1+2\mathrm{i}\nu}}$$

$$\int \mathrm{d}^2\mathbf{p}_\perp f(\mathbf{p}_\perp, \nu)\langle \psi_n(\mathbf{r}) \mid \exp(\mathrm{i}(\mathbf{p}_\perp - \mathbf{q}) \cdot \mathbf{r}) \mid \psi_0(\mathbf{r})\rangle$$

$$\times \frac{1}{(\mathbf{q} - \mathbf{p}_\perp)^2}\langle u_m(\boldsymbol{\rho}) \mid Z_A \exp(\mathrm{i}\mathbf{p}_\perp \cdot \boldsymbol{\rho}) - \exp(\mathrm{i}\mathbf{q} \cdot \boldsymbol{\rho}) \mid u_0(\boldsymbol{\rho})\rangle, \quad (3.17)$$

where the function $f(\mathbf{p}_\perp, \nu)$ is defined in Eq. (6.4).

In order to get some insight into the physical picture of the collision described by the amplitude (3.17) we rewrite the last line in (3.17) by using the completeness relation for the target states which yields

$$\langle u_m \mid Z_A \exp(\mathrm{i}\mathbf{p}_\perp \cdot \mathbf{r}) - \exp(\mathrm{i}\mathbf{q} \cdot \mathbf{r}) \mid u_0\rangle$$

$$= \sum_{m'}\langle u_m \mid Z_A - \exp(\mathrm{i}(\mathbf{q} - \mathbf{p}_\perp) \cdot \mathbf{r}) \mid u_{m'}\rangle\langle u_{m'} \mid \exp(\mathrm{i}\mathbf{p}_\perp \cdot \mathbf{r}) \mid u_0\rangle. \quad (3.18)$$

Here the sum runs over the complete set $\{u_{m'}\}$ of the atomic states. According to (3.18), the transition amplitude (3.17), due to the presence of the eikonal distortion factors, can be interpreted as taking into account the virtual excitation of the target by the field of the projectile nucleus. Correspondingly, the vector $\mathbf{p}_\perp$ can be thought of as the virtual momentum transfer at the intermediate stage of the collision process.

One should emphasize that, while both the inelastic ($m \neq 0$) and elastic ($m = 0$) target collision modes can be described by (3.17), the expression (3.17) is only valid under the assumption that the initial and final internal states of the projectile are *different*, i.e. $n \neq 0$. Indeed, it was the latter assumption which resulted in the absence of terms with derivatives in the integrands of (3.11) and (3.12), as well as in the integrand of (3.17). In contrast, terms with derivatives do contribute to the transition amplitude obtained in the symmetric eikonal approximation in the case of collisions with a projectile which either does not have atomic structure (see e.g. [37]) or when its atomic structure remains 'frozen' in the collision.

3.2.2 Symmetric Eikonal Model: 'Electrostatic' Approach

Now we shall very briefly discuss how the transition amplitude (3.17) can be derived by using another approach [41]. Compared to the 'standard' consideration which employs the transition amplitude in the form given by (3.7), the valuable merit of this approach is that it is suitable for a natural generalization to relativistic collision velocities.

We start with the following expression for the transition amplitude

$$a_{fi}(\mathbf{b}) = -i \int_{-\infty}^{+\infty} dt \int d^3 \mathbf{x} \varrho_A(\mathbf{x}, t) \, \varphi_I(\mathbf{x}, t). \tag{3.19}$$

Here $\varrho_A(\mathbf{x}, t)$ is the transition charge density created by the target at a time t and a space point $\mathbf{x}$ and $\varphi_I(\mathbf{x}, t)$ is the transition scalar potential generated by the projectile at the same t and $\mathbf{x}$[3].

The charge density of the target is given by

$$\varrho_A(\mathbf{x}, t) = \int d^3 r \phi_f^*(\mathbf{r}, t) \left[Z_A \delta^{(3)}(\mathbf{x}) - \delta^{(3)}(\mathbf{x} - \mathbf{r}) \right] \phi_i(\mathbf{r}, t), \tag{3.20}$$

where ϕ_i and ϕ_f are the initial and final states of the target, respectively, and $\delta^{(3)}$ is the 3-dimensional delta-function. According to the symmetric eikonal approximation we choose these states as

$$\phi_i(\mathbf{r}, t) = \exp(-i\nu \ln(vs + \mathbf{v} \cdot \mathbf{s}))\varphi_0(\mathbf{r}) \exp(-i\varepsilon_0 t),$$
$$\phi_f(\mathbf{r}, t) = \exp(+i\nu \ln(vs - \mathbf{v} \cdot \mathbf{s}))\varphi_n(\mathbf{r}) \exp(-i\varepsilon_n t). \tag{3.21}$$

Inserting the states (3.21) into (3.20) one can show (see [41]) that

$$\varrho_A(\mathbf{x}, t) = \exp(i(\varepsilon_n - \varepsilon_0)t) \int d^2 \mathbf{p}_\perp f(\mathbf{p}_\perp, \nu) \exp(-i\mathbf{p}_\perp \cdot \mathbf{b})$$
$$\times \int d^3 r \phi_f^*(\mathbf{r}) \exp(i\mathbf{p}_\perp \cdot \mathbf{r}) \phi_i(\mathbf{r}) \left[Z_A \delta^{(3)}(\mathbf{x}) - \delta^{(3)}(\mathbf{x} - \mathbf{r}) \right]. \tag{3.22}$$

The scalar potential, created by the projectile in the collision, is a solution of Poisson's equation

$$\Delta \varphi_I(\mathbf{x}, t) = -4\pi \varrho_I(\mathbf{x}, t), \tag{3.23}$$

where $\varrho_I(\mathbf{x}, t)$ is the transition charge density of the projectile. In contrast to the charge density of the target, the transition charge density of the projectile is evaluated with the undistorted electron states. Applying the 3-dimensional Fourier transformation to both sides of (3.23) we obtain that

$$\tilde{\varphi}_I(\mathbf{k}, t) = \frac{4\pi}{k^2} \tilde{\varrho}_I(\mathbf{k}, t), \tag{3.24}$$

where $\tilde{\varrho}_I(\mathbf{k}, t)$ and $\tilde{\varphi}_I(\mathbf{k}, t)$ are the Fourier transforms of the charge density and the scalar potential of the ion, respectively.

---

[3] One can take the transition amplitude in a fully equivalent form where the projectile charge density is coupled to the scalar potential of the target. Note also that (3.19) is similar in form to (2.1).

By expanding the charge density and the scalar potential in (3.19) into the Fourier integrals the transition amplitude is rewritten as

$$a_{fi}(\mathbf{b}) = -4\pi i \int_{-\infty}^{+\infty} dt \int d^3 k \, \tilde{\varrho}_A(\mathbf{k}, t) \frac{1}{k^2} \tilde{\varrho}_I(-\mathbf{k}, t), \qquad (3.25)$$

where $\tilde{\varrho}_A(\mathbf{k}, t)$ is the Fourier transform of the charge density of the atom.

It is not difficult to show that in the $\mathbf{k}$-space the charge density of the target is given by

$$\tilde{\varrho}_A(\mathbf{k}, t) = \frac{1}{(2\pi)^{3/2}} \int d^3 x \varrho_A(\mathbf{x}, t) \exp(-i\mathbf{k} \cdot \mathbf{x})$$

$$= \frac{\exp(i(\varepsilon_n - \varepsilon_0)t)}{(2\pi)^{3/2}} \int d^2 p_\perp f(p_\perp, \nu) \exp(-i\mathbf{p}_\perp \cdot \mathbf{b})$$

$$\times \langle \varphi_n(\mathbf{r}) \mid Z_A \exp(i\mathbf{p}_\perp \cdot \mathbf{r}) - \exp(i(\mathbf{p}_\perp - \mathbf{k}) \cdot \mathbf{r}) \mid \varphi_0(\mathbf{r}) \rangle \quad (3.26)$$

and the transition charge density of the projectile for the case $m \neq 0$ is given by

$$\tilde{\varrho}_I(\mathbf{k}, t) = \frac{1}{(2\pi)^{3/2}} \int d^3 x \varrho_I(\mathbf{x}, t) \exp(-i\mathbf{k} \cdot \mathbf{x})$$

$$= \frac{\exp(i(\epsilon_m - \epsilon_0)t) \exp(-i\mathbf{k} \cdot \mathbf{R}(t))}{(2\pi)^{3/2}} \langle \chi_m(\boldsymbol{\xi}) \mid \exp(-i\mathbf{k} \cdot \boldsymbol{\xi}) \mid \chi_0(\boldsymbol{\xi}) \rangle.$$

$$(3.27)$$

Using the above two expressions and taking into account the relation (3.13), the eikonal transition amplitude in the $\mathbf{q}_\perp$-space is obtained to be

$$S_{fi}(\mathbf{q}_\perp) = \frac{1}{2\pi} \int d^2 b \, a_{fi}(\mathbf{b}) \exp(i\mathbf{q}_\perp \cdot \mathbf{b})$$

$$= -\frac{2i}{v} \int d^2 p_\perp f(p_\perp, \nu) \langle \chi_m(\boldsymbol{\xi}) \mid \exp(i(\mathbf{p}_\perp - \mathbf{q}) \cdot \boldsymbol{\xi}) \mid \chi_0(\boldsymbol{\xi}) \rangle \frac{1}{(\mathbf{q} - \mathbf{p}_\perp)^2}$$

$$\times \langle \varphi_n(\mathbf{r}) \mid Z_A \exp(i\mathbf{p}_\perp \cdot \mathbf{r}) - \exp(i\mathbf{q} \cdot \mathbf{r}) \mid \varphi_0(\mathbf{r}) \rangle. \qquad (3.28)$$

The amplitude (3.28) is identical to that given by (3.17).

One has to add that the two approaches to derive the symmetric eikonal transition amplitude, which have been considered in the last subsections, lead to the same expression for the transition amplitude only in the case of collisions in which the projectile changes its internal state, $m \neq 0$. This peculiarity is not just the property of the symmetric eikonal model: the same relation between results of the 'standard' and 'electrostatic' approaches holds in any distorted-wave model in which the initial and final distortion factors are functions depending only on $\mathbf{s}$. Note also that within the first Born approximation the two approaches yield identical results independent of whether $m = 0$ or $m \neq 0$.

3.2.3 An Example of Applications: Electron Angular Distribution

Important information about the interactions governing the ion–atom collisions can be obtained by considering the cross section, $\mathrm{d}^2\sigma/\mathrm{d}\varepsilon\mathrm{d}\Omega$, which is differential in energy and angle of the electron emitted from the target. An example of the cross section $\mathrm{d}^2\sigma/\mathrm{d}\varepsilon\mathrm{d}\Omega$ is shown in Fig. 3.1 for the reaction $3.6\,\mathrm{MeV}\ \mathrm{C}^{5+}(1s)+\mathrm{He}(1s^2) \to \mathrm{C}^{6+} + \mathrm{He}^+(1s)+2e^-$ ($v = 12$ a.u.). In this figure the cross section is given for a fixed electron energy of $\varepsilon = k^2/2 = 40\,\mathrm{eV}$ as a function of the polar emission angle of the target electron, $\theta = \arccos(\mathbf{k}\cdot\mathbf{v}/kv)$, where $\mathbf{k}$ is the electron momentum with respect to the target nucleus and $k = |\mathbf{k}|$. Compared to the result of the first order consideration (dash curve) the eikonal theory shows very substantial differences, both in the magnitude and the shape of the cross section (see solid curve). In particular, it predicts the very strong enhancement of the target emission into the backward direction. This enhancement is due to the contribution of the two-center

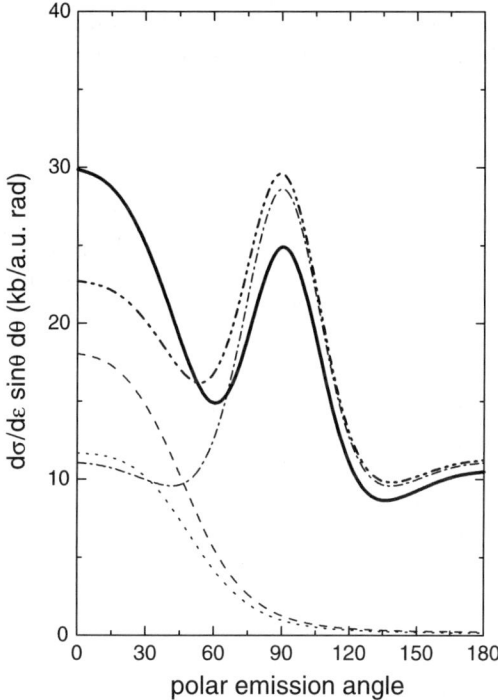

Fig. 3.1. Angular distribution of $40\,\mathrm{eV}$ electrons emitted from the target in $3.6\,\mathrm{MeV\,u}^{-1}$ $\mathrm{C}^{5+}(1s)+\mathrm{He}(1s^2) \to \mathrm{C}^{6+}+\mathrm{He}^+(1s) + 2e^-$ collisions. *Solid curve*: calculation using the amplitude (3.17). *Dash curve*: the first Born result. *Dot curve*: calculation with the amplitude (3.17) neglecting the two-center dielectronic interaction. *Dash–dot* curve: calculation with the amplitude (3.17) neglecting the interaction between the projectile electron and the target nucleus. *Dash–dot–dot* curve: the sum of the dot and dash–dot curves.

electron–nucleus interactions and reflects the fact that in this reaction channel the target electron does not get a large recoil in the forward direction since the momentum transfer necessary to remove the tightly bound projectile electron is provided by the target core. The angular distribution calculated with the amplitude (3.17) has rather an unusual shape which is due to the interplay between the two different reaction channels. One of the interesting features of this distribution is the clear interference between the contributions to the two-center dielectronic transitions arising due to the two-center electron–electron and electron–nucleus interactions (compare solid and dash–dot–dot curves in Fig. 3.1).

3.3 Coupled Channel Approach

In the previous section we have considered collisions between a highly charged hydrogen-like ion $(Z_I/v \lesssim 1)$ and a light atom $(Z_A/v \ll 1)$. We have seen that, despite the field of the atom acting on the electron of the ion is weak, the projectile-electron excitation and loss processes in general cannot be treated within the first order theory.

In the rest of this chapter we consider how the projectile-electron excitation and loss can be dealt with when the target atom is heavy enough to violate the condition $Z_A/v \ll 1$.

When, for a given collision velocity, the target atomic number substantially increases, the interaction between the electron of the projectile and the target becomes too strong, making first-order theories irrelevant. In reality, each transition from the initial electron state of the projectile would lead to a reduction of the population of this state making further transitions from this state less probable. Such a reduction is not taken into account in first-order theories. The latter ones do not preserve unitarity and, therefore, often result in strongly overestimated cross sections in the case of large perturbations. This may be especially true for the elastic mode, where the screened field of the target nucleus can become so strong in collisions with small impact parameters that first order calculations yield elastic cross sections which are an order of magnitude larger than the experimental total cross sections (see [42]).

We return to considering a collision between a projectile-ion with a nuclear charge Z_I, which initially has one electron, and a neutral target atom with atomic number Z_A. We shall again use the semi-classical approximation and describe the electronic system of the colliding particles containing $Z_A + 1$ electrons by the time-dependent Schrödinger equation of the form given by (2.13). The time-dependent wave function $\Psi^{el}(\mathbf{r}, \{\boldsymbol{\rho}\}, t)$ now describes the motion of $Z_A + 1$ electrons, the atomic Hamiltonian H_a^{el} depends on the $3Z_A$ coordinates of the atomic electrons $\{\boldsymbol{\rho}\} = \{\boldsymbol{\rho_1}, \boldsymbol{\rho_1}, ...\}$ with respect to the atomic nucleus and the interaction between the ion and atom reads

$$V = -\frac{Z_A}{|\,\mathbf{R}(t) + \mathbf{r}\,|} - \sum_{j=1}^{Z_A} \frac{Z_I}{|\,\mathbf{R}(t) - \boldsymbol{\rho}_j\,|} + \sum_{j=1}^{Z_A} \frac{1}{|\,\mathbf{R}(t) - \boldsymbol{\rho}_j + \mathbf{r}\,|}, \qquad (3.29)$$

where the nucleus–nucleus interaction has again been omitted.

One way to solve the time-dependent Schrödinger equation is to expand the wavefunction in a complete set, $\{\varphi_\alpha(\mathbf{r}, \{\boldsymbol{\rho}\})\}$, of internal wavefunctions of the noninteracting ion–atom system

$$\Psi^{\text{el}} = \sum_\alpha a_\alpha(t)\,\exp(-\mathrm{i}E_\alpha t)\,\varphi_\alpha, \qquad (3.30)$$

where E_α is the sum of internal electronic energies of the ion and atom. We assume that the electron translational factors as well as kinetic energies of the relative motion of the electrons are included in the wavefunctions φ_α. It is implied that the summation in (3.30) runs also over continuum states of the electron in the ion and those of the electrons in the atom. Inserting (3.30) into the Schrödinger equation one obtains for the unknown time-dependent coefficients a_α the following system of differential equations

$$\mathrm{i}\frac{\mathrm{d}a_\alpha}{\mathrm{d}t} = \sum_{\alpha'} V_{\alpha\alpha'}(t)\exp(\mathrm{i}(E_\alpha - E_{\alpha'})t)a_{\alpha'} \qquad (3.31)$$

with the initial conditions

$$a_0(t \to -\infty) = 1,$$
$$a_{\alpha' \neq 0}(t \to -\infty) = 0. \qquad (3.32)$$

At sufficiently large collision velocities transitions of the electrons between different centers (charge exchange channels) are suppressed because of the electron translational factors and can be neglected. For electron transitions at the same centers the translational factors for initial and final states as well as kinetic energies of the relative motion of the electrons in the corresponding exponents mutually cancel. Therefore, the remaining terms $V_{\alpha\alpha'} = \langle \varphi_\alpha \mid V(t) \mid \varphi_{\alpha'} \rangle$ can be rewritten as $V_{\alpha\alpha'} = \langle \chi_\alpha \mid V(t) \mid \chi_{\alpha'} \rangle$, where the functions χ_α describe the internal motion of the electrons within the colliding particles.

The system of the differential equations (3.31), in principle, is equivalent to the Schrödinger equation and, in this sense, is exact. However, it includes an infinite number of channels and, even after the implementation of the simplifications discussed in the previous paragraph, cannot be solved without making further approximations.

One way to solve approximately the system (3.31) would be to keep all the channels and to develop a perturbation series by using an iteration procedure. Coupled channel approaches represent an alternative. They consist of (i) a restriction of the number of channels considered and (ii) an exact (numerical) solution of the resulting finite set of the coupled equations. Coupled channel approaches preserve unitarity and, compared to first order treatments, are much better suited for considering strong perturbations.

For instance, coupled channel calculations for the electron loss from 0.25–1 MeV u^{-1} He$^{+}(1s)$ projectiles colliding with Ne, Ar and Kr atoms, which were performed in [43], enabled one to get a much better agreement with experiment than that obtained in the first order calculations. Coupled channel approaches turned out to be also rather powerful at lower impact energies, where the electron capture starts to play an important role [44–46].

3.4 Sudden Approximation

Equation (3.31) can be solved analytically if we assume that the exponents of the oscillating factors on the right hand side of (3.31) are small and can be neglected. This is the case if the effective collision time $T(b)$, when the interaction $V(t)$ reaches considerable magnitudes, is short compared to typical electron transition times $\tau_{\alpha\alpha'} \simeq |E_\alpha - E_{\alpha'}|^{-1}$, i.e. if $|E_\alpha - E_{\alpha'}| T \ll 1$. If we neglect these oscillating factors then the solution of the infinite set of equations (3.31) reads [47]

$$a^{\text{SA}}_{0 \to n, 0 \to m}(t) = < \chi_\alpha \mid \exp\left(-i \int_{-\infty}^{t} dt\, V(t)\right) \mid \chi_0 >$$

$$= \left\langle u_m\, \psi_n \left| \exp\left(-i \int_{-\infty}^{t} dt'\, V(t')\right) \right| \psi_0\, u_0 \right\rangle. \quad (3.33)$$

The corresponding cross section is given by

$$\sigma^{\text{SA}}_{0 \to n, 0 \to m} = \int d^2\mathbf{b} \left| \left\langle u_m\, \psi_n \left| \exp\left(-i \int_{-\infty}^{+\infty} dt\, V(t)\right) \right| \psi_0\, u_0 \right\rangle \right|^2. \quad (3.34)$$

The amplitude (3.33) has the familiar form of the transition amplitude obtained within the first order of the Magnus (or sudden) approximation [48]. For a discussion of this approximation and its applications to various processes see a review [49], this approximation is very briefly considered also in [50,51]. In the weak interaction limit the exponent in (3.33) can be expanded in series keeping only the first order term in V. The resulting amplitude coincides with the first order amplitude (2.16), provided one drops in the latter the oscillating factors $\exp(-i(\varepsilon_n + \epsilon_m - \varepsilon_0 - \epsilon_0)t)$.

The valuable merit of the sudden approximation is that it preserves unitarity. Using the completeness of the states of the ion and atom it is easy to show that the total probability to find the electronic system of the colliding particles in any of its possible states satisfies the condition

$$P^{\text{SA}}_{\text{tot}}(\mathbf{b}) = \sum_{n,m} \left| a^{\text{SA}}_{0 \to n, 0 \to m}(t) \right|^2 \equiv 1. \quad (3.35)$$

Working within the sudden approximation it is convenient to split the ion–atom interaction according to

$$V = V_{\text{scr}} + \Delta V, \tag{3.36}$$

where

$$V_{\text{scr}} = = -\frac{Z_A}{|\mathbf{R} + \mathbf{r}|} + \left\langle u_0 \left| \sum_{j=1}^{Z_A} \frac{1}{|\mathbf{R} - \boldsymbol{\rho}_j + \mathbf{r}|} \right| u_0 \right\rangle \tag{3.37}$$

and

$$\Delta V = -\sum_{j=1}^{Z_A} \frac{Z_I}{|\mathbf{R} - \boldsymbol{\rho}_j|} + \sum_{j=1}^{Z_A} \frac{1}{|\mathbf{R} - \boldsymbol{\rho}_j + \mathbf{r}|} - \left\langle u_0 \left| \sum_{j=1}^{Z_A} \frac{1}{|\mathbf{R} + \boldsymbol{\rho}_j - \mathbf{r}|} \right| u_0 \right\rangle. \tag{3.38}$$

Here, the term V_{scr} represents the interaction between the electron of the ion and the atom which occupies its initial state. It includes the interaction with the atomic nucleus as well as with the atomic electrons averaged over the atomic ground state u_0. The latter can be considered as an uncorrelated part of the two-center electron–electron interaction. In collisions with heavy atoms $(Z_A \gg 1)$ the interaction V_{scr} can be very strong.

The term ΔV contains the interaction of the electrons of the atom with the nucleus of the ion and the $\mathbf{r}$-dependent part of this term describes the correlated part of the interaction between the electron of the ion and the electrons of the atom.

3.4.1 Elastic Contribution from the Target

Using (3.33) and (3.36)–(3.38), the elastic amplitude $(m = 0)$ for the projectile-electron excitation or loss can be presented as

$$
\begin{aligned}
a_{0 \to n, 0 \to 0}^{\text{SA}} &= \left\langle \psi_n u_0 \left| \exp\left(-\mathrm{i} \int_{-\infty}^{+\infty} V(t) \mathrm{d}t \right) \right| \psi_0 u_0 \right\rangle \\
&= \left\langle \psi_n \left| \exp\left(-\mathrm{i} \int_{-\infty}^{+\infty} V_{\text{scr}}(t) \mathrm{d}t \right) G(\mathbf{r}) \right| \psi_0 \right\rangle,
\end{aligned} \tag{3.39}
$$

where

$$G(\mathbf{r}) = \int \prod_{j=1}^{Z_A} \mathrm{d}\boldsymbol{\rho}_j \, | u_0 |^2 \exp\left(-\mathrm{i} \int_{-\infty}^{+\infty} \Delta V \mathrm{d}t \right). \tag{3.40}$$

Considering the $\mathbf{r}$-dependent part of the interaction $\Delta V(\mathbf{r})$ as a weak perturbation acting on the electron of the projectile, one can show [52] that the elastic contribution to the excitation or loss probability, $P^{\text{SA}}_{0\to n, 0\to 0} = |a^{\text{SA}}_{0\to n, 0\to 0}|^2$, can be approximated by

$$
P^{\text{SA}}_{0\to n, 0\to 0} = \left| \langle \psi_n \left| \exp\left(-i \int_{-\infty}^{+\infty} V_{\text{scr}}(t) dt \right) \right| \psi_0 \rangle \right|^2
$$

$$
\times \left(1 - \sum_{m\neq 0} |\langle u_m | \int_{-\infty}^{+\infty} V_0(t) dt | u_0 \rangle|^2 \right), \qquad (3.41)
$$

where

$$
V_0(t) = \sum_{j=1}^{Z_{\text{A}}} \left(-\frac{Z_{\text{I}}}{|\mathbf{R}(t) - \boldsymbol{\rho}_j|} + \langle \psi_0 \left| \frac{1}{|\mathbf{R}(t) - \boldsymbol{\rho}_j + \mathbf{r}|} \right| \psi_0 \rangle \right) \qquad (3.42)
$$

is the net potential which the projectile in the ground state exerts on the target electrons.

Equation (3.42) has a very simple meaning: the elastic probability is equal to the probability for the projectile electron to make a transition in the field of the target atom in the ground state multiplied by the probability for the target atom to remain in this state during the collision. Within the sudden approximation (3.41) appears rather naturally. On the other hand, such a product of probabilities is often introduced in an ad hoc way, in connection with the independent electron model (see e.g. [2, 21]).

3.4.2 Total Contribution from the Target

Let us now turn to the consideration of the total contribution to the excitation or loss. With the 'partial' cross sections of the form (3.34) the summation over the final states of the atom can be done with the help of the closure relation which yields

$$
\sigma^{\text{SA}}_{0\to n} = \sum_m \sigma^{\text{SA}}_{nm} = 2\pi \int_0^\infty db\, b\, P^{\text{SA}}_{0\to n}(b), \qquad (3.43)
$$

where the total probability for the transition $\psi_0 \to \psi_n$ of the electron of the ion is given by

$$
P^{\text{SA}}_{0\to n}(b) = \int \prod_{j=1}^{Z_{\text{A}}} d\boldsymbol{\rho}_j\, |u_0|^2 \left| \int d\mathbf{r} \psi_n^*(\mathbf{r}) \exp\left(-i \int_{-\infty}^{+\infty} dt\, V \right) \psi_0(\mathbf{r}) \right|^2. \qquad (3.44)
$$

Compared to collisions with light atoms, in collisions with many-electron atoms the strength of the interaction V_{scr} strongly increases while the relative role of the two-center electron correlations diminishes. Therefore, in such a case one can try to omit in (3.44) the term ΔV. This results in the drastic simplification for the transition probability (3.44), where the integral over the coordinates of the atomic electrons reduces to

$$\int \prod_{j=1} d\boldsymbol{\rho}_j \mid u_0 \mid^2 = 1,$$

and from (3.43) and (3.44) we obtain

$$\sigma_{0\to n}^{SA} \simeq \sigma_{0\to n}^{scr} = 2\pi \int_0^\infty db\, b \mid < \psi_n \mid \exp\left(-\frac{i}{v} \int_{-\infty}^{+\infty} dZ\, V_{scr}(\mathbf{r})\right) \mid \psi_0 > \mid^2 .$$

$$(3.45)$$

Two points may be noted here. First, the cross section (3.45) might seem to look like the purely elastic contribution. Moreover, in the limit of a weak ion–atom interaction it indeed goes over into the first order elastic cross section. However, in the case of a strong interaction this cross section actually represents much more than just the elastic part of the projectile-electron excitation/loss cross section. Equation (3.45) describes the cross sections for the transitions of the projectile-electron caused by its interaction with the atom whose electronic density is 'frozen' in space during the effective collision time, which does not imply that the atom will finally remain in its initial state. Therefore, expression (3.45) can still be denoted as the screening cross section. However, one should keep in mind that now, in contrast to the first order treatment, the screening cross section is no longer the elastic cross section because it has been obtained by taking into account transitions to all final states of the atom.

Second, despite the summation performed over all states of the atom, in the limit of a weak ion–atom interaction (3.45) does not reproduce the inelastic contribution to the total cross section. That is in contrast to (3.43)–(3.44) which in this limit show both the elastic and inelastic contributions [52]. The obvious reason is that the term ΔV, which represents the two-center electron–electron correlations, has been neglected in (3.45).

3.5 Glauber Approximation

Collisions between two atomic systems both carrying electrons can be also described by the time-independent Schrödinger equation

$$(E - H_i - H_a - V)\Psi(\mathbf{R}, \mathbf{r}, \{\boldsymbol{\rho}\}) = 0. \qquad (3.46)$$

In (3.46), $\Psi(\mathbf{R}, \mathbf{r}, \{\boldsymbol{\rho}\})$ is the wavefunction describing both the nuclear and electronic motion of the colliding partners (for instance, in the center-of-mass frame), E is the total energy of the colliding particles, $\mathbf{R}$ is the internuclear coordinate which now does not depend explicitly on time, the quantities $\mathbf{r}$ and $\{\boldsymbol{\rho}\} = \{\boldsymbol{\rho}_1, \boldsymbol{\rho}_1, ...\}$ have the same meaning as in the previous subsection,

H_i and H_a are the total Hamiltonians of the free ion and atom, respectively. In (3.46) $V = V(\mathbf{R}, \mathbf{r}, \{\boldsymbol{\rho}\})$ is the total time-independent interaction between the ion and the atom.

The key approximations of the Glauber approach for collisions between composite atomic systems are [9, 54–55]

$$k_i a_0 \gg 1, \; E_i \gg \overline{V} \text{ and } \frac{a_0 \Delta \varepsilon}{v} \ll 1. \tag{3.47}$$

Here $k_i = Mv$ and E_i are the incident momentum and energy of the relative motion, respectively, M is the reduced mass of the relative motion. $\overline{V}$ is a typical potential strength in the problem, a_0 is a typical dimension of the domain, where the interaction occurs between the colliding particles (in our case $a_0 \sim 1$ a.u. is the dimension of the neutral atom), and $\Delta \varepsilon$ is the difference between final and initial internal electron energies of the colliding systems. The first two inequalities in (3.47) represent the "short wavelength" (semiclassical) condition and the "high-energy" requirement, respectively [9, 53]. Both these inequalities are well fulfilled for energetic heavy-particle collisions with velocities of several atomic units and higher. In contrast, the last inequality in (3.47) can be rather restrictive in this region of the collision velocity. This inequality implies that during the collision the inner electronic motion of the colliding particles can be viewed as 'frozen'. The effective collision time can be estimated as $T \sim a_0/v$ and the last inequality in (3.47) then reads: $T \ll \tau_0$, where $\tau_0 \sim \Delta \varepsilon^{-1}$ is an electron transition time in the colliding particles. The latter inequality is also the condition for the applicability of the sudden approximation.

Assuming that the minimum and maximum momenta transferred into the inner motion of the colliding particles are equal to 0 and ∞, the cross section for collisions, in which the ion and the atom make transitions $\psi_0 \to \psi_n$ and $u_0 \to u_m$, respectively, reads

$$\sigma_{nm}^{\text{GA}} = \int d^2\mathbf{b} \, |< \psi_n \, u_m \, | \exp\left(-\frac{i}{v} \int_{-\infty}^{+\infty} dZ \, V(\mathbf{R}, \mathbf{r}, \{\boldsymbol{\rho}\})\right) | \, \psi_0 \, u_0 >|^2. \tag{3.48}$$

Here $\mathbf{R} = \mathbf{b} + \frac{\mathbf{v}}{v}Z$, where $\mathbf{b}$ is interpreted as the impact parameter. In (3.48) the integration over b is implied to run from 0 to b_{max}. In general, the upper limit b_{max} should not be very large, $b_{\text{max}} \ll v/\Delta \varepsilon$, in order that the inner electron motion in the colliding particles can be considered as frozen during the collision and, thus, the Glauber approximation can be applied. Setting $Z = vt$ in the integrand of (3.48) and assuming that, since for ionic transitions ($n \neq 0$) the ion–atom interaction is effectively short ranged, one can take $b_{\text{max}} \to \infty$, we see that expression (3.48) is precisely the result (3.34) obtained in the sudden approximation. Such a coincidence is actually not very surprising because the last relation in (3.47) represents also the condition for the applicability of the sudden perturbation.

3.6 Classical Trajectory Monte Carlo Approach

Both purely quantum mechanical and semi-classical approaches to the projectile–target collisions in general become very complicated when more than two electrons actively participate in the collision. In such a case descriptions based on classical mechanics may represent an alternative.

The applications of classical mechanics to atomic collisions date back to J. Thomson and N. Bohr. With the advent of quantum mechanics the attraction of classical models for atomic collisions had waned but, nevertheless, had never completely disappeared. The renewed interest in descriptions of atomic collisions on the base of classical mechanics started when the Classical Trajectory Monte Carlo method (CTMC) was proposed for atomic collisions in [55]. Since then, a lot of papers has appeared which treat the motion of electrons classically, no matter how small are their de Broglie wave lengths compared to the other characteristic dimensions of the problem (for a brief review of the CTMC in atomic collisions see [56]).

In the simplest case of the projectile–target collisions, when two hydrogen-like systems interact with each other, the CTMC method consists of the following main ingredients. First, the initial condition at $t \to -\infty$ is given by simulating the noninteracting atomic systems by suitably chosen micro-canonical momentum distributions. Second, the dynamical description of the collision between the systems is given by the classical Hamilton equations:

$$\frac{\mathrm{d}\mathbf{r}_a}{\mathrm{d}t} = \frac{\partial H}{\partial \mathbf{p}_a}, \quad a = 1, 2,$$

$$\frac{\mathrm{d}\mathbf{p}_a}{\mathrm{d}t} = -\frac{\partial H}{\partial \mathbf{r}_a}, \quad a = 1, 2,$$

$$\frac{\mathrm{d}\mathbf{R}_a}{\mathrm{d}t} = \frac{\partial H}{\partial \mathbf{P}_a}, \quad a = 1, 2,$$

$$\frac{\mathrm{d}\mathbf{P}_a}{\mathrm{d}t} = -\frac{\partial H}{\partial \mathbf{R}_a}, \quad a = 1, 2. \tag{3.49}$$

Here, $a = 1, 2$ refer to the 'first' and the 'second' hydrogen-like systems, $\mathbf{r}_a$ and $\mathbf{p}_a$ are the coordinates and momenta of the ath electron, $\mathbf{R}_a$ and $\mathbf{P}_a$ are the coordinates and momenta of the ath nucleus and H is the Hamilton function,

$$H = \frac{\mathbf{p}_1^2}{2} + \frac{\mathbf{P}_1^2}{2} - \frac{Z_1}{|\mathbf{R}_1 - \mathbf{r}_1|} + \frac{\mathbf{p}_2^2}{2} + \frac{\mathbf{P}_1^2}{2} - \frac{Z_2}{|\mathbf{R}_2 - \mathbf{r}_2|}$$

$$+ \frac{Z_1 Z_2}{|\mathbf{R}_1 - \mathbf{R}_2|} + \frac{1}{|\mathbf{r}_1 - \mathbf{r}_2|} - \frac{Z_1}{|\mathbf{R}_1 - \mathbf{r}_2|} - \frac{Z_2}{|\mathbf{R}_2 - \mathbf{r}_1|}, \tag{3.50}$$

where Z_a is the charge of the ath nucleus. In energetic ion–atom collisions trajectories of the heavy nuclei are, to a very good approximations, straight lines. This can be explicitly taken into account in (3.49)–(3.50) leading to substantial simplifications.

When, for a given set of the initial conditions, the system of (3.49)–(3.50) has been solved, its solutions have to be sorted out in order to interpret the result in terms of ionization/loss and excitation. For instance, if at $t \to \infty$ it is found that both the electrons are far away from the nuclei, this event is interpreted as mutual projectile–target ionization.

Since in the CTMC one deals with microcanonical distributions of the initial electron orbits, very many individual collision trajectories must be considered in order to obtain sufficient statistics on various cross sections. Compared to the total cross sections, calculations of the differential cross sections demand even more trajectories to be considered.

The advantage of the CTMC approach is the relative simplicity of the classical equations (3.49)–(3.50). This, in particular, enables one to take directly into account all the pair-wise interactions between the colliding atomic particles which may have initially several electrons.

The disadvantages of this approach are of fundamental character and are directly related to the limited ability of classical physics to yield reasonable results when applied to treat atomic objects. For instance, it is well known that an hydrogen atom, according to classical physics, is unstable because of the emission of radiation by the electron moving around the nucleus. It is only if we 'forget' about the interaction with the radiation field, becomes it possible to model the atom classically.

Moreover, according to classical physics, any bound state of an atom with two and more electrons will not be stable also because of the Coulomb interaction between the electrons. Since the latter belongs to the same kind of interaction which also governs ion–atom collisions, it cannot simply be ignored. Therefore, certain 'tricks' are introduced in CTMC calculations in order to allow atoms with two and more electrons to 'survive' in the absence of the collision.

Nevertheless, after such tricks are implemented, applications of the CTMC approach may enable one to get a reasonable agreement with experiment in situations when quantum theories turn out to be too complicated to apply.

In the case of collisions between a bare nucleus and a hydrogen-like system the relationship between classical and quantum descriptions was studied in [57]. In particular, it has been shown in [57] that, as quite expected, the classical description fails to yield reasonable results for collisions with large impact parameters. When such collisions dominate in the process, as is the case at very high impact energies where the collision velocity greatly exceeds the typical electron velocity in the atomic ground state, the CTMC approach clearly fails to reproduce experimental data (see, for instance, [58,59]).

3.7 Projectile Electron Loss. Comparison with Experiment

3.7.1 Total Loss Cross Section

In Fig. 3.2 experimental and theoretical results are shown for the total cross section of the electron loss from 3.5 MeV $He^+(1s)$ projectiles colliding with different atomic targets ranging from Ne ($Z_A = 10$) to Xe ($Z_A = 54$). The impact energy corresponds to the collision velocity of 5.94 a.u. and, thus, the parameter Z_A/v varies between 1.68 and 9.1.

Three sets of theoretical results are shown in Fig. 3.2. Results shown by diamonds were obtained in [42] by using the first Born approximation and considering the elastic target contribution only.

The results of [43] are displayed by circles. These results were obtained by applying a coupled channel approach to the electron transitions in the projectile. It was assumed in [43] that the atom has a rigid structure, i.e. the

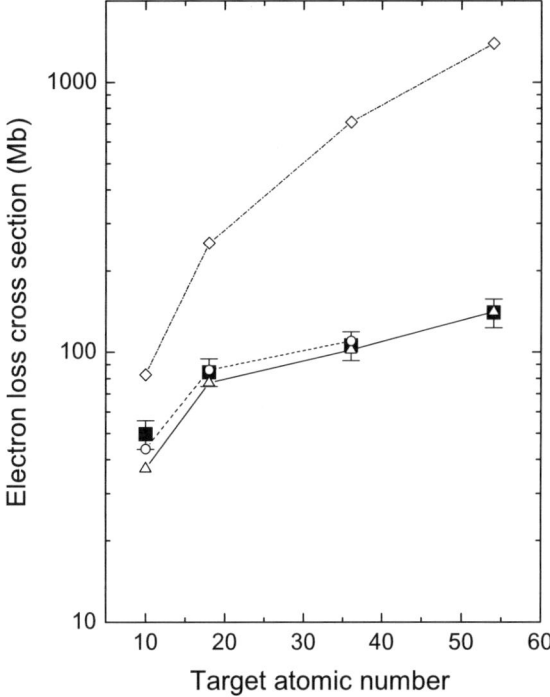

Fig. 3.2. The total cross section for the electron loss from incident 3.5 MeV $He^+(1s)$ projectiles colliding with Ne, Ar, Kr and Xe atoms. *Solid squares* with error bars are experimental data from [42]. Results of the coupled-channel calculations of [43] are shown by *circles*. Results of the calculations using (3.45) are displayed by triangles. Results of the first order calculations for the screening target mode are shown by diamonds. The lines connecting the theoretical results are just to guide the eye.

electrons of the atom do not undergo transitions and are always in the ground atomic state. This enabled the authors of [43] to replace in their calculations the atom by an external potential acting on the electron of the projectile in the collision.

The results of [47] are shown by triangles. In order to obtain the loss cross sections the authors of [47] used the sudden approximation and their calculations were based on (3.45).

All the three calculations have one common point: the contribution to the loss cross section arising from the 'active' behavior of the atomic electrons in the collision – the two-center electron–electron correlations – were neglected.

One immediately sees in Fig. 3.2 that the first order calculation strongly fails. Already in collisions with Ne the results of this calculation substantially overestimate the experimental data. When the atomic number of the target increases, these results become even more inaccurate: for collisions with Xe the first order result for the elastic target mode is by an order of magnitude larger than the experimental total loss cross section.

Compared to the first order results, the other two theoretical treatments certainly represent drastic improvements. Indeed, their results differ by no more than about 30% from the experimental data.

However, both these treatments fully ignore the 'active' role which the electrons of the atom might play in the projectile-electron loss process. This seems to be the main reason why the considerable differences between the experiment and theory exist in the case of the Ne target. Compared to the other target atoms Ne has the substantially smaller number of electrons and this increases the relative role of the two-center electron correlations in the loss process.

The similarity in the magnitudes for the loss cross sections obtained by using the sudden approximation and the coupled-channel approach deserves a brief discussion. At the first glance these two approaches seem to be quite different. Indeed, the cross section (3.45) was obtained in the sudden approximation by summing over all possible final target states[4] whereas the coupled channel approach of [43] assumed the rigid target structure which is not influenced by the collision and, hence, might seem to account only for the contribution of the elastic target mode.

However, one can obtain the same form for the sudden cross section (3.45) by considering that the target has a rigid inner structure, which cannot be changed by any collision. In this case the problem is reduced to the projectile-electron transitions in collisions with a structureless particle which creates a short-range potential V_{scr}. Then, if one applies the sudden approximation, the excitation or loss cross sections are given by precisely the same (3.45). If one now takes into account that the sudden approximation can be derived from the coupled-channel formalism (see Sects. 3.3 and 3.4), then the close values for the loss cross sections, obtained with the sudden and coupled-channel

---

[4] Although neglecting the two-center electron correlations.

approaches, are already looking quite natural. Moreover, the coupled-channel calculations of [43] can now be viewed as calculations for the total loss rather than calculations for the elastic target mode alone, as they were regarded in [43]. This point allows one to understand the unexpectedly good agreement between the results of the coupled-channel calculations for the "elastic" target mode and the experimental data on the total loss cross sections.

3.7.2 Loss Cross Section Resolved over the Final Charge States of the Target

The results in Fig. 3.2 show the total loss cross sections but do not provide any information about what happens with the target atom in such collisions. More insight into the dynamics of the ion–atom collision process is obtained when the projectile-electron loss cross section is resolved over the final charge states of the target. Such cross sections were studied, both experimentally and theoretically, in [60] for collisions of 1–4 MeV $He^+(1s)$ ions with Ne atoms and are shown in Fig. 3.3.

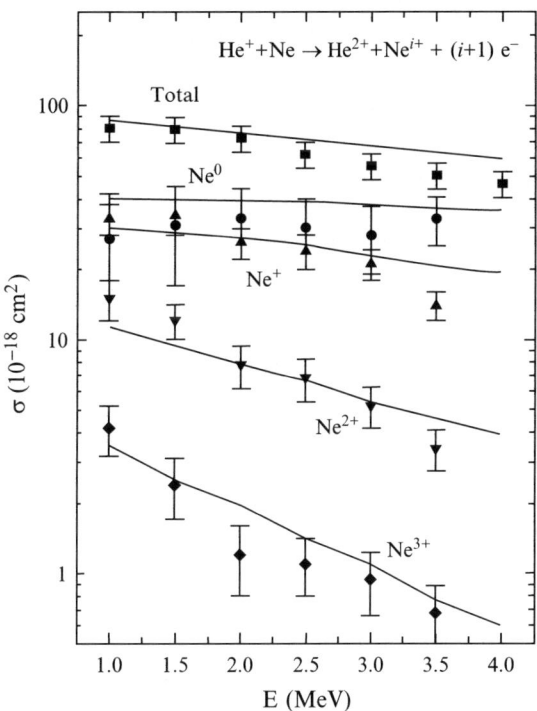

Fig. 3.3. The cross section for projectile-electron loss in $He^+(1s)$–Ne collisions. The cross section is given as a function of the impact energy and is resolved over the final charge states of the target in the interval Ne^0–Ne^{3+}. Symbols with error bars denote the experimental results. Curves show results of the (n)CTMC calculations. From [60].

According to the results of this figure, the main contribution to the projectile-electron loss is given by collisions in which the target atoms either keep the net charge zero or get singly ionized. Besides, at the lowest impact energies considered in [60], the noticeable contribution to the loss cross section arises also from collisions resulting in the double ionization of Ne atoms. However, when the impact energy increases, the relative role of this reaction channel rapidly diminishes. Results of the CTMC calculations performed by the authors of [60] are in surprisingly good agreement with their experimental data.

Compared to the exploration of the total cross sections, much more information about the various mechanisms, which govern the projectile–target collisions, can be obtained by studying differential cross sections. During the last two decades experimental techniques have reached quite a high level of sophistication [61,62]. In particular, in the study of ion–atom collisions these techniques enable one to detect in coincidence electrons emitted from the projectile and the target and the target recoil ions. In principle, experimental explorations of the projectile-electron excitation and loss processes have become possible even at the most basic level of the fully differential cross sections.

In the next subsections we shall consider some of the differential cross sections for the projectile-electron loss process which have been recently studied experimentally and theoretically.

3.7.3 Longitudinal Momentum Distribution of Target Recoil Ions

The electron loss from fast highly charged hydrogen-like projectiles, occurring simultaneously with the target ionization, was studied in [63,64], where the cross section differential in the longitudinal component $P_\parallel$ of the target recoil momentum, $d\sigma/dP_\parallel$, was measured.

In Fig. 3.4 experimental data from [64] for the reaction 75 MeV $O^{7+}(1s)$ $+He(1s^2) \to O^{8+} + He^+ + 2e^-$ are shown. This impact energy corresponds to a collision velocity of $v = 13.7$ a.u. The collision velocity is sufficiently high, $Z_A/v \ll 1$, and the atomic number of the target is much less than that of the projectile. This means that the total cross section for the projectile-electron loss is well described by the first order theory in the projectile–target interaction.

However, the latter strongly fails in an attempt to describe the loss cross section differential in the longitudinal component $P_\parallel$ of the target recoil momentum (see Fig. 3.4). Indeed, the first order theory predicts a single maximum in the spectrum of the recoil ions which is centered close to the zero momentum. This is in contrast to the experimental data which show not only the maximum at the small recoil momentum but also one more maximum at substantially larger values of $P_\parallel$.

The reason of the failure of the first order theory is that, according to this theory, the simultaneous transitions of the electrons in the projectile and the

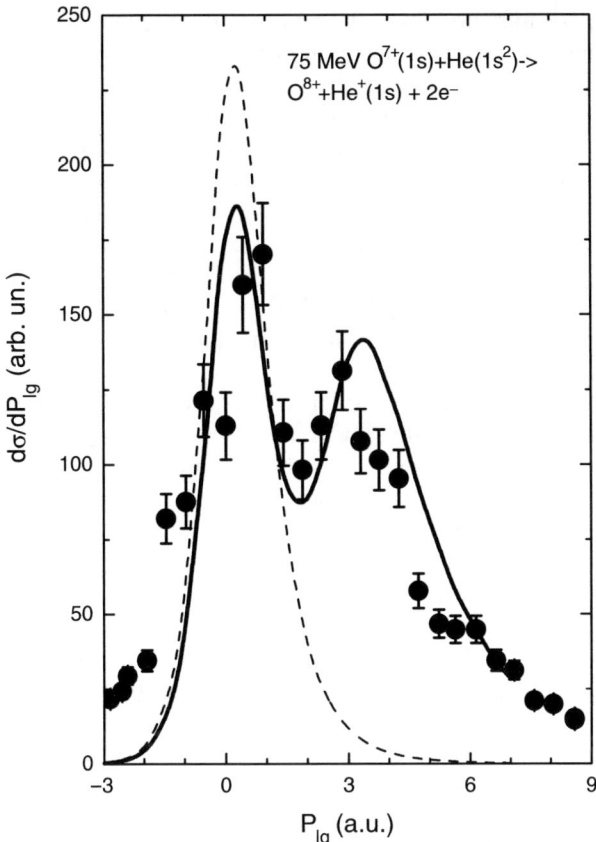

Fig. 3.4. Longitudinal momentum distribution of the recoil ions. *Solid curve*: calculation using the amplitude (3.17). *Dash curve*: first Born calculation. *Circles with error bars*: experimental results reported in [64]. From [40].

target may only occur via the two-center electron–electron interactions.[5] In order to remove the tightly bound electron from the projectile it is necessary to transfer to this electron a sufficiently large momentum. If this momentum is transferred via the two-center electron–electron interaction, then the electron of the target gets a recoil. On the target scale this recoil is strong enough in order that the target electron leaves the atom with the longitudinal momentum approximately equal to the longitudinal component of the total momentum transfer to the atom. In such a collision the residual target ion acts merely as a spectator carrying almost no longitudinal momentum. This is the basic

---

[5] Note that this point was already emphasized when we discussed the inelastic target mode (Sect. 2.1.2 of Chap. 2) and also when the distorted-wave approach was considered (see Sect. 3.2).

physics lying behind the one-peak structure of the first order prediction for the longitudinal momentum spectrum of the target recoil ions.

However, as was already discussed in Sects. 3.1 and 3.2, there is another mechanism which may also cause the simultaneous transitions of the projectile and target electrons. Within this mechanism the electron of the projectile is removed due to its interaction with the (screened) target nucleus whereas the electron of the target is transferred to the continuum via the interaction with the nucleus of the projectile. Such a mechanism is predicted by the second order theory in the projectile–target interaction. This mechanism also appears very naturally in the general distorted-wave model (see (3.14)–(3.14)) discussed in Sect. 3.2, as well as in its particular case – the symmetric eikonal model – described in Sect. 3.2.1.

Within this mechanism the origin of the second maximum in the recoil momentum spectrum is easily understood [64] as arising due to the large recoil transferred to the atomic core in the process of the removal of the tightly bound electron from the projectile via the direct interaction between the core and the projectile electron.

Results of calculations performed in [40] based on the symmetric eikonal amplitude (3.17) are shown in Fig. 3.4 by the solid curve. In these calculations the helium target was regarded as a hydrogen-like ion, where the 'active' electron moves in the field of the atomic core with an effective core charge of $Z_A = 1.69$. The loss of a tightly bound electron from $O^{7+}(1s)$ projectiles occurs in collisions with momentum transfers much larger than typical electron momenta in the ground state of helium. Therefore, the effective charge of the atomic core, which is 'seen' by the projectile electron in the collision, was set to 2.

It is seen in Fig. 3.4 that the calculations of [40] do show the two-peak structure and are in reasonably good agreement with the experimental data.

3.7.4 Two-Center Interactions in $3.6\,\mathrm{MeV\,u^{-1}}$ $C^{2+} + He \rightarrow C^{3+}$ $+ He^+ + 2e^-$ Collisions

In [65] the reaction $C^{2+} + He \rightarrow C^{3+} + He^+ + 2e^-$ was studied. The impact energy was $3.6\,\mathrm{MeV\,u^{-1}}$ corresponding to the collision velocity of 12 a.u. One of the main goals of [65] was to learn whether e-$2e$ processes on ions, $e^- + A^{n+} \rightarrow A^{(n+1)+} + 2\,e^-$, can be experimentally explored in ion–atom collisions using the atoms as a source of a very dense beam of 'quasi-free' electrons.

In particular, in [65] an attempt was made to separate the loss events, appearing due to the interaction of the projectile electron with the active electron of the target, from the loss events, which are caused by the interaction of the projectile electron with the atomic core. For the former events it was expected that their signature should be a strong correlation between the momenta of the electrons emitted from the projectile and target. In the latter events such a correlation was expected to be between the momenta of the electron of the projectile and the target recoil ion.

The conditions $|\mathbf{p}| > |\mathbf{P}|$ and $|\mathbf{p}| < |\mathbf{P}|$, where $\mathbf{p}$ is the momentum of the electron emitted from the target and $\mathbf{P}$ is the momentum of the target recoil ion, were chosen in [65] as approximate criteria for these two types of the collision events.

Perhaps, the simplest way to look into the correlations is to consider the projections of the momenta $\mathbf{p}$, $\mathbf{P}$ and $\mathbf{k}$, where the latter is the momentum of the electron emitted from the projectile, on a plane perpendicular to the collision velocity $\mathbf{v}$. Figure 3.5 shows experimental and theoretical results of [65] for the correlation between the azimuthal angles $\Phi(n,e) = \arccos(\mathbf{P}_\perp \cdot \mathbf{k}_\perp / P_\perp k_\perp)$ and $\Phi(e,e) = \arccos(\mathbf{p}_\perp \cdot \mathbf{k}_\perp / p_\perp k_\perp)$, where $\mathbf{a}_\perp = \mathbf{a} - (\mathbf{a} \cdot \mathbf{v})\mathbf{v}/v^2$.

It is seen in the figure that the conditions $|\mathbf{p}| > |\mathbf{P}|$ and $|\mathbf{p}| < |\mathbf{P}|$ indeed lead to two qualitatively different emission patterns. For the events, which correspond to $|\mathbf{p}| > |\mathbf{P}|$, the emission pattern along $\Phi(e,e)$ is concentrated in the narrow interval of angles around $\Phi(e,e) = 180°$, but is almost a constant

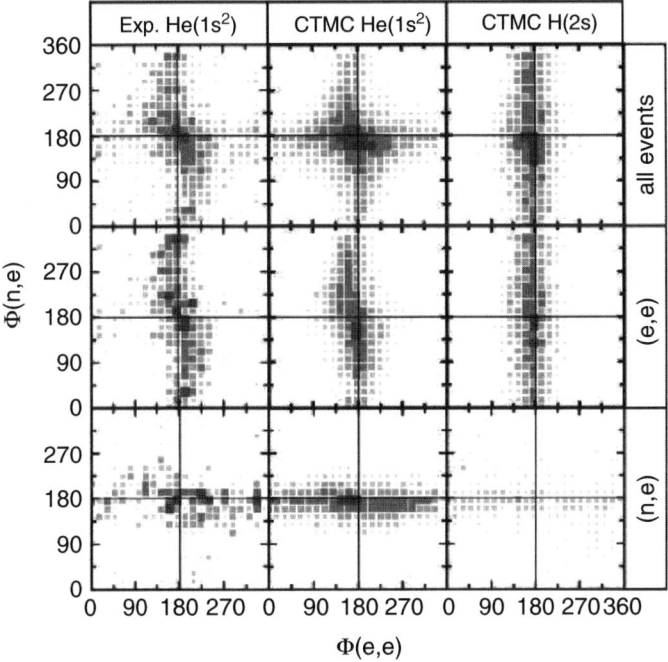

Fig. 3.5. $3.6\,\mathrm{MeV\,u^{-1}}$ $\mathrm{C^{2+}} + \mathrm{He} \rightarrow \mathrm{C^{3+}} + \mathrm{He^+} + e^- + e^-$ collisions. The azimuthal angle $\Phi(n,e)$ between the momenta of the electron emitted from the projectile and the target recoil ion $\mathrm{He^+}$ versus the azimuthal angle $\Phi(e,e)$ between the momenta of the electrons emitted from the projectile and the target. *Left column:* experiment. *Middle column:* CTMC. *Upper row:* all events. *Middle row:* events with $|\mathbf{p}| > |\mathbf{P}|$. *Lower row:* events with $|\mathbf{p}| < |\mathbf{P}|$. For more information the right column shows CTMC results for $3.6\,\mathrm{MeV\,u^{-1}}$ $\mathrm{C^{2+}} + \mathrm{H(2s)} \rightarrow \mathrm{C^{3+}} + p^+ + e^- + e^-$ collisions. From [65].

along $\Phi(n,e)$. In contrast, for the events selected by the condition $|\mathbf{p}| < |\mathbf{P}|$ we observe exactly the opposite. Now the emission pattern strongly depends on $\Phi(n,e)$, with most of the events concentrated close to $\Phi(n,e) = 180^0$, but is a very smooth function of $\Phi(e,e)$. We observe, roughly speaking, that compared to the previous case, the emission pattern in the plane $\Phi(n,e)$-$\Phi(e,e)$ was rotated by $90°$.

Because of the relatively high impact velocity and not very different values of the binding energies of the outer electrons in C^{2+} and the electrons in helium, the momentum transferred to the target is on overall not much larger than the typical momentum of the target electrons in the ground state. Therefore, in the case of the carbon-helium collisions the active electron of the target atom cannot be regarded as quasi-free and the experiment of [65] cannot be really viewed as exploring e^--2e^- processes on C^{2+}.

In this respect another collision system – 3.6 MeV u^{-1} C^{2+} + H(2s) → C^{3+} + p^+ + $2e^-$, which was considered in [65] using the CTMC approach – is much more suitable to mimic the e^--2e^- process on C^{2+} since the binding energy of the H(2s) target is much smaller. Indeed, it is seen in the right column of Fig. 3.5 that in this case the emission pattern is almost fully determined by the two-center electron–electron correlation whereas the target nucleus in practically all collisions acts merely as a spectator.

3.7.5 Mutual Electron Removal in 0.2 MeV H^- + He Collisions

Compared to neutral atoms and positively charged ions, negative ions represent an atomic system possessing qualitatively new properties [66]. The electron removal from negative ions by fast collisions with bare nuclei was studied, experimentally and theoretically, in a number of publications (see, for instance, [67–70] and references therein).

If a negative ion collides with an atom the electron detachment from the negative ion can proceed simultaneously with ionization of the atom. An example of such an detachment-ionization process was considered in [71], where the reaction 0.2 MeV H^- + He → H^0 + He^+ + $2\,e^-$ was investigated in some detail.

The collision system consists of six particles. In the past it had been found out that the CTMC method, which enables one to deal relatively easily with few-particle systems, was not successful when applied to collisions with negative ions [67].

In a quantum consideration it is difficult to explicitly include all these particles. Therefore, in an attempt to analyze the experimental data, it was assumed in [71] that both helium and negative hydrogen can be regarded as consisting of just two particles: the 'active' electron and the core. The interaction of the active electron with the core of its parent atomic system was described by using a model potential.

The impact energy corresponds to a collision velocity of about 2.8 a.u. This velocity value is certainly not sufficiently large in order that the sudden

approximation can be applied. The distorted-wave models, discussed in
Sect. 3.2, are also not expected to yield good results in this case. Therefore,
an attempt of theoretical analysis undertaken in [71] was based on two very
simplified models: (a) the first order approximation for H^-–He collisions and
(b) the continuum-distorted-wave-eikonal-initial-state approximation [33], assuming that the action of H^- on the active electron of He can in some cases
be approximated by that of an equivelocity antiproton.

Results of this model calculations are shown in Figs. 3.6 and 3.7, where
they are compared with the corresponding experimental data.

Figure 3.6 shows the spectrum of the electrons emitted from H^- and He
given as a function of the longitudinal component $p_{ez} = \mathbf{p}_e \cdot \mathbf{v}/(p_e v)$ of the electron momentum $\mathbf{p}_e$. The experimental data show that this spectrum consists
of two parts which are relatively well separated. It is, therefore, plausible to
interpret these parts as related mainly either to the emission from the target
($p_{ez} \lesssim 1$ a.u.) or from the projectile ($p_{ez} \gtrsim 1$ a.u.).

It is seen in this figure that the first order calculation reproduces rather
well the longitudinal spectrum of the electrons emitted from the projectile.

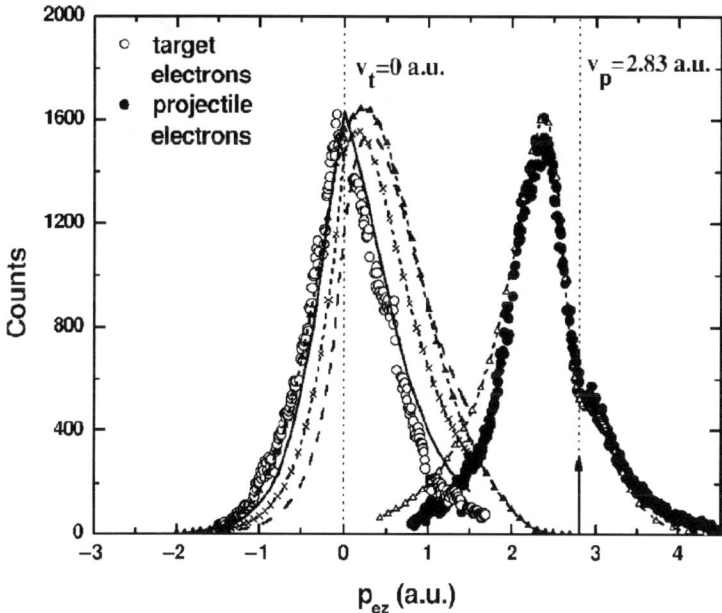

Fig. 3.6. Longitudinal momentum spectra of electrons emitted in 0.2 MeV H^- + He
$\rightarrow H^0 + He^+ + e^- + e^-$ collisions. *Open* and *solid circles* are the experimental data.
Triangles connected by *dash curves* are results of the first order calculation for the
H^- + He collisions. Besides, *solid* and *dash curves* show continuum-distorted-wave-
eikonal-initial-state results for the ionization of helium by the impact of 0.2 MeV
antiproton and proton, respectively, and *cross–dash* curve displays the first order
results for the last two systems. From [71].

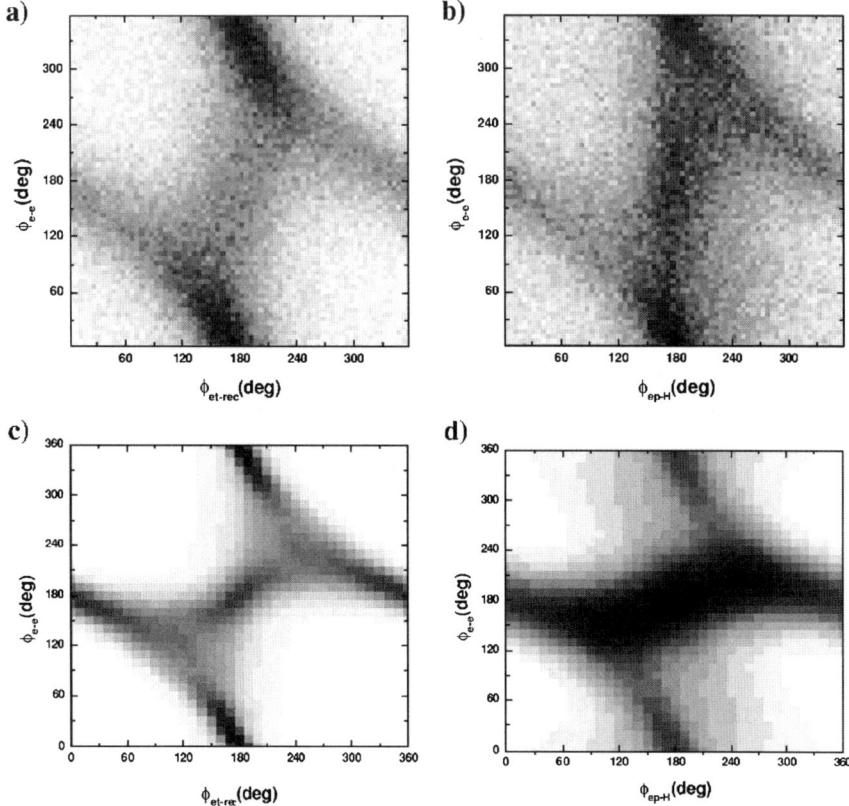

Fig. 3.7. $0.2\,\text{MeV}\ H^- + \text{He} \rightarrow H^0 + \text{He}^+ + 2\,e^-$ collisions. The angle between the ejected electrons ϕ_{e-e} versus the angle between the electron emitted from the target and the target recoil ion He^+ $\phi_{e_t-\text{rec}}$ [panel (a)] and versus the angle between the electron emitted from the projectile and the projectile core ϕ_{e_p-H} [panel (b)] in the azimuthal plane. Panels (c) and (d) show the corresponding theoretical spectra calculated in the first order approximation in the projectile–target interaction. From [71].

In particular, it describes the bump-like structure observed in the experimental spectrum at $p_{ez} \simeq v$. According to the analysis of [71] this structure is connected with a short-range character of the potential which binds the active electron in H^-.

However, the same first order calculation obviously fails to reproduce that part of the longitudinal spectrum which is related to the emission from the target. For that part of the spectrum a much better agreement between experiment and theory is obtained, if the action of the H^- on the active electron of helium in the collision is modeled by that of an equivelocity antiproton and the CDW-EIS approximation is applied.

It follows from Fig. 3.7 that the first order theory also leads to mixed results when applied to describe the relative strength of the correlations between the various particles by projecting the momenta of these particles onto a plane perpendicular to the collision velocity (azimuthal plane).

It is seen in this figure that the first order results are rather similar to the experimental data when ϕ_{e-e} and ϕ_{e_t-rec} variables are considered, where ϕ_{e-e} is the angle between the ejected electrons in the azimuthal plane and ϕ_{e_t-rec} is the angle in this plane between the electron emitted from the target and the target recoil ion He^+. However, the same first order model strongly fails to describe the experiment in the case when ϕ_{e-e} and ϕ_{e_p-H} are chosen as variables, where ϕ_{e_p-H} is the angle between the electron emitted from the projectile and the projectile core in the azimuthal plane.

The possible reasons for the failure of the first order approximation are discussed in [71], where the interested reader is referred to.

Part II

Relativistic Collisions

4

Introduction to Relativistic Collisions

The space and time used to describe nonrelativistic collisions are the classical space and absolute time; transitions between different inertial reference frames are mediated by a well known Galilean transformation. The space and time, in which relativistic collisions take place, is the four-dimensional flat space–time of the special theory of relativity.

The dynamics of nonrelativistic ion–atom collisions is fully described by the corresponding Schrödinger equation. This wave equation is of Hamiltonian type and explicitly contains all the interactions between the elementary particles (electrons and nuclei) the ion and atom are composed of, including interactions acting both between and within the colliding atomic particles.

In relativistic collisions such a 'single-equation' description is possible only if the fields acting on the colliding particles can be regarded as external perturbations, which themselves are not influenced by the collision. For instance, in collisions between a bare nucleus and an hydrogen-like ion the behavior of the electron can be treated by using the Dirac equation in which the interactions between the electron and the nuclei are taken as independent of the electron motion.

In ion–atom collisions, in which both the ion and atom carry electrons, the fields acting on the colliding particles in general cannot be regarded as external. If the latter is the case, the description of ion–atom collisions cannot be based merely on the Dirac equation. Instead, it has to include self-consistent considerations for charged particles and the electromagnetic field.

In this chapter we shall very briefly consider the special theory of relativity and the Maxwell and Dirac equations.

4.1 Elements of the Special Theory of Relativity

In this section we shall remind of some basic facts related to the special theory of relativity.

4.1.1 The Lorentz Transformation

Let us consider two inertial reference frames, K and K', which move with respect to each other with a constant velocity. Let O be the origin of the frame K and the coordinate axes in this frame be (X, Y, Z). Similarly, O' and (X', Y', Z') are the origin and the coordinate axes in the frame K'. Both these frames are also equipped with clocks which measure the time in these frames.

For simplicity we shall assume that the coordinate axes of the frames K and K' are chosen in a such way that the axes Z and Z' coincide and the axes X and Y are parallel to the axes X' and Y', respectively. With respect to the frame K the frame K' moves with a constant velocity $\mathbf{V} = (0, 0, V)$ along the Z-axis of the frame K. The clocks in both frames are set to zero when the origins O and O' coincide.

Let (x, y, z, t) be the space coordinates and time of some physical event observed in the frame K. Let (x', y', z', t') be the space coordinates and the time of the same event but viewed in the frame K'. According to the special theory of relativity the coordinates and time of the event in the frames K and K' are related by (see e.g. [7,8])

$$
\begin{aligned}
x' &= x, \\
y' &= y, \\
z' &= \gamma_V (z - Vt), \\
t' &= \gamma_V (t - Vz/c^2),
\end{aligned}
\tag{4.1}
$$

where c is the speed of light, which is independent of a reference frame, and $\gamma_V = 1/\sqrt{1 - V^2/c^2}$. The transformation (4.1) is referred to as the Lorentz transformation. The inverse Lorentz transformation is obtained from (4.1) by replacing $V \to -V$

$$
\begin{aligned}
x &= x', \\
y &= y', \\
z &= \gamma_V (z' + Vt'), \\
t &= \gamma_V (t' + Vz'/c^2).
\end{aligned}
\tag{4.2}
$$

Equations (4.1) and (4.2) represent the simplest case of the Lorentz transformation. If the axes in the frames K and K' are chosen as before, but the components of the velocity $\mathbf{V}$ of the frame K' in the frame K are all nonzero, the corresponding (more general) Lorentz transformation is given by (see [7])

$$
\begin{aligned}
\mathbf{r}' &= \mathbf{r} + (\gamma_V - 1)\mathbf{V}\, \mathbf{V} \cdot \mathbf{r} - \gamma_V \mathbf{V} t, \\
t' &= \gamma_V \left(t - \mathbf{V} \cdot \mathbf{r}/c^2\right).
\end{aligned}
\tag{4.3}
$$

In order to distinguish the transformations (4.1) and (4.3) the former is often referred to as a Lorentz boost.

4.1.2 Four-Dimensional Space and Four-Vectors

The fact of fundamental importance is that the Lorentz transformation leaves unaltered the quadratic form $s^2 = c^2t^2 - x^2 - y^2 - z^2$: $c^2t^2 - x^2 - y^2 - z^2 = c^2t'^2 - x'^2 - y'^2 - z'^2$. Similarly to a vector $\mathbf{r} = (x, y, z)$ in the three-dimensional (Euclidean) space, whose length r is obtained from $r^2 = x^2 + y^2 + z^2$, the quadratic form $s^2 = c^2t^2 - x^2 - y^2 - z^2$ can be viewed as the squared length of the four-vector (ct, x, y, z) in the four-dimensional space–time. Since rotations do not vary the length of a vector, Lorentz transformations (4.1)–(4.3) can be regarded as rotations in the four-dimensional space–time.

In contrast to the square of the length in the Euclidean space, given by $x^2 + y^2 + z^2$, the quadratic form $s^2 = c^2t^2 - x^2 - y^2 - z^2$ contains terms with both positive and negative signs.[1] In order to account for the latter, two different forms of a given four-vector are introduced: the contravariant vector, defined by $x^\mu = (x^0, x^1, x^2, x^3) = (ct, x, y, z)$, and the covariant vector $x_\mu = (x_0, x_1, x_2, x_3) = (ct, -x, -y, -z)$. With these two vectors the quadratic form is given by $s^2 = x_\mu x^\mu = x^\mu x_\mu$, where $\mu = 0, 1, 2, 3$ and the summation over the repeated Greek indices is implied, and can be interpreted as a four-scalar product $x \cdot x$.

According to the special theory of relativity the space and the time together form a four-dimensional flat space–time continuum. The geometrical properties of this space–time are described by the metric tensor g which is a symmetric tensor of rank 2. The covariant and contravariant versions of this tensor coincide, $g_{\mu\nu} = g^{\mu\nu}$, and their elements presented in a matrix form read

$$g_{\mu\nu} = g^{\mu\nu} = \begin{pmatrix} 1 & 0 & 0 & 0 \\ 0 & -1 & 0 & 0 \\ 0 & 0 & -1 & 0 \\ 0 & 0 & 0 & -1 \end{pmatrix}. \tag{4.4}$$

The metric tensor with mixed components is simply equal to the Kronecker' δ-symbol: $g_\mu{}^\nu = g^\mu{}_\nu = 1$ for $\mu = \nu$ and $g_\mu{}^\nu = g^\mu{}_\nu = 0$ for $\mu \neq \nu$.

The metric tensor enables one to calculate the scalar product of two four-vectors: $x \cdot y = g_{\mu\nu} x^\mu y^\nu = g^{\mu\nu} x_\mu y_\nu$. In particular, with the help of the metric tensor the element of length in the four-dimensional space is expressed as $ds^2 = dx_\mu dx^\mu = g_{\mu\nu} dx^\mu dx^\nu = g^{\mu\nu} dx_\mu dx_\nu$, where $dx_0 = dx^0 = cdt$ and $dx_j = -dx^j$ with $j = 1, 2, 3$.

The element ds^2 is invariant under transformations, which combine (a) rotations in the three-dimensional space and in the space–time and (b) shifts in the space–time (translations):

$$x'^\alpha = \Lambda^\alpha{}_\beta x^\beta + y^\beta. \tag{4.5}$$

---

[1] Note that this is not the case if the four-vector is defined as (x, y, z, ict) and, thus, $s^2 = x^2 + y^2 + z^2 + (ict)^2$ In this book, however, such a definition will not be used.

The transformations (4.5) form a group. The matrix Λ in (4.5) satisfies the orthogonality conditions

$$\Lambda^{\mathrm{T}} g \Lambda = g, \tag{4.6}$$

where Λ^{T} is the transposed matrix. Note, that for a Lorentz boost in the Z-direction one has [7]

$$\Lambda_{\mu}{}^{\nu} = \begin{pmatrix} \gamma_V & 0 & 0 & -\frac{V}{c}\gamma_V \\ 0 & 1 & 0 & 0 \\ 0 & 0 & 1 & 0 \\ -\frac{V}{c}\gamma_V & 0 & 0 & \gamma_V \end{pmatrix}. \tag{4.7}$$

Obviously, that the four-vector formed using the space and time coordinates of a physical point is not the only four-vector of physical significance. There are other quantities which under the transformation from one reference frame to another behave like the four-vector (ct, x, y, z). In general, if a set of four quantities (a^0, a^1, a^2, a^3) behaves under the Lorentz transformation like the space–time vector (ct, x, y, z), i.e.

$$a'^{\alpha} = \frac{\partial x'^{\alpha}}{\partial x^{\beta}} a^{\beta},$$
$$a'_{\alpha} = \frac{\partial x'^{\beta}}{\partial x^{\alpha}} a_{\beta}, \tag{4.8}$$

these four quantities form a four-vector. By multiplying the left and right hand parts of the above two equations we see that the length of the vector is invariant: $a'_{\alpha} a'^{\alpha} = a_{\alpha} a^{\alpha}$.

For instance, the energy ε and the three-momentum $\mathbf{p} = (p_x, p_y, p_z)$ of a particle can be combined into the energy–momentum four-vector $p^{\alpha} = (\varepsilon/c, \mathbf{p})$ ($p_{\alpha} = (\varepsilon/c, -\mathbf{p})$). If the particle moves with a velocity $\mathbf{u} = (u_x, u_y, u_z)$ with respect to a frame K its momenta and energy in this frame are given by

$$\mathbf{p} = m_0 \gamma_u \mathbf{u},$$
$$\varepsilon = m_0 \gamma_u c^2, \tag{4.9}$$

where $\gamma_u = 1/\sqrt{1 - u^2/c^2}$ and m_0 is the rest mass of the particle. For the squared length of the four-momentum one obtains $p_{\alpha} p^{\alpha} = \varepsilon^2/c^2 - \mathbf{p}^2 = m_0^2 c^2$.

One more example of a four-vector, which is of great physical importance, is the four-current which is built using the charge density ρ and the current density $\mathbf{j}$:

$$j^{\alpha} = (c\rho, \mathbf{j}),$$
$$j_{\alpha} = (c\rho, -\mathbf{j}). \tag{4.10}$$

Partial derivatives in the three-dimensional space and time can also be combined into a four-vector. Indeed, since the differential of a four-scalar function $F(x)$, given by

$$dF = \frac{\partial F}{\partial x^\alpha}dx^\alpha = \frac{\partial F}{\partial x_\alpha}dx_\alpha,$$

is also a four-scalar, it is not difficult to convince oneself that

$$\partial^\alpha \equiv \frac{\partial}{\partial x_\alpha} = \left(\frac{\partial}{c\partial t}, -\nabla\right),$$

$$\partial_\alpha \equiv \frac{\partial}{\partial x^\alpha} = \left(\frac{\partial}{c\partial t}, \nabla\right) \qquad (4.11)$$

represent contravariant and covariant four-vectors, respectively.

4.1.3 Relativistic Addition of Velocities

Let a point move in the frame K' with a velocity $\mathbf{u}'$. Using the transformation (4.1) one can easily calculate the velocity of this point in the frame K in which the frame K' moves with a velocity V along the Z-axis. The result is

$$u_z = \frac{dz}{dt} = \frac{\gamma_V(dz' + Vdt')}{\gamma_V(dt' + Vdz'/c^2)} = \frac{u_z' + V}{1 + Vu_z'/c^2},$$

$$u_x = \frac{dx}{dt} = \frac{u_x'}{\gamma_V(1 + Vu_z'/c^2)},$$

$$u_y = \frac{dy}{dt} = \frac{u_y'}{\gamma_V(1 + Vu_z'/c^2)}. \qquad (4.12)$$

4.1.4 Transformation of Energy–Momentum

As was already mentioned, the energy and three-momentum of a particle form a four-vector which is referred to as the four-momentum of the particle. Let in the frame K the four-momentum be given by $p^\mu = (\varepsilon/c, \mathbf{p})$. This four-momentum can be recalculated to any other inertial frame according to the Lorentz transformation. For instance, in the frame K', which moves in the frame K with a velocity $\mathbf{V} = (0, 0, V)$, this four-momentum is represented by $p'^\mu = (\varepsilon'/c, \mathbf{p}')$ with the components given by

$$p_x' = p_x,$$
$$p_y' = p_y,$$
$$p_z' = \gamma_V(p_z - V\varepsilon/c^2),$$
$$\varepsilon'/c = \gamma_V(\varepsilon/c - Vp_z/c). \qquad (4.13)$$

Taking into account that $\mathbf{p} = (p_x, p_y, p_z) = (p\sin\theta\cos\phi, p\sin\theta\sin\phi, p\cos\theta)$ and $\mathbf{p}' = (p_x', p_y', p_z') = (p'\sin\theta'\cos\phi', p'\sin\theta'\sin\phi', p'\cos\theta')$, where θ, ϕ and θ', ϕ' are the polar and azimuthal angles of the three-momentum in the frames K and K, respectively, (4.13) can be also rewritten as

$$p' \sin \theta' = p \sin \theta,$$
$$p' \cos \theta' = \gamma_V (p \cos \theta - V\varepsilon/c^2),$$
$$\varepsilon'/c = \gamma_V (\varepsilon/c - Vp \cos \theta/c). \tag{4.14}$$

A particular case of the transformations (4.13)–(4.14) is encountered when one considers the energy–momentum of a photon. Let in the frame K the photon have a frequency ω and propagate under the angle θ with respect to the Z-axis. Then, using (4.14) it is not difficult to show that in the frame K' the frequency ω' and the propagation angle θ' of the photon are given by

$$\omega' = \gamma_V \omega \left(1 - \frac{V}{c} \cos \theta \right),$$

$$\tan \theta' = \frac{\sin \theta}{\gamma \left(\cos \theta - \frac{V}{c} \right)}. \tag{4.15}$$

The first equation in (4.15) describes the relativistic Doppler effect.

4.1.5 Transformations of Cross Sections

When exploring ion–atom collisions one needs to be able to relate cross sections calculated in one reference frame to the same cross sections but considered in another reference frame. For instance, it is often more convenient to perform calculations for cross sections of the projectile-electron loss in the rest frame of the projectile nucleus. However, since in an experiment cross sections are not, of course, measured in the rest frame of the projectile, it is necessary to transform these cross sections from the latter frame to the laboratory frame.

Let us consider two inertial reference frames, K and K'. As before, we choose the Z and Z' coordinate axes of these frames to coincide and the X and Y axes to be parallel to X' and Y', respectively. We also suppose that the frame K' moves with respect to K with a velocity $\mathbf{V} = (0, 0, V)$.

Let $N'(\varepsilon', \theta', \phi')$ be the density of particles (for instance, electrons or photons) emitted in ion–atom collisions which are observed in the frame K'. In this frame the particles have energies in the interval $(\varepsilon', \varepsilon' + \mathrm{d}\varepsilon')$ and propagate within the solid angle $\mathrm{d}\Omega' = \sin \theta' \mathrm{d}\theta' \mathrm{d}\phi'$ around the polar angle θ' and the azimuth angle ϕ'. Let $N(\varepsilon, \theta, \phi)$ be the density of the same particles, but observed in the frame K in which they have energies in the interval $(\varepsilon, \varepsilon + \mathrm{d}\varepsilon)$ and move within the solid angle $\mathrm{d}\Omega = \sin \theta \mathrm{d}\theta \mathrm{d}\phi$.

The number of emitted particles is a quantity which is naturally conserved under the transformation of a reference frame. Therefore, one has

$$N(\varepsilon, \theta, \phi)|\mathrm{d}\varepsilon \mathrm{d}\Omega| = N'(\varepsilon', \theta', \phi')|\mathrm{d}\varepsilon' \mathrm{d}\Omega'| \tag{4.16}$$

or

$$\frac{\mathrm{d}^2 \sigma}{\mathrm{d}\varepsilon \mathrm{d}\Omega} |\mathrm{d}\varepsilon \mathrm{d}\Omega| = \frac{\mathrm{d}^2 \sigma'}{\mathrm{d}\varepsilon' \mathrm{d}\Omega'} |\mathrm{d}\varepsilon' \mathrm{d}\Omega'|. \tag{4.17}$$

In the above equation $d^2\sigma/d\varepsilon d\Omega$ and $d^2\sigma'/d\varepsilon' d\Omega'$ are the doubly differential cross sections given in the frame K and K', respectively.

The energy and angle differentials are related by

$$d\varepsilon d\Omega = \frac{\partial(\varepsilon, \cos\theta)}{\partial(\varepsilon', \cos\theta')} d\varepsilon' d\Omega' \qquad (4.18)$$

with the Jakobian

$$\frac{\partial(\varepsilon, \cos\theta)}{\partial(\varepsilon', \cos\theta')} = \begin{vmatrix} \frac{\partial\varepsilon}{\partial\varepsilon'} & \frac{\partial\cos\theta}{\partial\varepsilon'} \\ \frac{\partial\varepsilon}{\partial\cos\theta'} & \frac{\partial\cos\theta}{\partial\cos\theta'} \end{vmatrix} = \frac{p'}{p} = \frac{\sin\theta}{\sin\theta'}. \qquad (4.19)$$

Taking into account (4.17) and (4.19) one obtains that the cross section differential in the emission energy and angle is transformed according to

$$\frac{1}{p} \frac{d^2\sigma}{d\varepsilon d\Omega} = \frac{1}{p'} \frac{d^2\sigma'}{d\varepsilon' d\Omega'}. \qquad (4.20)$$

The above expression and the transformations (4.14) contain all necessary information enabling one to recalculate the emission spectra from one reference frame to another.[2]

4.2 The Electromagnetic Field

4.2.1 The Maxwell Equations and the Conservation of Electric Charge

In ion–atom collisions the interaction between the atomic particles is transmitted by the electromagnetic field. An arbitrary electromagnetic field is described by the Maxwell equations which read (see e.g. [7])

$$\mathbf{\nabla} \cdot \mathbf{E} = 4\pi\rho,$$
$$\mathbf{\nabla} \times \mathbf{E} = -\frac{1}{c} \frac{\partial\mathbf{B}}{\partial t},$$
$$\mathbf{\nabla} \cdot \mathbf{B} = 0,$$
$$\mathbf{\nabla} \times \mathbf{B} = -\frac{4\pi}{c}\mathbf{j} - \frac{1}{c} \frac{\partial\mathbf{E}}{\partial t}. \qquad (4.21)$$

Here, $\mathbf{E}$ is the electric field strength, $\mathbf{B}$ is the magnetic field strength, ρ is the density of the electric charge and $\mathbf{j}$ is the density of the electric current.

Taking the partial time derivative of both sides in the first equation in (4.21), acting with the operator $\mathbf{\nabla}$ on both sides of the fourth equation in

---

[2] One should add that (4.20) is valid only provided the pairs of variables $(\varepsilon, \cos\theta)$ and $(\varepsilon', \cos\theta')$ are in a one-to-one relationship. A more careful analysis should be undertaken when a multivalued relationship between these pairs occurs (see [72]).

(4.21) and taking into account that $\operatorname{div}\operatorname{curl} \equiv 0$ we arrive at the so called continuity equation

$$\frac{\partial \rho}{\partial t} + \nabla \cdot \mathbf{j} = 0, \tag{4.22}$$

which expresses the local conservation of electric charge. With the help of (4.10) and (4.11), (4.22) can be rewritten in the four-dimensional form $\partial_\alpha j^\alpha = 0$ (or $\partial^\alpha j_\alpha = 0$) which demonstrates that the continuity equation is manifestly covariant.

4.2.2 Potentials of the Electromagnetic Field. Gauge Transformations

Since $\operatorname{div}\operatorname{curl} \equiv 0$, the third equation in (4.21) will be automatically fulfilled if we set

$$\mathbf{B} = \nabla \times \mathbf{A}. \tag{4.23}$$

The three-dimensional vector $\mathbf{A}$ is called the vector potential. Inserting the right hand side of (4.23) into the second line of (4.21) we obtain

$$\nabla \times \left(\mathbf{E} + \frac{1}{c}\frac{\partial \mathbf{A}}{\partial t} \right) = 0. \tag{4.24}$$

Taking into account that $\operatorname{curl}\operatorname{grad} \equiv 0$, the electric field can be expressed as

$$\mathbf{E} = -\frac{1}{c}\frac{\partial \mathbf{A}}{\partial t} - \nabla \Phi, \tag{4.25}$$

where Φ is called the scalar potential.

In (4.21) the electromagnetic field is represented by six quantities: the three components of the electric field and the three components of the magnetic field. If the scalar and vector potentials are known, the strengths of the magnetic and electric fields are unambiguously obtained from (4.23) and (4.25). Thus, the electromagnetic field can be described by introducing just four quantities, the vector potential $\mathbf{A}$ and the scalar potential Φ.

Moreover, there still remains a certain freedom in choosing these potentials. It is easy to check that if these potentials are transformed according to

$$\begin{aligned} \Phi \to \Phi' &= \Phi - \frac{1}{c}\frac{\partial f}{\partial t}, \\ \mathbf{A} \to \mathbf{A}' &= \mathbf{A} + \nabla f, \end{aligned} \tag{4.26}$$

where f is an arbitrary function (whose space and time derivatives exist), the fields $\mathbf{E}$ and $\mathbf{B}$ remain unchanged. The transformation (4.26) is called a gauge transformation for the field potentials and f is a gauge function.

In Classical Physics the introduction of the electromagnetic potentials is in essence just a matter of convenience. However, in quantum theory this introduction becomes mandatory since these potentials enter the quantum wave equations. In particular, the fact, that the electromagnetic field can be fully described by introducing the vector and scalar potentials which are not uniquely defined and maybe subjected to a gauge transformation, plays one of the key roles in quantum theory of electromagnetic processes.

4.2.3 Maxwell Equations for the Field Potentials

The electromagnetic potentials were introduced by using the second and third equations in (4.21). The rest two lines in (4.21) can be used to obtain the differential equations which these potentials have to obey to. Inserting (4.23) and (4.25) into the first and fourth equations in (4.21) and making use of the identity $\operatorname{curl}\operatorname{curl} \equiv \operatorname{grad}\operatorname{div} -\Delta$, we get

$$\Delta\Phi + \frac{1}{c}\frac{\partial}{\partial t}\mathbf{\nabla}\cdot\mathbf{A} = -4\pi\rho,$$

$$\Delta\mathbf{A} - \frac{1}{c^2}\frac{\partial^2\mathbf{A}}{\partial t^2} - \mathbf{\nabla}\left(\mathbf{\nabla}\cdot\mathbf{A} + \frac{1}{c}\frac{\partial\Phi}{\partial t}\right) = -\frac{4\pi}{c}\mathbf{j}. \tag{4.27}$$

By introducing the four-potential

$$A^\nu = (\Phi, \mathbf{A}), \tag{4.28}$$

which may be shown to be a contravariant four-vector, and taking into account that the differential operators $\partial/(c\partial t)$ and $-\mathbf{\nabla}$ form the components of a four-vector ∂^μ (see (4.11)), (4.26) and (4.27) can be written in a four-dimensional form:

$$\Box A^\mu + \partial^\mu\partial_\nu A^\nu = \frac{4\pi}{c}j^\mu \tag{4.29}$$

and

$$A^\mu \rightarrow A'^\mu = A^\mu - \partial^\mu f. \tag{4.30}$$

In (4.29)

$$\Box = \frac{\partial^2}{c^2\partial t^2} - \Delta \tag{4.31}$$

is the D'Alembert operator and the four-current j^μ is given by the first line in (4.10).

The equations for the field potentials can be simplified using the freedom to choose an appropriate gauge. For instance, the Lorentz condition, which is given by

$$\partial_\nu A^\nu = \mathbf{\nabla} \cdot \mathbf{A} + \frac{1}{c}\frac{\partial \varPhi}{\partial t} = 0,$$

leads to the equations

$$\varDelta\varPhi - \frac{1}{c^2}\frac{\partial^2 \varPhi}{\partial t^2} = -4\pi\rho,$$

$$\varDelta\mathbf{A} - \frac{1}{c^2}\frac{\partial^2 \mathbf{A}}{\partial t^2} = -\frac{4\pi}{c}\mathbf{j}, \tag{4.32}$$

in which the scalar and vector potentials are fully disentangled. The Lorentz condition is obviously invariant under a Lorentz transformation.

Another 'popular' gauge condition, $\mathbf{\nabla} \cdot \mathbf{A} = 0$ (the Coulomb gauge), results in

$$\varDelta\varPhi = -4\pi\rho$$

$$\varDelta\mathbf{A} - \frac{1}{c^2}\frac{\partial^2 \mathbf{A}}{\partial t^2} = -\frac{1}{c}\left(4\pi\mathbf{j} - \mathbf{\nabla}\frac{\partial\varPhi}{\partial t}\right). \tag{4.33}$$

In atomic collision theory the Coulomb gauge, in which the scalar potential remains exactly the same as if the speed of light were infinity, is convenient to analyze relativistic effects caused by the relative ion–atom motion.

4.3 The Dirac Equation

4.3.1 The Hamiltonian Form

The Hamiltonian form of the Dirac equation for an electron in the presence of an external electromagnetic field reads

$$i\frac{\partial\psi(\mathbf{r},t)}{\partial t} = \hat{H}\psi(\mathbf{r},t). \tag{4.34}$$

Here, $\mathbf{r}$ is the electron coordinate, t is the time. The Dirac Hamiltonian is given by

$$\hat{H} = c\boldsymbol{\alpha} \cdot \frac{\mathbf{\nabla}}{i} + c^2\beta - \varPhi + \boldsymbol{\alpha} \cdot \mathbf{A}, \tag{4.35}$$

where $\varPhi$ and $\mathbf{A}$ are the scalar and vector potentials, respectively, of the external field, and $\boldsymbol{\alpha} = (\alpha_x, \alpha_y, \alpha_z)$ and β are the Dirac matrices. These matrices satisfy the anticommutation relations:

$$\alpha_j\alpha_k + \alpha_k\alpha_j = 2\delta_{jk}I_d \ (j,k = 1,2,3),$$

$$\alpha_j\beta + \beta\alpha_j = 0_d,$$

$$\beta^2 = \alpha_j^2 = I_d, \tag{4.36}$$

where δ_{jk} is the Kronecker's delta and 0_d and I_d denote, respectively, the zero and unit matrices of a dimension d. The smallest dimension, in which matrices fulfilling (4.36) can be found, is $d = 4$ and, therefore, the Dirac matrices are normally represented by 4×4 matrices. Consequently, a state ψ corresponding to such a choice of matrices is a four-component column:

$$\psi = \begin{pmatrix} \psi_1 \\ \psi_2 \\ \psi_3 \\ \psi_4 \end{pmatrix}, \tag{4.37}$$

which is called the Dirac spinor (or four-spinor).

One of the representations of the Dirac matrices, which is commonly used in the theory of relativistic atomic collisions, is provided by (see e.g. [3, 10])

$$\boldsymbol{\alpha} = \begin{pmatrix} 0_2 & \boldsymbol{\sigma} \\ \boldsymbol{\sigma} & 0_2 \end{pmatrix},$$

$$\beta = \begin{pmatrix} I_2 & 0_2 \\ 0_2 & I_2 \end{pmatrix}, \tag{4.38}$$

where 0_2 and I_2 are the two-dimensional zero and unit matrices, respectively, and $\boldsymbol{\sigma} = (\sigma_x, \sigma_y, \sigma_z)$ are the set of the two-dimensional Pauli spin matrices. These matrices are normally taken as

$$\sigma_x = \begin{pmatrix} 0 & 1 \\ 1 & 0 \end{pmatrix}, \quad \sigma_y = \begin{pmatrix} 0 & i \\ -i & 0 \end{pmatrix}, \quad \sigma_z = \begin{pmatrix} 1 & 0 \\ 0 & -1 \end{pmatrix} \tag{4.39}$$

and possess the following basic properties

$$\sigma_x^2 = \sigma_y^2 = \sigma_z^2 = I_2,$$
$$\sigma_y \sigma_z = i\sigma_x, \quad \sigma_z \sigma_x = i\sigma_y, \quad \sigma_x \sigma_y = i\sigma_z,$$
$$\sigma_j \sigma_k + \sigma_k \sigma_j = 2\delta_{jk} I_2. \tag{4.40}$$

The Hermitian conjugate of the spinor (4.37) is $\psi^\dagger$, which is defined by

$$\psi^\dagger = (\psi_1^*, \psi_2^*, \psi_3^*, \psi_4^*) \tag{4.41}$$

and satisfies the equation

$$-i\frac{\partial \psi^\dagger}{\partial t} = ic(\boldsymbol{\nabla}\psi^\dagger) \cdot \boldsymbol{\alpha} + c^2 \psi^\dagger \beta - \Phi\psi^\dagger + \psi^\dagger \boldsymbol{\alpha} \cdot \mathbf{A}. \tag{4.42}$$

Multiplying both sides of (4.34) by $\psi^\dagger$ (from the left) and both sides of (4.42) by ψ (from the right) and subtracting the resulting equations from each other we obtain the continuity equation

$$\frac{\partial \rho}{\partial t} + \boldsymbol{\nabla} \cdot \mathbf{j} = 0, \tag{4.43}$$

where

$$\rho = \psi^\dagger \psi,$$
$$\mathbf{j} = c\psi^\dagger \boldsymbol{\alpha} \psi. \tag{4.44}$$

4.3.2 Gauge Invariance of the Dirac Equation

As was briefly discussed in the previous section, the electromagnetic field is
not changed when the field potentials Φ and $\mathbf{A}$ are subjected to a gauge
transformation given by (4.26). It is easy to check that, under the gauge
transformation (4.26) of the potentials, the form of the Dirac equation (4.34)
will not be altered if the wave function ψ is simultaneously subjected to the
transformation:

$$\psi \rightarrow \psi' = \exp(-\mathrm{i}f/c)\psi, \tag{4.45}$$

where f is the same gauge function as in (4.26).

4.3.3 The Covariant Form

If we multiply both sides of (4.34) by the matrix β and introduce matrices γ^ν
$(\nu = 0, 1, 2, 3)$ according to

$$\gamma^0 = \beta,$$
$$\gamma^j = \beta\alpha^j \ (j = 1, 2, 3), \tag{4.46}$$

we obtain the following equation

$$\left(\mathrm{i}\gamma^\nu \partial_\nu + \frac{1}{c}\gamma^\nu A_\nu - c\right)\psi = 0, \tag{4.47}$$

where $A_\nu = (\Phi, -\mathbf{A})$ is the four-potential. Equation (4.47) is referred to as
the covariant form of the Dirac equation. Equation (4.47) is a convenient
starting point to establish the covariance of the Dirac equation under a Lorentz
transformation (see, for instance, [13]). Note also that the matrices γ^ν, as it
follows from their definition (4.46), satisfy the anticommutation relations:

$$\gamma^\nu \gamma^\mu + \gamma^\mu \gamma^\nu = 2g^{\nu\mu} I_4, \tag{4.48}$$

where I_4 is the 4×4 unit matrix.

4.3.4 Classification of States in a Spherical Potential

At $t \rightarrow \pm\infty$, i.e. long before and after a collision between an ion and an atom
occurs, these atomic particles are well separated and the interaction between
them can be neglected. Then the internal motion of the electron(s) in the ion
and in the atom can be described by considering a time-independent Dirac
equation. A relatively simple but often arising problem is the motion of an
electron in a spherically symmetric external field which is described by the
equation:

$$\varepsilon\chi = \hat{H}_0\chi, \tag{4.49}$$

where

$$\hat{H}_0 = c\boldsymbol{\alpha} \cdot \frac{\boldsymbol{\nabla}}{i} + c^2\beta - V(r). \tag{4.50}$$

In these equations $-V$ is the potential energy of the electron in the external field, ε is the total electron energy and χ is the electron wave function.

The total angular momentum of a relativistic electron is given by

$$\mathbf{J} = \mathbf{L} + \frac{1}{2}\boldsymbol{\Sigma}, \tag{4.51}$$

where $\mathbf{L}$ is the orbital electron momentum[3] and

$$\boldsymbol{\Sigma} = \begin{pmatrix} \boldsymbol{\sigma} & 0_2 \\ 0_2 & \boldsymbol{\sigma} \end{pmatrix} \tag{4.52}$$

are the 4×4 spin matrices representing the spin part of the angular momentum.

One can show that the Hamiltonian $\hat{H}_0$ commutes with $\mathbf{J}^2$ and one of the components of $\mathbf{J}$ (say, J_z). Therefore, eigenstates of $\mathbf{J}^2$ and J_z,

$$\mathbf{J}^2 \chi_{j\,m_j} = \begin{pmatrix} \left(\mathbf{L}+\frac{1}{2}\boldsymbol{\sigma}\right)^2 & 0_2 \\ 0_2 & \left(\mathbf{L}+\frac{1}{2}\boldsymbol{\sigma}\right)^2 \end{pmatrix} \chi_{j\,m_j} = j(j+1)\,\chi_{j\,m_j},$$

$$J_z \chi_{j\,m_j} = \begin{pmatrix} L_z+\frac{1}{2}\sigma_z & 0_2 \\ 0_2 & L_z+\frac{1}{2}\sigma_z \end{pmatrix} \chi_{j\,m_j} = m_j\,\chi_{j\,m_j} \tag{4.53}$$

with the total momentum j and its projection m_j are simultaneously eigenstates of the Hamiltonian $\hat{H}_0$.

Besides, one can show (see e.g. [50]) that the operator Ξ, defined by

$$\Xi = \gamma^0 \left(\boldsymbol{\Sigma}\cdot\mathbf{L} + I_4\right) = \begin{pmatrix} (\boldsymbol{\sigma}\cdot\mathbf{L}+I_2) & 0_2 \\ 0_2 & -(\boldsymbol{\sigma}\cdot\mathbf{L}+I_2) \end{pmatrix}, \tag{4.54}$$

also commutes with $\hat{H}_0$ (as well as with $\mathbf{J}^2$ and J_z, see (4.59)). Note that it is customary to denote an eigenvalue of Ξ by $-\kappa$,

$$\Xi\chi = -\kappa\chi. \tag{4.55}$$

Taking into account what has been said in the above two paragraphs, quantum states in a spherically symmetric potential can be classified according to their energy and quantum numbers j, m_j and $-\kappa$.

From the basic properties of the Pauli matrices, given by (4.40), it follows that (see e.g. [9])

$$(\boldsymbol{\sigma}\cdot\mathbf{a})(\boldsymbol{\sigma}\cdot\mathbf{b}) = \mathbf{a}\cdot\mathbf{b} + i\boldsymbol{\sigma}\cdot[\mathbf{a}\times\mathbf{b}], \tag{4.56}$$

---

[3] In (4.51) and below it is assumed that the operator $\mathbf{L}$ is multiplied by 4 × 4 (or 2 × 2) unit matrix where necessary.

where $\mathbf{a}$ and $\mathbf{b}$ are two arbitrary vectors. Using (4.56) it can be shown that

$$\Xi^2 = \mathbf{L}^2 + \boldsymbol{\Sigma} \cdot \mathbf{L} + I_4. \tag{4.57}$$

Comparing the above expression with

$$\mathbf{J}^2 = \mathbf{L}^2 + \boldsymbol{\Sigma} \cdot \mathbf{L} + \frac{3}{4} I_4, \tag{4.58}$$

we see that

$$\Xi^2 = \mathbf{J}^2 + \frac{1}{4} I_4 \tag{4.59}$$

and, therefore,

$$\kappa = \pm \left(j + \frac{1}{2} \right). \tag{4.60}$$

By decomposing the four-spinor χ into the upper and lower components,

$$\chi = \begin{pmatrix} \chi_u \\ \chi_d \end{pmatrix}, \tag{4.61}$$

the eigenvalue problem of (4.55) reduces to the two eigenvalue equations

$$(\boldsymbol{\sigma} \cdot \mathbf{L} + I_2) \chi_u = -\kappa \chi_u,$$
$$(\boldsymbol{\sigma} \cdot \mathbf{L} + I_2) \chi_d = \kappa \chi_d. \tag{4.62}$$

Besides, in terms of the upper and lower components, (4.53) can be rewritten as

$$\left(\mathbf{L} + \frac{1}{2} \boldsymbol{\sigma} \right)^2 \chi_{u(d)} = j(j+1) \chi_{u(d)},$$
$$\left(L_z + \frac{1}{2} \sigma_z \right) \chi_{u(d)} = m_j \chi_{u(d)}. \tag{4.63}$$

Comparing the first of the equations in (4.63) with equations in (4.62), we see that, since $\boldsymbol{\sigma}^2 = 3$, the two-component spinors χ_u and χ_d are also eigenstates of the operator $\mathbf{L}^2$. The latter can be expressed as

$$\mathbf{L}^2 \chi_u = l_u (l_u + 1) \chi_u,$$
$$\mathbf{L}^2 \chi_d = l_d (l_d + 1) \chi_d, \tag{4.64}$$

where l_u and l_d are the orbital angular momenta of the upper and lower components, respectively. Using (4.62) and (4.64) and the equations in the first line of (4.63) we obtain that

$$l_u (l_u + 1) = j (j + 1) + \kappa + \frac{1}{4},$$
$$l_d (l_d + 1) = j (j + 1) - \kappa + \frac{1}{4}. \tag{4.65}$$

Considering the quantum number κ in the above equations as a given value and taking into account that for any κ we have $j = |\kappa| - 1/2$ (see (4.60)), the solutions of these equations are

$$l_u = \kappa,$$
$$l_d = \kappa - 1 \qquad (4.66)$$

for positive values of κ and

$$l_u = |\kappa| - 1,$$
$$l_d = |\kappa| \qquad (4.67)$$

for negative values of κ.

As was already mentioned, quantum states of a relativistic electron moving in a spherically symmetric potential are fully specified by their energies (or the principal quantum number) and the angular quantum numbers j, m_j and κ. However, since the upper, χ_u, and lower, χ_d, components of the relativistic four-spinor χ are eigenstates of the orbital angular momentum and since in the nonrelativistic limit one has $|\chi_d| \ll |\chi_u|$, relativistic wave functions are normally denoted by specifying the total angular momentum and the orbital angular momentum of the upper component χ_u. For instance, the notation $2p_{3/2}(1/2)$ corresponds to a state with the principal quantum number $n = 2$, the orbital momentum $l_u = 1$ (a p-state), the total momentum $j = 3/2$ and its projection $m_j = 1/2$.[4] Such a convention for notations will also be used in this book.

A particularly important case of a central field is represented by a Coulomb potential, $-Z/r$. Detailed considerations of solutions of the Dirac equation for an electron moving in a Coulomb potential can be found in [10,12,73,74] and also in [3].

---

[4] Using (4.60), (4.66) and (4.67) one can obtain that in this state $\kappa = -2$.

5

Descriptions of Collisions Within the First Order Approximation in the Projectile–Target Interaction

5.1 Preliminary Remarks

The processes of ionization and excitation of atoms in relativistic collisions with bare nuclei have been theoretically studied for several decades (for reviews see [3–6] as well as [75–77]). Theoretical studies of the projectile-electron transitions in relativistic collisions with neutral atoms represent a comparatively new field of research which does not yet have a long history and for which only a few review papers are available [78,79].

The first attempt to formulate a theory for relativistic collisions of an ion and an atom, which both initially have electrons actively or passively participating in the collision process, was undertaken in [80,81]. The approach of [80] to this problem was based on the first order perturbative treatment of ionization in relativistic collisions with structureless point-like charges [82,83]. In order to take into account the fundamental difference between the actions of a point-like charge and a neutral atom in the collision, results for the projectile-electron loss in nonrelativistic collisions with neutral atoms were employed and some intuitive assumptions were introduced to adapt the nonrelativistic results to relativistic collisions. The most complete set of results for the loss process in relativistic collisions, obtained in this way, was presented in [84]. In that paper the electron loss in ultrarelativistic collisions was considered for a variety of projectile–target pairs for collision energies up to those corresponding to $\gamma \leq 1,000$, where γ is the collisional Lorentz factor. The key finding of [84] was that the loss cross section for any projectile–target pair can be well approximated for the range $10 \lesssim \gamma \lesssim 1,000$ by the following simple formula: $\sigma_{\text{loss}} = A + B \ln \gamma$, where the parameters A and B depend on the projectile–target pair but are γ-independent.

One should note that the structure of the above expression closely resembles that of the cross section for single ionization of atoms (e.g. K-shell ionization) by point-like charged particles moving at relativistic velocities. In particular, the above loss cross section includes the term $\ln \gamma$ which is well known to appear in the cross section for atomic ionization by relativistic charged particles. In the loss cross section such a term would arise if collisions with large impact parameters $b_{max} \sim v\gamma/\omega_{fi}$, where v is the collision velocity and ω_{fi} is the energy transfer to the electron of the projectile (in the rest frame of the projectile), would substantially contribute to the loss process.

Elementary estimates show, however, that even in the case when one considers the electron loss from the heaviest hydrogen-like projectiles, for which the energy transfers ω_{fi} reach the largest values (and, therefore, b_{max} are smaller than those for lighter projectiles), the impact parameters $b_{max} \sim v\gamma/\omega_{fi}$ can substantially exceed the size of a neutral atom already in collisions where the Lorentz factor is still far below $1,000$. Therefore, it is rather obvious that the simple expression for the loss cross section, suggested in [84], in general cannot be valid for ultrarelativistic collisions. In particular, as the same estimates suggest, this expression should not be applied to evaluate cross sections for the electron loss from very heavy ions at $\gamma \simeq 30$ and higher and its applicability may become even more questionable in cases of the electron loss from lighter ions (for illustrations see Figs. 5.1 and 5.2).

The main reason for this is that the theory of [84] does not account for important peculiarities in the screening effect of the atomic electrons in collisions with large γ. However, this shortcoming had not been revealed for

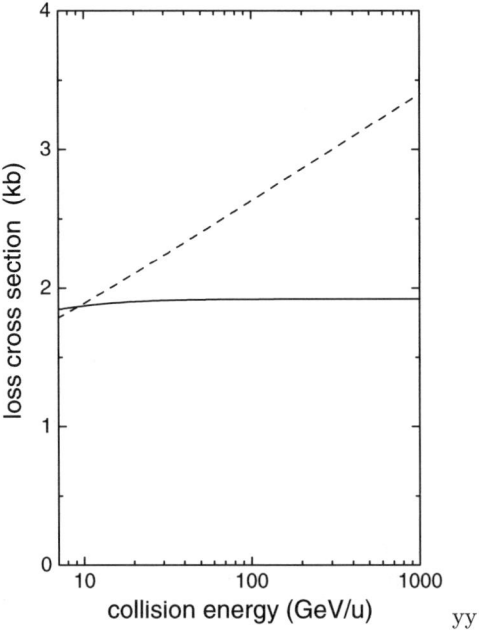

Fig. 5.1. Total cross section for the electron loss from 7–$1,000\,\text{GeV}\,\text{u}^{-1}\,\text{Sn}^{49+}$ ions ($Z_I = 50$) colliding with Ne atoms ($Z_A = 10$). The *dash line* was obtained by using the results of [84]. The *solid line* shows the correct results.

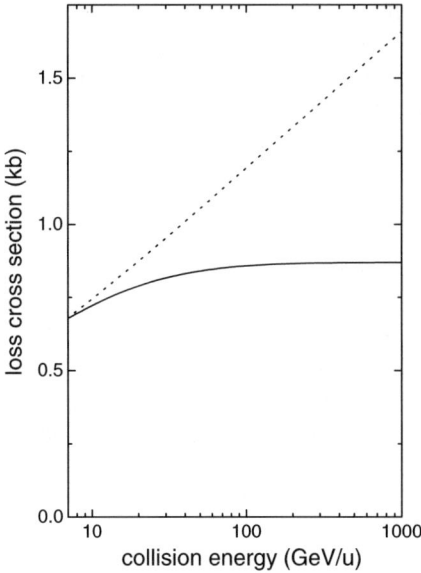

Fig. 5.2. Same as in Fig. 5.1 but for the electron loss from Hg^{79+} ($Z_I = 80$).

more than 10 years until an experiment [85] on the electron loss from ultra-relativistic hydrogen-like Pb ions unveiled the considerable difference between the predictions of the theory of [84] and the experimental observations.

It was, for the first time, pointed out in [86], that the loss cross section, obtained in [84], does not correctly describe screening effects in ultrarelativistic collisions. As an alternative to the theory of [84], a simple model was suggested in [86] to describe, within the framework of the first order perturbation approach, the elastic target mode of the electron loss cross section.[1] Within this model the contribution of the elastic target mode to the loss cross section is separated into contributions from 'close' and 'distant' collisions. The dividing distance between the 'close' and 'distant' collisions was chosen in [86] to be essentially the radius of the electron bound state in the projectile. The close-collision contribution was evaluated by regarding the projectile electron as free and by assuming that the action of the neutral atom on the projectile electron is equivalent to that of the atomic nucleus whereas the atomic electrons play no role. The distant-collision contribution was estimated by using the method of equivalent photons [88, 89].[2] The contribution to the loss cross section given by the inelastic target mode cannot be treated within such an approach. Therefore, the total loss cross section σ_t was estimated as $\sigma_t = (1 + 1/Z_A)\sigma_{el}$ where σ_{el} is that part of the loss cross section, which is given by the elastic target mode, and the rest accounts for the incoherent action of Z_A 'active' atomic electrons.

---

[1] The model of [86] is also briefly discussed in a review article [87].

[2] This method is also described in a number of books, see for instance [7].

The model briefly described above is appealing by its simplicity. However, it has a number of obvious and serious shortcomings. For example, although the result for the total loss cross section in the model is dependent on the impact parameter, which separates 'close' and 'distant' collisions, the latter is not strictly defined. Further, the projectile electron can be treated as (quasi-) free only in collisions where the momentum transfer to the electron is much larger than its typical momentum in the initial bound state of the projectile (all the momenta are considered in the projectile frame). However, even for 'close' impact parameters this is not the case for the overwhelming majority of the collisions. In addition, there is also an arbitrariness in estimating the contribution arising from the 'distant' collisions.

In 'practical' terms the main deficiency of the oversimplified model of [86] is that the accuracy of its predictions is difficult to estimate and, therefore, it should in general be used merely to provide first rough estimates for the loss cross section.

From the theoretical point of view the shortcomings of this model are much less tolerable. In particular, this model by no means can serve as a substitute for more regular and rigorous theoretical methods. It is, therefore, the main goal of this and the next chapters to discuss some of such methods.

5.2 Simplified Semi-Classical Consideration

We start with a simplified semi-classical first-order consideration for the contribution of the elastic target mode to the projectile-electron excitation or loss in relativistic collisions with an atomic target [90]. It is convenient to consider the collision in the projectile-ion frame. The nucleus of the ion with charge Z_I is assumed to be at rest and taken as the origin. The translational motion of the neutral atom is treated classically. The nucleus of the incident neutral atom with atomic number Z_A moves on a straight-line trajectory $\mathbf{R}(t) = \mathbf{b} + \mathbf{v}t$, where $\mathbf{b}$ is the impact parameter and $\mathbf{v} = (0, 0, v)$ the collision velocity. This nucleus is 'dressed' by the electrons, the positions of which are assumed to be 'frozen' with respect to the nucleus. The coordinates of the electrons with respect to the origin are $\mathbf{r}_j = \mathbf{R}(t) + \boldsymbol{\eta}_j$, where $\boldsymbol{\eta}_j$ are the coordinates of the electrons with respect to the nucleus of the atom.

The fields created by the incident atom are described by the scalar potential $\Phi(\mathbf{r}, t)$ and the vector potential $\mathbf{A}(\mathbf{r}, t)$ obeying the Maxwell equations which in the Lorentz gauge[3] read

---

[3] As is well known, the Lorentz condition $\operatorname{div}\mathbf{A} + \frac{1}{c}\frac{\partial \Phi}{\partial t} = 0$, which enables one to get the Maxwell equations in the form (5.1), is not sufficient to uniquely define the potentials. In fact there exists an infinite number of gauges for which the Lorentz condition is fulfilled. In this book, whenever words 'Lorentz gauge' are used, that particular gauge of the Lorentz family of gauges is implied which in the case of a point-like charge would yield the Lienard–Wiechert potentials (see (5.86)).

$$\Delta\Phi(\mathbf{r},t) - \frac{1}{c^2}\frac{\partial^2\Phi(\mathbf{r},t)}{\partial t^2} = -4\pi\rho(\mathbf{r},t),$$

$$\Delta\mathbf{A}(\mathbf{r},t) - \frac{1}{c^2}\frac{\partial^2\mathbf{A}(\mathbf{r},t)}{\partial t^2} = -\frac{4\pi}{c}\mathbf{J}(\mathbf{r},t), \tag{5.1}$$

where $c \simeq 137$ a.u. is the speed of light. Considering for the moment that the incident atom is represented by a beam of point-like classically moving charges, which all have the same velocity $\mathbf{v}$, the charge and current densities of the incident atom are simply given by

$$\rho(\mathbf{r},t) = Z_A\delta(\mathbf{r} - \mathbf{R}(t)) - \sum_j^{N_A}\delta(\mathbf{r} - \mathbf{R}(t) - \boldsymbol{\eta}_j),$$

$$\mathbf{J}(\mathbf{r},t) = \rho(\mathbf{r},t)\mathbf{v}, \tag{5.2}$$

where the sum runs over all atomic electrons ($N_A = Z_A$ for a neutral atom). Equation (5.1) can be solved by using Fourier transformations:

$$\Phi(\mathbf{r},t) = \frac{1}{4\pi^2}\int d^3q\int_{-\infty}^{+\infty}d\omega F(\mathbf{q},\omega)\exp(i(\mathbf{q}\cdot\mathbf{r} - \omega t))$$

$$\rho(\mathbf{r},t) = \frac{1}{8\pi^3}\int d^3q\int_{-\infty}^{+\infty}d\omega\delta(\omega - \mathbf{q}\cdot\mathbf{v})\exp(i(\mathbf{q}\cdot(\mathbf{r} - \mathbf{b}) - \omega t))$$

$$\times\left(Z_A - \sum_j^{N_A}\exp(-i\mathbf{q}\cdot\boldsymbol{\eta}_j)\right). \tag{5.3}$$

Inserting (5.3) into (5.1) we get for the Fourier transform $F(\mathbf{q},\omega)$

$$F(\mathbf{q},\omega) = \frac{2\,\delta(\omega - \mathbf{q}\cdot\mathbf{v})\exp(-i\mathbf{q}\cdot\mathbf{b})}{q^2 - \frac{\omega^2}{c^2}}\left(Z_A - \sum_j^{N_A}\exp(-i\mathbf{q}\cdot\boldsymbol{\eta}_j)\right) \tag{5.4}$$

and obtain the integral representation for the scalar potential:

$$\Phi(\mathbf{r},t) = \frac{1}{2\pi^2}\int d^3q\,\frac{\exp(i\mathbf{q}\cdot(\mathbf{r} - \mathbf{b} - \mathbf{v}t))}{q^2 - \frac{(\mathbf{q}\mathbf{v})^2}{c^2}}\left(Z_A - \sum_j^{N_A}\exp(-i\mathbf{q}\cdot\boldsymbol{\eta}_j)\right). \tag{5.5}$$

In the Lorentz gauge the vector potential is very simply related to the scalar one by

$$\mathbf{A}(\mathbf{r},t) = \frac{\mathbf{v}}{c}\Phi(\mathbf{r},t). \tag{5.6}$$

The quantum nature of the 'frozen' electrons of the incident atom can be explicitly taken into account by using the following replacement:

$$\sum_{j}^{N_A} \exp(-i\mathbf{q} \cdot \boldsymbol{\eta}_j) \rightarrow \int d^3\boldsymbol{\eta}_1 \int d^3\boldsymbol{\eta}_2 \cdots \int d^3\boldsymbol{\eta}_N \varphi_0^{\dagger}(\boldsymbol{\zeta}_N)\varphi_0(\boldsymbol{\zeta}_N)$$

$$\times (\exp(-i\mathbf{q} \cdot \boldsymbol{\eta}_1) + \exp(-i\mathbf{q} \cdot \boldsymbol{\eta}_2) + \cdots + \exp(-i\mathbf{q} \cdot \boldsymbol{\eta}_N))$$

$$= \langle \varphi_0 | \sum_{j}^{N_A} \exp(-i\mathbf{q} \cdot \boldsymbol{\eta}_j) | \varphi_0 \rangle. \tag{5.7}$$

In (5.7) $\varphi_0(\boldsymbol{\zeta}_N)$ is the electronic wavefunction of the ground state of the incident atom, transformed into the rest frame of the ion, $\boldsymbol{\zeta}_N = \{\boldsymbol{\eta}_1, \boldsymbol{\eta}_2, ..., \boldsymbol{\eta}_N\}$ is the 3-N dimensional vector representing the coordinates of the N electrons of the incident atom with respect to the nucleus of the atom ($\boldsymbol{\zeta}_N$ is given in the rest frame of the ion).

We will assume that the ion has a single electron. In the ion frame the interaction of this electron with the incident atom is given by

$$V_{\text{int}}(\mathbf{r}, t) = -\Phi(\mathbf{r}, t)\left(1 - \frac{1}{c}\mathbf{v} \cdot \boldsymbol{\alpha}\right), \tag{5.8}$$

where $\mathbf{r}$ are the coordinates of the electron, $\boldsymbol{\alpha} = (\alpha_x, \alpha_y, \alpha_z)$ are the Dirac matrices (see (4.38)) and the scalar potential Φ is given by (5.5) with the replacement (5.7).

Within the first order perturbation theory the electron transition amplitude $a_{0 \rightarrow n}$ is given by

$$a_{0 \rightarrow n} = \frac{i}{2\pi^2} \int_{-\infty}^{+\infty} dt \exp(i\omega_{n0}t) \int d^3\mathbf{q} \, \frac{< \psi_n \mid \exp(i\mathbf{q} \cdot \mathbf{r})\left(1 - \frac{1}{c}\mathbf{v} \cdot \boldsymbol{\alpha}\right) \mid \psi_0 >}{q^2 - \frac{\omega_{n0}^2}{c^2}}$$

$$\times Z_{\text{A,eff}}(\mathbf{q}) \exp(-i\mathbf{q} \cdot (\mathbf{b} + \mathbf{v}t)), \tag{5.9}$$

where $\psi_0(\mathbf{r})e^{-i\varepsilon_0 t}$ and $\psi_n(\mathbf{r})e^{-i\varepsilon_n t}$ are the initial and final electron states of the ion, $\mathbf{r}$ is the electron coordinate with respect to the nucleus of the ion and $\omega_{n0} = \varepsilon_n - \varepsilon_0$ is the transition frequency of the electron of the ion. The indices 0 and n denote all quantum numbers of the corresponding states including spin.

In (5.9) the quantity

$$Z_{\text{A,eff}}(\mathbf{q}) = Z_A - \langle \varphi_0 | \sum_{j}^{N_A} \exp(-i\mathbf{q} \cdot \boldsymbol{\eta}_j) | \varphi_0 \rangle \tag{5.10}$$

represents an 'effective charge' of the incident atom which is 'seen' by the electron of the ion. The magnitude of this charge depends on the momentum $\mathbf{q}$.

Integration over time in (5.9) gives the factor $2\pi\delta(\mathbf{qv} + \varepsilon_0 - \varepsilon_n)$ which allows one to integrate easily over the longitudinal component, $q_z = \mathbf{qv}/v$, of the momentum transfer $\mathbf{q}$. The result of these two integrations is

$$a_{0 \to n} = \frac{i}{\pi v} \int d^2 \mathbf{q}_\perp \frac{< \psi_n \mid \exp(i\mathbf{q} \cdot \mathbf{r}) \left(1 - \frac{1}{c}\mathbf{v} \cdot \boldsymbol{\alpha}\right) \mid \psi_0 >}{q_\perp^2 + \frac{\omega_{n0}^2}{v^2 \gamma^2}}$$

$$\times Z_{A,\text{eff}}(\mathbf{q}) \exp(-i\mathbf{q}_\perp \cdot \mathbf{b}). \tag{5.11}$$

Here, $\gamma = \frac{1}{\sqrt{1 - v^2/c^2}}$ and now $\mathbf{q} = (\mathbf{q}_\perp, q_z)$ has a fixed z-component, $q_z = \varepsilon_n - \varepsilon_0/v$, which represents the minimum momentum transfer to the ion in the ion frame. In (5.11) the integration runs over the transverse momentum transfer $\mathbf{q}_\perp$ ($0 \le q_\perp < \infty$) and $\mathbf{q}_\perp \cdot \mathbf{v} = 0$.

The corresponding cross section, $\sigma_{0 \to n}$, is given by

$$\sigma_{0 \to n} = \int d^2 \mathbf{b} \mid a_{0 \to n} \mid^2 . \tag{5.12}$$

Inserting the transition amplitude (5.11) into (5.12) and performing the integration over the impact parameter in (5.12) we finally obtain

$$\sigma_{0 \to n} = \frac{4}{v^2} \int d^2 \mathbf{q}_\perp \, Z_{A,\text{eff}}^2(\mathbf{q}) \frac{\mid < \psi_n \mid \exp(i\mathbf{q} \cdot \mathbf{r}) \left(1 - \frac{1}{c}\mathbf{v} \cdot \boldsymbol{\alpha}\right) \mid \psi_0 >\mid^2}{\left(q_\perp^2 + \frac{\omega_{n0}^2}{v^2 \gamma^2}\right)^2}. \tag{5.13}$$

The above discussed semi-classical approach for the projectile-electron excitation and loss in relativistic collisions is rather simple. However, this approach is not appropriate to treat the inelastic atomic mode of the collisions. Besides, it is not quite clear how the assumption, that the atomic electrons are 'frozen' and do not represent a source of the charge current in the atomic rest frame, could influence the result obtained for the elastic mode. Keeping these two points in mind, we now proceed to discuss more general first order theories.

5.3 Plane-Wave Born Approximation

In this subsection we consider the first order quantum treatment for relativistic collisions of two atomic particles, which both carry (active) electrons [91]. The general form of the transition S-matrix element which describes collisions of atomic particles, interacting via the electromagnetic field, is given by (see e.g. [13])

$$S_{fi} = \left(-\frac{i}{c} \int d^4x \, J_\mu^I(x) \, A_A^\mu(x)\right)_{fi} . \tag{5.14}$$

This formula represents the natural generalization of the nonrelativistic expression (2.1) for the relativistic case. Here, $J_\mu^I(x)$ ($\mu = 0, 1, 2, 3$) is the electromagnetic four-current of the projectile-ion at a space–time point x and $A_A^\mu(x)$ is the four-potential of the electromagnetic field created by the target-atom at the same point x. In (5.14) and below the summation over repeated Greek indices is implied.

The four-potential of the target-atom obeys the Maxwell equations which, in the Lorentz gauge, read

$$\Box A_{\mathrm{A}}^{\mu}(x) = +\frac{4\pi}{c} J_{\mathrm{A}}^{\mu}(x), \tag{5.15}$$

where $J_{\mathrm{A}}^{\mu}(x)$ is the four-current of the target-atom and $\Box$ denotes the D'Alembert operator given by (4.31).

Since the nuclear and atomic scales are very different, Coulomb collisions between the ion and the atom, resulting in excitation of nuclear degrees of freedom, are normally of negligible importance for cross sections of electron transitions. Therefore, the nuclei of the atom and the ion can be regarded as point-like unstructured charges.

Simple estimates show that in a reference frame, where the atom or the ion is initially at rest, its typical recoil velocity after the collision is not only nonrelativistic but is also orders of magnitude less than the Bohr velocity $v_0 = 1$ a.u.

Taking into account the two points mentioned above, the transition matrix element (5.14) can be calculated as follows. First, the ion current $J_{\mu}^{\mathrm{I}}(x)$ is evaluated in the reference frame K_{I}, where the ion is initially at rest and where this current is denoted by $J_{\mu}^{\mathrm{I}}(x_{\mathrm{I}})$ with $x_{\mathrm{I}} = (ct_{\mathrm{I}}, \mathbf{x}_{\mathrm{I}})$ being the coordinates of the space–time point x in K_{I}. Second, the atom current $J_{\mathrm{A}}''^{\mu}(x_{\mathrm{A}})$ is calculated in the reference frame K_{A}, where the atom is initially at rest, and then the atom potential $A_{\mathrm{A}}''^{\mu}(x_{\mathrm{A}})$ is evaluated in this frame. Finally, this potential is transformed to the frame K_{I} in order to calculate the transition matrix elements and corresponding cross sections in K_{I}.

Assuming that the ion carries only one electron the transition four-current J_{μ}^{I} of the ion in the frame K_{I} is written as

$$J_0^{\mathrm{I}}(x_{\mathrm{I}}) = c \int \mathrm{d}^3 \mathbf{R}_{\mathrm{I}} \int \mathrm{d}^3 \mathbf{r} \Psi_{\mathrm{f}}^{\dagger}(\mathbf{R}_{\mathrm{I}}, \mathbf{r}, t_{\mathrm{I}})$$
$$\times \left(Z_{\mathrm{I}} \delta^{(3)}(\mathbf{x}_{\mathrm{I}} - \mathbf{R}_{\mathrm{I}}) - \delta^{(3)}(\mathbf{x}_{\mathrm{I}} - \mathbf{R}_{\mathrm{I}} - \mathbf{r}) \right) \Psi_{\mathrm{i}}(\mathbf{R}_{\mathrm{I}}, \mathbf{r}, t_{\mathrm{I}}),$$
$$J_l^{\mathrm{I}}(x_{\mathrm{I}}) = c \int \mathrm{d}^3 \mathbf{R}_{\mathrm{I}} \int \mathrm{d}^3 \mathbf{r} \Psi_{\mathrm{f}}^{\dagger}(\mathbf{R}_{\mathrm{I}}, \mathbf{r}, t_{\mathrm{I}}) \alpha_l$$
$$\times \delta^{(3)}(\mathbf{x}_{\mathrm{I}} - \mathbf{R}_{\mathrm{I}} - \mathbf{r}) \Psi_{\mathrm{i}}(\mathbf{R}_{\mathrm{I}}, \mathbf{r}, t_{\mathrm{I}}); \; l = 1, 2, 3. \tag{5.16}$$

In (5.16) Z_{I} is the atomic number of the ion, $\mathbf{R}_{\mathrm{I}}$ is the coordinate of the ion nucleus, $\mathbf{r}$ is the coordinate of the electron of the ion with respect to the ion nucleus, α_l are the Dirac matrixes for the electron of the ion, and $\delta^{(3)}$ is the three-dimensional delta-function. The mass of the nucleus is much larger than that of the electron. Therefore, in the frame K_{I} the three-velocity of the ion nucleus is negligible compared to that of the electron and we have neglected in the second line in (5.16) the contribution to the ion current due to the motion of the nucleus. The large mass of the nucleus also permitted us to omit that part of the ion current, which is connected with the spin degrees of the nucleus of the ion. Further, in a first order treatment the initial and final states in (5.16) are just unperturbed states of the ion and are given by

$$\Psi_j(\mathbf{R}_I, \mathbf{r}, t_I) = \frac{1}{\sqrt{V_I}} \exp(i\mathbf{P}_j^I \cdot \mathbf{R}_I - iE_j^I \, t_I)\psi_{0,n}(\mathbf{r}). \tag{5.17}$$

Here the symbol j stands for both i and f, which refer to the initial and final states of the ion, respectively. $\mathbf{P}_i^I$ and $\mathbf{P}_f^I$ are the total three-momenta ($\mathbf{P}_i^I = 0$), E_i^I and E_f^I are the total energies (including the rest energies) of the ion, ψ_0 and ψ_n are the initial and final internal states of the ion and V_I is a normalization volume for the plane wave describing a free motion of the ion before and after the collision. The ansatz (5.17) represents a common form of a wavefunction for a free atomic system moving with a nonrelativistic velocity, where we have neglected the spin of the nucleus and the difference between the coordinate of the nucleus of the ion and the coordinate of the center of mass of the ion. The justification of both approximations lies in the extremely large difference between the masses of nuclei and of electrons.

As we mentioned, in this book we are not interested in discussing collisions where the ion remains in its initial internal state and, therefore, in what follows consider only $n \neq 0$. Inserting (5.17) into (5.16) and integrating over $\mathbf{R}_I$ we obtain

$$J_\mu^I(x_I) = c \, \frac{F_\mu^I\left(n0; \mathbf{P}_f^I - \mathbf{P}_i^I\right)}{V_I} \exp(i\left(\mathbf{P}_i^I - \mathbf{P}_f^I\right) \cdot \mathbf{x}_I - i\left(E_i^I - E_f^I\right)t_I). \tag{5.18}$$

We will refer to the four-component quantity $F_\mu^I(n0; \mathbf{Q})$ with components

$$F_0^I(n0; \mathbf{Q}) = - \int d^3\mathbf{r} \, \psi_n^\dagger(\mathbf{r}) \, \exp(i\, \mathbf{Q} \cdot \mathbf{r}) \, \psi_0(\mathbf{r}),$$

$$F_l^I(n0; \mathbf{Q}) = \int d^3\mathbf{r} \, \psi_n^\dagger(\mathbf{r}) \, \exp(i\, \mathbf{Q} \cdot \mathbf{r}) \, \alpha_l \psi_0(\mathbf{r}) \tag{5.19}$$

as to the inelastic form-factor of the ion.

Now we turn to the evaluation of the potential $A''^\mu_A(x_A)$, which describes the field created by the atom in the frame K_A where the atom is initially at rest. Here, $x_A = (ct_A, \mathbf{x}_A)$ is the space–time four-vector in K_A. In a way similar to that used to get the ion current (5.18), one can show that, within the first-order consideration, the transition four-current of the atom in the frame K_A reads

$$J_A'^{\,\mu}(x_A) = c \, \frac{F_A^\mu\left(m0; \mathbf{P'}_f^A - \mathbf{P'}_i^A\right)}{V_A'}$$
$$\times \exp\left(i\left(\mathbf{P'}_i^A - \mathbf{P'}_f^A\right) \cdot \mathbf{x}_A - i\left(E'_i^A - E'_f^A\right)t_A\right). \tag{5.20}$$

In (5.20) $\mathbf{P'}_{i,f}^A$ $\left(\mathbf{P'}_i^A = 0\right)$ are the three-momenta and $E'_{i,f}^A$ the total energies (including the rest energies) of the atom in the initial and final states, respectively, and V_A' is a normalization volume for the atom in the frame K_A.

The components of the form-factor of the atom $F_\mu^{\mathrm{A}}(m0; \mathbf{Q})$ are defined by

$$F_0^{\mathrm{A}}(m0; \mathbf{Q}) = Z_{\mathrm{A}}\delta_{m0} - \int \prod_{i=1}^{N_{\mathrm{A}}} \mathrm{d}^3\boldsymbol{\xi}_i \, u_m^\dagger(\boldsymbol{\tau}_{N_{\mathrm{A}}}) \sum_{j=1}^{N_{\mathrm{A}}} \exp(\mathrm{i}\,\mathbf{Q}\cdot\boldsymbol{\xi}_j)\, u_0(\boldsymbol{\tau}_{N_{\mathrm{A}}}),$$

$$F_l^{\mathrm{A}}(m0; \mathbf{Q}) = \int \prod_{i=1}^{N_{\mathrm{A}}} \mathrm{d}^3\boldsymbol{\xi}_i \, u_m^\dagger(\boldsymbol{\tau}_{N_{\mathrm{A}}}) \sum_{j=1}^{N_{\mathrm{A}}} \alpha_{l(j)} \, \exp(\mathrm{i}\,\mathbf{Q}\cdot\boldsymbol{\xi}_j)\, u_0(\boldsymbol{\tau}_{N_{\mathrm{A}}}), \quad (5.21)$$

where Z_{A} is the atomic number, N_{A} is the number of electrons of the atom, $\alpha_{l(j)}$ are the Dirac matrices for the jth electron, $u_{0,m}$ are the initial and final internal states of the atom, $\boldsymbol{\tau}_{N_{\mathrm{A}}} = \{\boldsymbol{\xi}_1, \boldsymbol{\xi}_2, \ldots, \boldsymbol{\xi}_{N_{\mathrm{A}}}\}$ represents the coordinates of the N_{A} atomic electrons with respect to the atomic nucleus.

The potential $A_\mu'^{\mathrm{A}}(x_{\mathrm{A}})$ of the atom in the frame K_{A}, is to be calculated from the Maxwell equations

$$\left(\boldsymbol{\Delta}_{\mathrm{A}} - \frac{\partial^2}{c^2\partial t_{\mathrm{A}}^2}\right) A_\mu'^{\mathrm{A}}(x_{\mathrm{A}}) = -\frac{4\pi}{c} J_\mu'^{\mathrm{A}}(x_{\mathrm{A}}), \quad (5.22)$$

where the transition four-current of the atom, $J_\mu'^{\mathrm{A}}(x_{\mathrm{A}})$, is defined by (5.20) and (5.21). Equation (5.22) can be solved by using four-dimensional Fourier transformations for the potential and the current

$$A_{\mathrm{A}}'^\mu(x_{\mathrm{A}}) = \frac{1}{(2\pi)^2}\int \mathrm{d}^4k \, B_{\mathrm{A}}^\mu(k) \, \exp(\mathrm{i}kx_{\mathrm{A}}),$$

$$J_{\mathrm{A}}'^{\,\mu}(x_{\mathrm{A}}) = \frac{c}{V_{\mathrm{A}}'}\int \mathrm{d}^4k \, \exp(\mathrm{i}kx_{\mathrm{A}}) \, \delta^{(4)}(k + P_{\mathrm{f}}'^{\mathrm{A}} - P_{\mathrm{i}}'^{\mathrm{A}}) F_{\mathrm{A}}^\mu(m0; -\mathbf{k}), \quad (5.23)$$

where $P_{\mathrm{i,f}}'^{\mathrm{A}}$ are the four-momenta of the atom in the frame K_{A}, $\mathbf{k}$ is the 'spatial' part of k and $k\,x_{\mathrm{A}} = k_\mu x_{\mathrm{A}}^\mu$. Inserting (5.23) into the Maxwell equation, the Fourier transform $B_{\mathrm{A}}^\mu(k)$ is found to be

$$B_{\mathrm{A}}^\mu(k) = 4\pi \frac{(2\pi)^2 \, \delta^{(4)}(k + P_{\mathrm{f}}'^{\mathrm{A}} - P_{\mathrm{i}}'^{\mathrm{A}}) \, F_{\mathrm{A}}^\mu(m0; -\mathbf{k})}{k^2 - \mathrm{i}\varsigma} \cdot \frac{1}{V_{\mathrm{A}}'}. \quad (5.24)$$

Correspondingly, the four-potential is given by

$$A_{\mathrm{A}}'^\mu(x_{\mathrm{A}}) = 4\pi \frac{\exp\left(\mathrm{i}\left(P_{\mathrm{i}}'^{\mathrm{A}} - P_{\mathrm{f}}'^{\mathrm{A}}\right)x_{\mathrm{A}}\right)}{\left(P_{\mathrm{i}}'^{\mathrm{A}} - P_{\mathrm{f}}'^{\mathrm{A}}\right)^2 - \mathrm{i}\varsigma} \cdot \frac{F_{\mathrm{A}}^\mu\left(m0; \mathbf{P}_{\mathrm{f}}'^{\mathrm{A}} - \mathbf{P}_{\mathrm{i}}'^{\mathrm{A}}\right)}{V_{\mathrm{A}}'}. \quad (5.25)$$

In (5.24) and (5.25) the term $-\mathrm{i}\varsigma$ with $\varsigma \to +0$ gives a prescription to handle the singularity.

If we denote by $\Lambda_{\mu\nu}$ the matrix for the Lorentz transformation from the frame K_{A} to the frame K_{I}, then the potential of the atom in the frame K_{I} is given by

$$A_A^\mu(x_I) = \Lambda^\mu_{\ \nu} A_A'^\nu(\Lambda^{-1} x_I)$$

$$= 4\pi \frac{\exp\left(i\left(P_i^A - P_f^A\right)x\right)}{\left(P_i^A - P_f^A\right)^2 - i\varsigma} \Lambda^\mu_{\ \nu} \frac{F_A^\nu\left(m0; \mathbf{P}_f'^A - \mathbf{P}_i'^A\right)}{\gamma V_A}. \quad (5.26)$$

In (5.26) P_i^A and P_f^A are the initial and final four-momentum of the atom in the frame K_I, $V_A = V_A'/\gamma$ is the normalization volume for the atom in K_I, $\gamma = 1/\sqrt{1 - \frac{v^2}{c^2}}$ is the Lorentz factor and $\mathbf{v} = (0, 0, v)$ the velocity of the incident atom in K_I.

The three-momentum transfer to the atom $\mathbf{q}_A = \mathbf{P}_f'^A - \mathbf{P}_i'^A$ in the frame K_A can be rewritten in terms of the atom momentum in the frame K_I and the atomic initial and final internal energies, given in the frame K_A,

$$\mathbf{q}_A = \left(\mathbf{P}_{f\perp}^A - \mathbf{P}_{i\perp}^A, \frac{1}{\gamma}\left(P_{f\parallel}^A - P_{i\parallel}^A\right) + \frac{v}{c^2}\left(E_i'^A - E_f'^A\right)\right)$$

$$= \left(\mathbf{P}_{f\perp}^A - \mathbf{P}_{i\perp}^A, \frac{1}{\gamma}\left(P_{f\parallel}^A - P_{i\parallel}^A\right) + \frac{v}{c^2}(\epsilon_0 - \epsilon_m)\right). \quad (5.27)$$

Here $\mathbf{P}_\perp^A$ and $P_\parallel^A$ are the parts of the three-momentum $\mathbf{P}^A$ of the atom in the frame K_I, which are perpendicular and parallel to the velocity $\mathbf{v}$, respectively. Further, ϵ_0 and ϵ_m are the initial and final electron energies of the atom given in the atomic frame. In the second line of (5.27) the recoil energy of the atom in the frame K_A has been neglected because it is negligible due to very large atomic mass.

By inserting the right hand sides of (5.18) and (5.26) into (5.14) and performing there the integration over d^4x we obtain

$$S_{fi} = -i\frac{4\pi}{V_I V_A}(2\pi)^4 \delta^{(4)}\left(P_i^I + P_i^A - P_f^I - P_f^A\right) G_{fi}, \quad (5.28)$$

where

$$G_{fi} = \frac{F_\mu^I(n0; \mathbf{q}_I)\gamma^{-1}\Lambda^\mu_{\ \nu} F_A^\nu(m0; \mathbf{q}_A)}{\left(P_i^A - P_f^A\right)^2 - i\varsigma} \quad (5.29)$$

and $\mathbf{q}_I = \mathbf{P}_f^I - \mathbf{P}_i^I = \mathbf{P}_i^A - \mathbf{P}_f^A$ is the three-dimensional momentum transfer to the ion in the frame K_I. Having derived the transition S-matrix, one can now obtain the cross section for a process where the electron of the ion and those of the atom make a transition $\psi_0 \to \psi_n$ and $u_0 \to u_m$, respectively. This cross section reads [91]

$$\frac{\mathrm{d}^2\sigma_{0\to n}^{0\to m}}{\mathrm{d}^2\mathbf{q}_\perp} = \frac{4}{v^2}\frac{E_f^A}{E_i^A}\mid G_{fi}\mid^2 = \frac{4}{v^2}\frac{E_f^A}{E_i^A}$$

$$\times \frac{\left|F_\mu^I\left(n0; \mathbf{q}_\perp, q_{min}^I\right)\gamma^{-1}\Lambda^\mu_\nu F_A^\nu\left(m0; -\mathbf{q}_\perp, -\frac{q_{min}^I}{\gamma} - \frac{v}{c^2}(\epsilon_m - \epsilon_0)\right)\right|^2}{\left(q_\perp^2 + \left(q_{min}^I\right)^2 - \left(E_f^I - E_i^I\right)^2/c^2\right)^2 + \varsigma^2}.$$

$$(5.30)$$

Here E_i^A and $E_f^A = E_i^A + E_i^I - E_f^I$ are the initial and final total energies of the atom given in the ion frame and $\mathbf{q}_\perp$ is the transverse part of the momentum transfer $\mathbf{q}_I$, which is perpendicular to the initial momentum $\mathbf{P}_i^A$ of the incident atom. Note that the factor E_f^A/E_i^A in (5.30) can be approximated by unity. Neglecting the recoil energy of the ion in the frame K_I, the difference between the total ion energies in that frame, $E_f^I - E_i^I$, is replaced by $\varepsilon_n - \varepsilon_0$, where ε_0 and ε_n are the energies of the electron of the ion in the initial and final internal states, ψ_0 and ψ_n, respectively. The component q_{min}^I of the momentum transfer $\mathbf{q}_I$, which is parallel to the initial momentum of the incident atom and represents the minimum momentum transfer to the ion in the frame K_I, is determined from the energy conservation in the collision and is given by [91]

$$q_{min}^I = \frac{\varepsilon_n - \varepsilon_0}{v} + \frac{\epsilon_m - \epsilon_0}{v\gamma}. \tag{5.31}$$

The quantity

$$\begin{aligned}
q_{min}^A &= \frac{q_{min}^I}{\gamma} + \frac{v}{c^2}(\epsilon_m - \epsilon_0) \\
&= \frac{\epsilon_m - \epsilon_0}{v} + \frac{\varepsilon_n - \varepsilon_0}{v\gamma},
\end{aligned} \tag{5.32}$$

represents the absolute value of the minimum momentum transfer to the atom in the frame K_A.

If we choose the corresponding pairs of the coordinate axes in the frames K_I (x_I, y_I, z_I) and K_A (x_A, y_A, z_A) to point in the same direction and assume that in the frame K_I the atom moves along the Z_I-axis in the positive direction the total momenta, which are transferred to the ion in the frame K_I and to the atom in the frame K_A, will be given by

$$\begin{aligned}
\mathbf{q}_I &= (q_{I,x}, q_{I,y}, q_{I,z}) = \left(\mathbf{q}_\perp; q_{min}^I\right), \\
\mathbf{q}_A &= (q_{A,x}, q_{A,y}, q_{A,z}) = \left(-\mathbf{q}_\perp; -q_{min}^A\right).
\end{aligned} \tag{5.33}$$

Note that in contrast to the nonrelativistic consideration the absolute values of the momenta q_{min}^A and q_{min}^I are no longer equal. Instead, in the relativistic treatment they are related according to

$$\left(q_{min}^A\right)^2 - \frac{(\varepsilon_n - \varepsilon_0)^2}{c^2} = \left(q_{min}^I\right)^2 - \frac{(\epsilon_m - \epsilon_0)^2}{c^2}. \tag{5.34}$$

The above equality just expresses the fact that the quantity $k_3^2 - k_0^2$, where k_0 and k_3 are the components of the virtual photon four-momentum k_μ, is invariant under a Lorentz transformation.

Using (5.32) the cross section (5.30) can be rewritten in quite a symmetric form:

$$\frac{\mathrm{d}^2\sigma_{0\to n}^{0\to m}}{\mathrm{d}^2\mathbf{q}_\perp} = \frac{4}{v^2}\frac{\mid F_\mu^{\mathrm{I}}\left(n0;\mathbf{q}_{\mathrm{I}}\right)\gamma^{-1}\Lambda_\nu^\mu\,F_{\mathrm{A}}^\nu\left(m0;\mathbf{q}_{\mathrm{A}}\right)\mid^2}{\left(\mathbf{q}_{\mathrm{I}}^2 - \frac{(\varepsilon_n - \varepsilon_0)^2}{c^2}\right)^2 + \varsigma^2}$$

$$= \frac{4}{v^2}\frac{\mid F_\mu^{\mathrm{I}}\left(n0;\mathbf{q}_{\mathrm{I}}\right)\gamma^{-1}\Lambda_\nu^\mu\,F_{\mathrm{A}}^\nu\left(m0;\mathbf{q}_{\mathrm{A}}\right)\mid^2}{\left(\mathbf{q}_{\mathrm{A}}^2 - \frac{(\epsilon_m - \epsilon_0)^2}{c^2}\right)^2 + \varsigma^2}$$

$$= \frac{4}{v^2}\frac{\mid F_\mu^{\mathrm{I}}\left(n0;\mathbf{q}_{\mathrm{I}}\right)\gamma^{-1}\Lambda_\nu^\mu\,F_{\mathrm{A}}^\nu\left(m0;\mathbf{q}_{\mathrm{A}}\right)\mid^2}{\left(q_\perp^2 + \frac{(\varepsilon_n - \varepsilon_0 + \epsilon_m - \epsilon_0)^2}{v^2\gamma^2} + 2(\gamma - 1)\frac{(\varepsilon_n - \varepsilon_0)(\epsilon_m - \epsilon_0)}{v^2\gamma^2}\right)^2 + \varsigma^2}.$$

$$(5.35)$$

5.3.1 The Form-Factor Coupling

The nonzero elements of the Lorentz transformation matrix Λ_ν^μ are (see (4.7)): $\Lambda_0^0 = \Lambda_3^3 = \gamma$, $\Lambda_1^1 = \Lambda_2^2 = 1$, and $\Lambda_3^0 = \Lambda_0^3 = -\frac{v}{c}\gamma$. Taking this into account, the explicit form of the relativistic coupling of the form-factors in (5.48) is given by

$$F_\mu^{\mathrm{I}}\gamma^{-1}\Lambda_\nu^\mu\,F_{\mathrm{A}}^\nu = \left(F_0^{\mathrm{I}} + \frac{v}{c}F_3^{\mathrm{I}}\right)\left(F_{\mathrm{A}}^0 + \frac{v}{c}F_{\mathrm{A}}^3\right) + \frac{F_3^{\mathrm{I}}F_{\mathrm{A}}^3}{\gamma^2} + \frac{F_1^{\mathrm{I}}F_{\mathrm{A}}^1 + F_2^{\mathrm{I}}F_{\mathrm{A}}^2}{\gamma}. \quad (5.36)$$

5.4 Semi-Classical Approximation

Additional important information about physics of the ion–atom collisions can be obtained by considering impact parameter dependencies of the projectile-electron excitation and loss. Such dependencies can be studied within the semi-classical approximation where both nuclei are regarded as classical particles. In Sect. 5.2 we have discussed the simplified version of the semi-classical approximation. That treatment, however, is not appropriate for considering the inelastic atomic mode. In addition, even for the elastic mode $m = 0$ the first order cross section (5.35) contains the more complicated coupling between the form-factors of the ion and atom and in general does not coincide with the cross section (5.13) obtained within the simplified semi-classical approximation. Therefore, we now will turn to the consideration of the general version of the first order semi-classical approximation for the projectile-electron excitation and loss in relativistic collision [92].

The starting expression for the semi-classical transition amplitude is formally the same as that used to develop the first order plane-wave approximation in the previous section

$$a_{fi} = -\frac{\mathrm{i}}{c}\int \mathrm{d}^4x\,J_\mu^{\mathrm{I}}(x)\,A_{\mathrm{A}}^\mu(x). \quad (5.37)$$

As before, $J_\mu^{\mathrm{I}}(x)$ denotes the electromagnetic transition four-current of the ion at a space–time point x and $A_{\mathrm{A}}^\mu(x)$ is the four-potential of the electromagnetic field created by the atom at the same point x. Now, however, these quantities and the transition amplitude (5.37) have to be evaluated within the first order perturbation theory where only the electrons are treated quantum mechanically whereas the nuclei are regarded as classical particles and their relative motion is assumed to be a straight-line.

The evaluation of the semi-classical matrix element (5.37) can be split into steps exactly similar to those used to obtain the first order results in the previous subsection. Namely, the ion current $J_\mu^I(x)$ is evaluated in the reference frame K_I, where the ion nucleus is at rest. The current $J'^\mu_A(x_A)$ and potential $A'^A_\mu(x_A)$ of the atom are calculated in the reference frame K_A, where the atomic nucleus is at rest. Then the atom potential is transformed to the frame K_I, where the transition matrix elements and corresponding probabilities are evaluated.

Assuming that the nucleus of the ion in the frame K_I rests at the origin, the transition four-current J_μ^I of the ion in this frame reads

$$J_0^I(x_I) = c \int d^3r \Psi_n^\dagger(\mathbf{r}, t_I) \left(Z_I \delta^{(3)}(\mathbf{x}_I) - \delta^{(3)}(\mathbf{x}_I - \mathbf{r}) \right) \Psi_0(\mathbf{r}, t_I),$$

$$J_l^I(x_I) = c \int d^3r \Psi_n^\dagger(\mathbf{r}, t_I) \alpha_l \, \delta^{(3)}(\mathbf{x}_I - \mathbf{r}) \Psi_0(\mathbf{r}, t_I); \; l = 1, 2, 3. \quad (5.38)$$

In (5.38) Z_I is the atomic number of the ion, $\mathbf{r}$ is the coordinate of the electron of the ion with respect to the ion nucleus, α_l are the Dirac matrices for the electron of the ion and $\delta^{(3)}$ is the three-dimensional delta-function. Further, $\Psi_{0,n}(\mathbf{r}, t) = \psi_{0,n}(\mathbf{r}) \exp(-i\varepsilon_{0,n}t)$ are the initial and final electronic states of the ion with the energies $\varepsilon_{0,n}$. As in the previous sections we will be interested only in collisions where the internal state of the ion is changed: $n \neq 0$.

It is convenient to rewrite the four-current (5.38) using the integral representation

$$\delta^{(3)}(\mathbf{x}) = \frac{1}{(2\pi)^3} \int d^3k \exp(-i\mathbf{k} \cdot \mathbf{x}) \quad (5.39)$$

for the δ-functions in (5.38). This yields

$$J_\mu^I(x_I) = \frac{c}{(2\pi)^3} \int d^3k \exp\left(i(\varepsilon_n - \varepsilon_0)t_I - i\mathbf{k} \cdot \mathbf{x}_I\right) F_\mu^I(n0; \mathbf{k}). \quad (5.40)$$

In (5.40) the four components $F_\mu^I(n0; \mathbf{k})$ of the inelastic form-factor of the ion are the same as in (5.19).

Similarly, for the transition four-current of the atom in the atom rest frame K_A one obtains

$$J'^A_\mu(x_A) = \frac{c}{(2\pi)^3} \int d^3k \exp\left(i(\epsilon_m - \epsilon_0)t_A - i\mathbf{k} \cdot \mathbf{x}_A\right) F_\mu^A(m0; \mathbf{k})$$

$$= \frac{c}{(2\pi)^3} \int d^4k \exp(ikx_A) F_\mu^A(m0; \mathbf{k}) \delta\left(\frac{\omega + \epsilon_0 - \epsilon_m}{c}\right). \quad (5.41)$$

In (5.41) $x_A = (ct_A, \mathbf{x}_A)$, $k = (\omega/c, \mathbf{k})$ and $kx_A = \omega t_A - \mathbf{k} \cdot \mathbf{x}_A$ and the components of the atomic form-factor F_μ^A are given in (5.21).

The four-potential, $A'^A_\mu(x_A)$, of the atom in its rest frame is obtained from the Maxwell equations with the source terms given by the transition four-current $J_\mu^{'A}(x_A)$ of the atom which is defined by (5.41) and (5.21). Like in the quantum consideration, in the semi-classical treatment the Maxwell equations can be easily solved with the help of a four-dimensional Fourier transformation. The result is

$$A'^A_\mu(x_A) = \frac{\exp(\mathrm{i}(\epsilon_m - \epsilon_0)t_A)}{2\pi^2} \int d^3\mathbf{k} \exp(-\mathrm{i}\mathbf{k} \cdot \mathbf{x}_A) \frac{F_\mu^A(m0; \mathbf{k})}{\mathbf{k}^2 - \frac{(\epsilon_m - \epsilon_0)^2}{c^2} - \mathrm{i}\varsigma}. \quad (5.42)$$

Note that the solutions for the four-potentials in the quantum and semi-classical considerations, given by (5.25) and (5.42), of course, do not coincide.

Let the atom move in the frame K_I along a straight-line trajectory with velocity $\mathbf{v} = (0, 0, v)$ and impact parameter $\mathbf{b} = (b_1, b_2, 0)$. Let $\Lambda_{\mu\nu}$ be the Lorentz transformation matrix from the frame K_A to the frame K_I. Then, taking into account that $t_A = \gamma(t_I - \frac{v}{c^2}x_{I3})$, $x_{A1} = x_{I1} - b_1$, $x_{A2} = x_{I2} - b_2$ and $x_{A3} = \gamma(x_{I3} - vt_I)$, the atomic four-potential in the frame K_I is given by

$$A^\mu_A(x_{I1}, x_{I2}, x_{I3}, t_I) = \frac{1}{2\pi^2} \exp\left(\mathrm{i}(E_m - E_0)t_I + \mathrm{i}(p_i - p_f)x_{I3}\right)$$

$$\times \int d^3\mathbf{k} \exp\left(-\mathrm{i}\mathbf{k}_\perp \cdot (\mathbf{x}_{I,\perp} - \mathbf{b}) - \mathrm{i}k_3\gamma(x_{I3} - vt)\right)$$

$$\times \frac{\Lambda^\mu_\nu F_A^\nu(m0; \mathbf{k})}{\mathbf{k}^2 - \frac{(\epsilon_m - \epsilon_0)^2}{c^2} - \mathrm{i}\varsigma}. \quad (5.43)$$

Here, $\mathbf{x}_{I,\perp} = (x_{I1}, x_{I2}, 0)$, $E_0 = \gamma\epsilon_0$ and $E_m = \gamma\epsilon_m$ are the total energies of the atomic electrons in the initial and final states, respectively, given in the frame K_I. Further, $p_i = \frac{v}{c^2}E_0$ and $p_f = \frac{v}{c^2}E_m$. In (5.43) the component k_3 is parallel and $\mathbf{k}_\perp$ is perpendicular to the collision velocity. Introducing the vector $\mathbf{q} = (\mathbf{q}_\perp, q_3) = (\mathbf{k}_\perp, \gamma k_3)$ (5.43) is rewritten as

$$A^\mu_A(x_I) = \frac{1}{2\pi^2} \int d^3\mathbf{q} \frac{\Lambda^\mu_\nu \gamma^{-1} F_A^\nu(m0; \mathbf{q}_\perp, \gamma^{-1}q_3)}{\mathbf{q}_\perp^2 + \frac{q_3^2}{\gamma^2} - \frac{(\epsilon_m - \epsilon_0)^2}{c^2} - \mathrm{i}\varsigma}$$

$$\times \exp\left(-\mathrm{i}\mathbf{q}_\perp(\mathbf{x}_\perp - \mathbf{b}) - \mathrm{i}(q_3 - p_i + p_f)x_{I3}\right)$$

$$\times \exp\left(\mathrm{i}(E_m - E_0 + q_3v)t_I\right). \quad (5.44)$$

Inserting (5.40) and (5.44) into (5.37), we obtain for the transition amplitude

$$a_{fi}(\mathbf{b}) = a_{0\to n}^{0\to m}(\mathbf{b}) =$$

$$-\frac{i}{2\pi^2(2\pi)^3} \int d^4 x_I \int d^3\mathbf{k} \int d^3\mathbf{q} \frac{F_\mu^I(n0; \mathbf{k}_\perp, k_3) \Lambda_\nu^\mu \gamma^{-1} F_A^\nu(m0; \mathbf{q}_\perp, \gamma^{-1} q_3)}{\mathbf{q}_\perp^2 + \frac{q_3^2}{\gamma^2} - \frac{(\epsilon_m - \epsilon_0)^2}{c^2} - i\varsigma}$$

$$\times \exp\left(-i(\mathbf{k}_\perp + \mathbf{q}_\perp) \cdot (\mathbf{x}_{I,\perp} - \mathbf{b}) - i(q_3 + k_3 - p_i + p_f)x_{I3}\right)$$

$$\times \exp\left(i(E_m + \varepsilon_n - E_0 - \varepsilon_0 + q_3 v)t_I\right). \tag{5.45}$$

This transition amplitude describes the collision process where the electron of the ion makes a transition $0 \to n$ and the electrons of the atom make a transition $0 \to m$. After performing the eightfold integration over x, $\mathbf{k}$ and q_3 in (5.45) and re-denoting $\mathbf{q}_\perp = -\mathbf{q}_\perp$ one arrives at the following expression for the transition amplitude [92]:

$$a_{0\to n}^{0\to m}(\mathbf{b}) = -\frac{i}{\pi v} \int d^2\mathbf{q}_\perp \exp(-i\mathbf{q}_\perp \cdot \mathbf{b})$$

$$\times \frac{F_\mu^I\left(n0; \mathbf{q}_\perp, q_{\min}^I\right) \gamma^{-1} \Lambda_\nu^\mu F_A^\nu\left(m0; -\mathbf{q}_\perp, -q_{\min}^A\right)}{q_\perp^2 + \frac{(\varepsilon_n - \varepsilon_0 + \epsilon_m - \epsilon_0)^2}{v^2\gamma^2} + 2(\gamma - 1)\frac{(\varepsilon_n - \varepsilon_0)(\epsilon_m - \epsilon_0)}{v^2\gamma^2} - i\varsigma}. \tag{5.46}$$

In (5.46) the integration runs over the two-dimensional vector $\mathbf{q}_\perp$ ($0 \le q_\perp < \infty$), which is perpendicular to the collision velocity. The minimum momentum transfers $q_{\min}^I$ and $q_{\min}^A$ are defined by (5.31) and (5.32), respectively.

By comparing the semi-classical transition amplitudes, given by (5.46) and (5.11), one can draw two main conclusions. First, in contrast to the simplified semi-classical treatment the more general version of the semi-classical approximation allows one to consider also collisions in which both projectile and target electrons make transitions. Second, even for the elastic target mode the transition amplitude (5.11) is, in general, not equivalent to that given by (5.46) since the latter includes a more complicated coupling between the form-factors of the ion and atom.

5.4.1 Equivalence of the Semi-Classical and the Plane-Wave Born Treatments

Very often it is more convenient to use in calculations the transition amplitude written in the momentum space $S_{fi}(\mathbf{q}_\perp)$. The latter is related to the amplitude in the impact parameter space (5.46) by the two-dimensional Fourier transformation (3.13); $\mathbf{q}_\perp$ has the physical meaning of the transverse part of the total three-momentum transfer in the collision. Taking into account (5.46) and (5.47) we obtain

$$S_{fi}(\mathbf{q}_\perp) = -\frac{2i}{v} \frac{F_\mu^I(n0; \mathbf{q}_I) \gamma^{-1} \Lambda_\nu^\mu F_A^\nu(m0; \mathbf{q}_A)}{q_\perp^2 + \frac{(\varepsilon_n - \varepsilon_0 + \epsilon_m - \epsilon_0)^2}{v^2\gamma^2} + 2(\gamma - 1)\frac{(\varepsilon_n - \varepsilon_0)(\epsilon_m - \epsilon_0)}{v^2\gamma^2} - i\varsigma}. \tag{5.47}$$

For a process, where the electron of the ion makes a transition $\psi_0 \to \psi_n$ and those of the atom make a transition $u_0 \to u_m$, the semi-classical cross section differential in the transverse momentum transfer is then given by

$$\frac{d^2\sigma_{0 \to n}^{0 \to m}}{d^2\mathbf{q}_\perp} = |\, S_{fi}(\mathbf{q}_\perp)\,|^2$$

$$= \frac{4}{v^2} \frac{|\, F_\mu^I\,(n0;\mathbf{q}_I)\,\gamma^{-1}\Lambda_\nu^\mu F_A^\nu\,(m0;\mathbf{q}_A)\,|^2}{\left(q_\perp^2 + \frac{(\varepsilon_n - \varepsilon_0 + \epsilon_m - \epsilon_0)^2}{v^2\gamma^2} + 2(\gamma - 1)\frac{(\varepsilon_n - \varepsilon_0)(\epsilon_m - \epsilon_0)}{v^2\gamma^2}\right)^2 + \varsigma^2}. \quad (5.48)$$

The cross section (5.48), which was obtained assuming that the relative projectile–target motion is described by a classical straight-line trajectory, is identical to that given by (5.35). The latter was obtained within the plane-wave Born treatment under the usual assumptions that the recoils of the nuclei can be neglected in the rest frames of these nuclei and that the maximum momentum transfer in the collision can be set to infinity.

The cross section (5.48), the amplitude (5.47) as well as the integrand of the amplitude (5.46) contain in their denominators the square $q_\mu q^\mu = (\varepsilon_n - \varepsilon_0)^2/c^2 - \mathbf{q}_I^2 \equiv (\epsilon_m - \epsilon_0)^2/c^2 - \mathbf{q}_A^2$ of the four-momentum of the virtual photon which transmits the ion–atom interaction. If this photon is on mass shell, $q_\mu q^\mu = 0$, the singularity might appear.

However, if the condition $(\varepsilon_n - \varepsilon_0)(\epsilon_m - \epsilon_0) \geq 0$ is fulfilled in the collision, one can have $q_\mu q^\mu = 0$ only if the three quantities $q_\perp$, $(\epsilon_m - \epsilon_0)$ and $(\varepsilon_n - \varepsilon_0)$ are all simultaneously equal to zero. The latter would merely mean that the collision simply does not occur and such a situation is, of course, of no interest. From the physical point of view it means that in the case when $(\varepsilon_n - \varepsilon_0)(\epsilon_m - \epsilon_0) \geq 0$ is fulfilled, the restrictions, imposed by the energy-momentum conservation in the collision, do not permit the electromagnetic interaction between the systems to be transmitted by an on-mass-shell photon.

For the moment we assume that $(\varepsilon_n - \varepsilon_0)(\epsilon_m - \epsilon_0) \geq 0$ and the terms $i\varsigma$ and ς^2 may be omitted in the amplitudes and the cross section. The case when the singularity is present and the physics corresponding to the singularity will be discussed in detail in Sect. 5.14.2.

5.5 Relativistic Features and the Nonrelativistic Limit

Compared to the nonrelativistic cross section given by (2.4) and (2.6), the cross section (5.48), in which the explicit form of the coupling of the form-factors is given by (5.36), contains two types of relativistic effects. The first type depends on the collision velocity v and disappears when $v/c \ll 1$. In detail it includes the following:

1. The so called retardation effect described by the term ω_{fi}^2/c^2. This effect leads to the appearance of the Lorentz factor in the denominator in the integrand of (5.48) and decreases its value, thus, tending to increase the cross section.

2. In contrast to the nonrelativistic consideration, the minimum values of the momenta q_{min}^I and q_{min}^A, transferred to the ion and to the atom in the corresponding reference frames, are no longer equal (compare (5.31) and (5.32)). Because of the presence of the Lorentz factor now these momenta depend differently on the transition energies $\varepsilon_n - \varepsilon_0$ and $\epsilon_m - \epsilon_0$. As a result, the dependences of the form-factors of the ion and of the atom on these transition energies are also different which has important consequences for the shielding effects in relativistic collisions.

   At the extreme relativistic impact energies, $\gamma \to \infty$, one has $q_{min}^I = (\varepsilon_n - \varepsilon_0)/c$ and $q_{min}^A = (\epsilon_m - \epsilon_0)/c$. This means that the collision kinematics for the ion and atom along the **v**-direction completely decouple from each other: the colliding particles no longer 'know' about the final internal states of their collision partner.

3. The coupling between the zeroth and third components of the form-factors in (5.36).

The second type is due to relativistic effects in the inner motions of the electron of the ion and those of the atom. Since the cross section (5.48) not only accounts for the effects related to the closeness of the collision velocity to the speed of light but also describes relativistic effects connected with the inner motion of electrons within each of the colliding centers this type of the effects does not vanish when $v/c \ll 1$ and is reflected in the cross section by the coupling between the space components of the corresponding form-factors.[4] The relativistic effects in the inner motion become less important for atomic systems having low atomic numbers. Therefore, in the case of relatively light projectiles and targets the influence of these terms on the cross section (5.48) is very weak at any collision velocity.

In the full nonrelativistic limit $c \to \infty$ both types of the relativistic effects vanish and the cross section (5.48) coincides with the corresponding nonrelativistic result obtained from (2.4) by differentiating in the transverse momentum transfer.

5.6 Consideration on the Base of Quantum Electrodynamics

The semi-classical transition amplitudes given by (5.46) and (5.47) and the cross section (5.48) have been derived using the classical description of the electromagnetic field. Besides, the electrons were described without using the second quantization. One can show that the same amplitudes and cross section are obtained if both the electromagnetic field and the electrons are treated using Quantum Electrodynamics. Below we shall briefly

---

[4] For instance, the second and third terms in (5.36) vanish at $\gamma \to \infty$ but do not disappear when $v/c \ll 1$.

discuss how the projectile–target collisions can be treated using Quantum Electrodynamics.

We start our consideration with the standard picture of fast atomic collision in a rarefied gaseous medium which involves just one particle from the beam and one particle from the target and shall consider only those collisions in which both colliding particles change their internal states. In the target frame the projectile nucleus is assumed to move along a straight-line classical trajectory $\mathbf{R}(t) = \mathbf{b} + \mathbf{v}t$, where $\mathbf{v} = (0, 0, v)$ is the collision velocity and $\mathbf{b} = (b_1; b_2; 0)$ the impact parameter with respect to the target nucleus which is at rest and taken as the origin. At $v \gg 1$ the ion–atom interaction in collisions, in which both these particles change their internal states, basically reduces to the two-center dielectronic interaction. The latter directly couples two electrons initially bound to the different colliding centers and occurs predominantly via the single photon exchange. Thus, such ion–atom collisions are in essence collisions of two electrons, which are bound to the heavy centers moving at a given relative velocity, and can be treated in the framework of the Quantum Electrodynamics.

According to Quantum Electrodynamics the general S-matrix operator for a scattering process is given by (see e.g. [73–94])

$$\hat{S} = \hat{T}\left\{\exp\left(-i\int d^4x \hat{H}_I(x)\right)\right\}, \qquad (5.49)$$

where $\hat{T}\{\}$ denotes the time-ordered product. The density of the interaction Hamiltonian $\hat{H}_I$ at a space–time point x, defined in the interaction picture, is given by the normal product, $\hat{H}_I(x) = N(\hat{j}^\mu(x)\hat{A}_\mu(x))$, $\mu = 0, 1, 2, 3$. Further, $\hat{j}^\mu = (\hat{j}^0, \hat{\mathbf{j}})$, where $\hat{j}^0(x) = \hat{\psi}^\dagger(x)\,\hat{\psi}(x)$ and $\hat{\mathbf{j}}(x) = \hat{\psi}^\dagger(x)\,\boldsymbol{\alpha}\,\hat{\psi}(x)$ are the operators for the lepton four-current density, $\hat{\psi}^\dagger(x)$ and $\hat{\psi}(x)$ are the secondly quantized lepton fields and $\hat{A}_\alpha$ is the quantized four-potential of the electromagnetic field.

In our case the initial, $|\text{ in}\rangle$, and final, $|\text{ out}\rangle$, states of the total 'projectile+target+electromagnetic field' system are the following. At $t \to -\infty$ the target is in its initial internal state which in the target rest frame is described by a four-spinor φ_0 with an energy ϵ_0. The initial internal state of the projectile in the projectile rest frame is given by a four-spinor ψ_0 and has in this frame an energy ε_0. At $t \to +\infty$ the internal states of the target and projectile in the corresponding rest frames are given by four-spinors φ_m (with an energy ϵ_m) and ψ_n (ε_n), respectively. At both $t \to -\infty$ and $t \to +\infty$ the electromagnetic field subsystem occupies its lowest possible state, the vacuum state $|0_{\text{ph}}\rangle$.

The second order term in the expansion of the S-matrix operator (5.49),

$$\hat{S}^{(2)} = -\frac{1}{2}\int d^4x \int d^4y \, T\left\{\hat{H}_I(x)\hat{H}_I(y)\right\}, \qquad (5.50)$$

is the lowest order term yielding a nonzero contribution to the transition amplitude in the case under consideration. By restricting our treatment to the lowest order term the transition amplitude is given by

$$a_{0 \to n}^{0 \to m}(\mathbf{b}) = \langle \text{out} \mid \hat{S}^{(2)} \mid \text{in} \rangle. \tag{5.51}$$

The transition amplitude (5.51) contains the so called direct and exchange parts. The exchange part describes the scattering process in which the electrons are exchanged between the binding centers during the collision process. In high-velocity collisions such an exchange is unlikely and by neglecting it in (5.51) the transition amplitude is obtained to be

$$a_{0 \to n}^{0 \to m}(\mathbf{b}) = \lim_{\varsigma \to +0} \frac{-\mathrm{i}}{\pi v} \int \mathrm{d}^2 \mathbf{q}_\perp \exp(-\mathrm{i}\mathbf{q}_\perp \cdot \mathbf{b}) \frac{G}{q^2 + \mathrm{i}\varsigma}. \tag{5.52}$$

The integration in (5.52) runs over the two-dimensional transverse momentum transfer $\mathbf{q}_\perp$ ($\mathbf{q}_\perp \cdot \mathbf{v} = 0$) in the collision.

The denominator in the integrand of (5.52) depends on the square $q^2 = q_\mu q^\mu$ of the four-momentum transfer q_μ in the collision. This term arises from the Fourier transform of the photon vacuum expectation value

$$\left\langle 0_{\text{ph}} | T \left\{ \hat{A}^\lambda(x) \hat{A}^\eta(y) \right\} | 0_{\text{ph}} \right\rangle$$

and is thus directly related to the Stückelberg–Feynman propagator $D^{\lambda\eta}$ in the four-momentum space

$$D^{\lambda\eta}(q) \propto \lim_{\varsigma \to +0} \frac{g^{\lambda\eta}}{q^2 + \mathrm{i}\varsigma}, \tag{5.53}$$

where $g^{\lambda\eta}$ is the metric tensor.

One can also show that the factor G represents the coupling between the form-factors of the projectile and the target and its form simply coincides with the form-factor coupling given by (5.36). Then, comparing (5.52) and (5.46), we see that the transition amplitude (5.52) obtained within the formalism of Quantum Electrodynamics is equivalent to the amplitude (5.46) which was derived describing the electromagnetic field classically (and without using the second quantization for the electron field).

5.7 Gauge Independence and the Continuity Equation

The electromagnetic four-potentials A_μ directly enter both the quantum and semi-classical transition amplitudes (5.14) and (5.37). However, the four-potential is not uniquely defined and depend on a choice of gauge. Namely, as was discussed in Sect. 4.2.2, any transformation of the form $A^\mu \to A'^\mu = A^\mu - \partial^\mu f$, where f is an arbitrary scalar function of x, leaves the electromagnetic field unchanged.

Formally the gauge independence of the transition matrix elements (5.14) and (5.37) can be proven by noting that one has

$$\int d^4x \, J_\mu(x) \, (A^\mu(x) - \partial^\mu f(x))$$

$$= \int d^4x \, J_\mu(x) \, A^\mu(x) - \int d^4x \, \partial^\mu \left(J_\mu(x) f(x) \right) + \int d^4x \, f(x) \partial^\mu J_\mu(x)$$

$$= \int d^4x \, J_\mu(x) \, A^\mu(x), \tag{5.54}$$

where we used the charge conservation condition, expressed by the continuity equation,

$$\partial^\mu J_\mu(x) = 0 \tag{5.55}$$

and the usual assumption that the terms $J_\mu(x)f(x)$ vanish on the four-dimensional hyper-sphere of infinite radius surrounding the charges. Using (5.18)–(5.21) (or their semi-classical counterparts) and the wave equations for the electron of the ion and those of the atom one can easily show that the first-order transition currents of the ion and atom obey the continuity equation (5.55) provided exact electronic states of the ion and atom are used.

Expanding the four-current $J_\mu(x)$ into a four-dimensional Fourier integral one can show that in the momentum space the continuity equation for the current is given by

$$q_\mu F^\mu = \frac{\omega_{fi}}{c} F^0(fi; \mathbf{q}) - q_x F^1(fi; \mathbf{q}) - q_y F^2(fi; \mathbf{q}) - q_z F^3(fi; \mathbf{q}) = 0. \tag{5.56}$$

Here, $q^\mu = \left(\frac{\omega_{fi}}{c}, q_x, q_y, q_z \right)$ is the four-momentum transfer to the particle (ion or atom) in the collision where the particle makes a transition $i \to f$ between its internal states and $F^\mu(fi; \mathbf{q})$ are the form-factors of the particle. Both q^μ and $F^\mu(fi; \mathbf{q})$ are given in the rest frame of the particle. Equation (5.56) represents a very useful relationship between the components of the form-factors.

In actual calculations one is often forced to use some approximations, e.g. for initial and final electron states of the colliding particles. In such a case the charge current in general will not be conserved and calculated results will not be gauge-independent. Since the Lorentz gauge is manifestly covariant, it is especially suited for a general consideration. However, in actual calculations this gauge may not always represent the best possible choice. For example, as is well known in the theory of ionization of atoms (or electron removal from ions) by collisions with point-like charges (see [83, 103, 104]), special care must be taken when treating ultrarelativistic collisions in the Lorentz gauge in the case when approximate electronic states of the atom/ion are used. The origin of the difficulties with the application of the Lorentz gauge is the near cancellation occurring in this gauge between the contributions from the scalar and vector potentials of an ultrarelativistically moving charge to that component of its electric field which is parallel to the velocity of the charge. Calculations with approximate states in general are not capable of dealing properly with this delicate point and yield much larger values for the field component which leads to very large errors in calculated cross sections [104].

Such a delicate interrelation between the potentials and the field is absent in other gauges which can make them more attractive in practical calculations.

5.8 Calculations in the Coulomb Gauge

As is known [83, 103, 104], results of calculations, performed for the ionization of atoms by point-like charges in the so called Coulomb gauge, are not so crucially sensitive to the accuracy of the electronic states as those performed in the Lorentz gauge. Therefore, in this section we briefly discuss the semi-classical transition amplitude and the cross section which are obtained by using the Coulomb gauge for the description of the field of the incident atom in *the ion rest frame* K_{I}.[5]

In the Coulomb gauge the potentials of the electromagnetic field are calibrated by the condition of transversality of the vector potential,

$$\mathrm{div}\,\mathbf{A}(\mathbf{x}, t) = 0, \qquad (5.57)$$

and the Maxwell equations for the field potentials in this gauge are given by (4.33). Using a method very similar to that discussed in Sect. 5.4 one can show that in the Coulomb gauge the first order semi-classical transition amplitude, written in the impact parameter space, is given by

$$a_{0 \to n}^{0 \to m}(\mathbf{b}) = -\frac{i}{\pi v} \int d^2 \mathbf{q}_\perp \exp(-i\mathbf{q}_\perp \cdot \mathbf{b})$$

$$\times \left\{ \frac{F_0^{\mathrm{I}}(n0; \mathbf{q}_{\mathrm{I}}) L_0}{\mathbf{q}_{\mathrm{I}}^2} + \frac{1}{\mathbf{q}_{\mathrm{I}}^2 - \frac{\omega_{n0}^2}{c^2}} \sum_{s=1}^3 F_s^{\mathrm{I}}(n0; \mathbf{q}_{\mathrm{I}}) L_s \right\}, \quad (5.58)$$

where the inelastic form-factors of the ion are defined in (5.19), $\omega_{n0} = \varepsilon_n - \varepsilon_0$ and $\mathbf{q}_{\mathrm{I}} = (\mathbf{q}_\perp, q_{\min}^{\mathrm{I}})$ are the energy and momentum transfer to the ion in the ion frame. Further,

$$L_0 = F_0^{\mathrm{A}}(m0; \mathbf{q}_{\mathrm{A}}) - \frac{v}{c} F_3^{\mathrm{A}}(m0; \mathbf{q}_{\mathrm{A}})$$

$$L_{1(2)} = -\frac{1}{\gamma} F_{1(2)}^{\mathrm{A}}(m0; \mathbf{q}_{\mathrm{A}}) - \frac{\omega_{n0}}{c} \frac{q_{\mathrm{I},1(2)}}{\mathbf{q}_{\mathrm{I}}^2} \left(F_0^{\mathrm{A}}(m0; \mathbf{q}_{\mathrm{A}}) - \frac{v}{c} F_3^{\mathrm{A}}(m0; \mathbf{q}_{\mathrm{I}}) \right)$$

$$L_3 = -F_3^{\mathrm{A}}(m0; \mathbf{q}_{\mathrm{A}}) \left(1 - \frac{v \omega_{n0}}{c^2} \frac{q_{\min}^{\mathrm{I}}}{\mathbf{q}_{\mathrm{I}}^2} \right) + F_0^{\mathrm{A}}(m0; \mathbf{q}_{\mathrm{A}}) \left(\frac{v}{c} - \frac{\omega_{n0}}{c} \frac{q_{\min}^{\mathrm{I}}}{\mathbf{q}_{\mathrm{I}}^2} \right),$$

$$(5.59)$$

---

[5] We remind that the Coulomb gauge is not covariant.

where the form-factors of the atom are defined in (5.21), $\mathbf{q}_A = (-\mathbf{q}_\perp, -q_{min}^A)$, as before, is the momentum transfer to the atom in its rest frame. The minimum momentum transfers q_{min}^I and q_{min}^A are given by (5.31) and (5.32). Provided exact states for the electron of the ion are used, the transition amplitudes (5.46) and (5.58) are identical.

Using the analogy with ionization–excitation processes in collisions with point-like charges, the first and second terms on the right-hand side of (5.58) can be termed as *longitudinal* and *transverse*, respectively. The first term represents the contribution to the transition amplitude which is due to the interaction of the electron of the ion with the scalar potential of the incident atom. In the Coulomb gauge the latter is the instantaneous (nonrelativistic) Coulomb potential that is reflected in (5.58) by the absence of the retardation correction $-\omega_{n0}^2/c^2$ in the photon propagator $\mathbf{q}_I^{-2}$. The transverse contribution arises due to the interaction with the vector potential of the incident atom. In the Coulomb gauge this interaction can be regarded as transmitted by a virtual photon with polarization vector perpendicular to the photon momentum $\mathbf{q}_I$, i.e. by a photon with transverse polarization. Indeed, one can show that the following condition holds

$$\mathbf{q}_I \cdot \mathbf{L} = 0, \tag{5.60}$$

where $\mathbf{L} = (L_1, L_2, L_3)$. Such a condition is inherent to a transverse photon with polarization vector $\propto \mathbf{L}$. Note that (5.60) is just the consequence of the continuity equation for the charge and current densities of the atom. Therefore, the condition given by (5.60) may be not fulfilled if any approximation for these densities are used (for example, if these densities are calculated with approximate electronic states).

The cross section, which corresponds to the transition amplitude (5.58), reads

$$\sigma_{0\to n}^{0\to m} = \frac{4}{v^2} \int d^2\mathbf{q}_\perp$$

$$\times \left| \frac{\langle \psi_n | L_0 \exp(i\mathbf{q}_I \cdot \mathbf{r}) | \psi_0 \rangle}{q_I^2} - \frac{\langle \psi_n | \mathbf{L} \cdot \boldsymbol{\alpha} \exp(i\mathbf{q}_I \cdot \mathbf{r}) | \psi_0 \rangle}{q_I^2 - \frac{\omega_{n0}^2}{c^2}} \right|^2, \tag{5.61}$$

where the form-factors $F_0^I(n0; \mathbf{q})$ and $F_s^I(n0; \mathbf{q})$ of the ion have been expressed using (5.19).

5.8.1 The Longitudinal and Transverse Contributions to the Loss Cross Section

In some cases the terms in the integrand of (5.61), which are proportional to $1/q_I^2$ and $1/(q_I^2 - \omega_{n0}^2/c^2)$, can be squared separately. This is possible, for example, if one calculates the total cross section for the electron loss from an unpolarized initial state ψ_0. In this case one can choose the quantization axis for the initial and final electron states of the ion to be along the total momentum transfer $\mathbf{q}_I$. In the case of a transverse photon with linear momentum $\mathbf{q}_I$ the projection of its angular momentum on the direction of $\mathbf{q}_I$ may take values ± 1. However, for the case of a longitudinal photon such a projection is zero. Therefore, for electron states ψ_0 and ψ_n, which are quantized along $\mathbf{q}_I$ and characterized by definite values of the magnetic quantum number, the matrix elements $\langle \psi_n | \exp(i\mathbf{q}_I \cdot \mathbf{r}) | \psi_0 \rangle$ and $\langle \psi_n | \mathbf{L} \cdot \boldsymbol{\alpha} \exp(i\mathbf{q}_I \cdot \mathbf{r}) | \psi_0 \rangle$ will satisfy different selection rules and the corresponding terms in the integrand of (5.61) can be squared separately. This will result in splitting the total loss cross section into two parts. These parts, where the corresponding integrands contain the terms proportional to $1/\mathbf{q}_I^4$ and $1/(\mathbf{q}_I^2 - \omega_{n0}^2/c^2)^2$, can be called $longitudinal$ and $transverse$ contributions, respectively, to the loss cross section.

Note that a similar separation of the ionization cross section into the longitudinal and transverse contributions[6] has been widely used in the theory of atomic K-shell ionization by relativistic collisions with point-like charges [76]. In the latter case such a separation was introduced in a paper by Fano [99] whose results in turn were based on the consideration of relativistic collisions given by Bethe and Fermi [100]. Note also that the separation of the ionization cross sections into the longitudinal and transverse contributions was a subject of certain confusion [95–98], for a brief discussion of its roots see [78].

5.9 Simplification of the Atomic Transition Four-Current: The 'Nonrelativistic Atom' Approximation

In general, the full relativistic coupling (see (5.36) and (5.58)–(5.59)) of the form-factors of the ion and atom is rather complicated. It is not only much more involved than its nonrelativistic limit but is also substantially more complicated even compared to the corresponding coupling obtained in the simplified version of the semi-classical approximation for the screening target mode in relativistic collisions. Although the keeping of the full relativistic coupling is still possible (and sometimes even necessary) when collisions with very simple (one- or two-electron) targets are treated, this becomes impractical if the projectile-electron excitation or loss occurs in collisions with multi- or many-electron targets.

In order to get simpler equations for the projectile-electron excitation and loss cross sections it was suggested in [91] to neglect the space components

---

[6] Sometimes the spin-flip part of the transverse contribution is separated and the ionization cross section is regarded as a sum of the longitudinal, transverse and spin-flip terms [83]. It is, however, more consistent to speak about just the longitudinal and transverse terms of the cross section since the spin-flip term is a natural part of the coupling of the electron to the transverse photon.

of the atomic form-factor, i.e. to disregard the three-current of the atom in the atom rest frame.[7] This step would break the symmetry between the descriptions of the ion and the atom in our consideration because the space components of the current of the ion in the ion frame are kept. However, one could immediately argue that, since we are interested in the study of the electron excitation (loss) processes mainly in (from) heavy and very heavy ions colliding with neutral atoms, this symmetry breaking in most cases should not be important because the electron of a highly charged ion and those of a neutral atom are not expected to behave similarly in the collision.

Let us first make some rough estimates for the atomic form-factors in (5.21). The component $F_A^0(m0; \mathbf{q}_A)$ of the atomic form-factor is connected with the charge distribution inside the atom. The components $F_A^l(m0; \mathbf{q}_A)$ are connected with the current, created by the motion of the electrons inside the atom in the rest frame of the atom. One can estimate roughly the magnitude of $F_A^l(m0; \mathbf{q}_A)$ as $F_A^l(m0; \mathbf{q}_A) \sim \frac{v_e}{c} F_A^0(m0; \mathbf{q}_A)$ where v_e is a characteristic velocity of the atomic electrons. For light and not too heavy atoms one has $v_e \ll c$ for all atomic electrons and the absolute values of all three components $F_A^l(m0; \mathbf{q}_A)$ are much smaller compared to that of $F_A^0(m0; \mathbf{q}_A)$. In heavy atoms the very inner electrons can have relativistic velocities. However, because the number of these electrons is relatively small compared to the total number of atomic electrons they are not expected to increase considerably the absolute value of $F_A^l(m0; \mathbf{q}_A)$. Therefore, the neglect of $F_A^l(m0; \mathbf{q}_A)$ seems to be approximately justified also for heavy atoms. In [91] the neglect of the space components of the atomic form-factor was termed 'the nonrelativistic atom (NRA) approximation'.

In the elastic target mode the electron of the ion makes a transition whereas the atomic electrons do not, and the symmetry between the highly charged ion and the neutral atom in the consideration becomes even more formal. Therefore, it seems to be obvious that the NRA approximation should be better suited for the elastic mode. Indeed, the analysis of the elastic atomic form-factors (see Sect. A.1 in the Appendix) and test calculations for the elastic mode suggest that in some cases the space components of the elastic atomic form-factor vanish *per se* and, thus, the NRA 'approximation' may actually even become exact.

In general, more care should be taken when using the NRA approximation for the inelastic mode. In the very rough estimates given above typical electron velocities in the free atom were chosen to draw conclusions about the relative importance of the form-factor components of the atom. However, in collisions with heavy projectile-ions the minimum momentum transfer $q_{min}^A = \frac{\varepsilon_n - \varepsilon_0}{v\gamma} + \frac{\epsilon_m - \epsilon_0}{v}$ can be large compared to the typical electron momenta in the atom. Because of this, the atomic electrons can acquire

---

[7] It is not very surprising that for the screening mode this would reduce the transition amplitudes and the corresponding cross sections exactly to the results following from the simplified semi-classical consideration of the Sect. 5.2.

velocities $\sim q_A = \sqrt{\mathbf{q}_\perp^2 + \left(q_{min}^A\right)^2}$, which are considerably higher than the typical electron velocities in the atomic ground state. Since it has been assumed, that the atomic electrons are nonrelativistic in the collisions, it means that the condition $q_A \ll c$ has to be fulfilled. The contributions of the atomic currents J'^1_A and J'^2_A to the transition matrix elements are suppressed by a factor of γ (see (5.36)) and the range of relatively large perpendicular momentum transfers $q_\perp \sim Z_I$ is not of great importance for the collision process. Therefore, the condition $q_A \ll c$ can be replaced by $q_{min}^A \ll c$. Further, if we estimate the energy difference $\varepsilon_n - \varepsilon_0$ as $\sim Z_I^2$ then the following condition

$$\gamma \gg \frac{Z_I^2}{vc}$$

is obtained for the use of the NRA approximation in relativistic collisions. It is certainly fulfilled for collisions with, say, $\gamma > 4$ for any heavy ion.

Another limitation for the application of the NRA approximation for the inelastic mode is also expected. As was already mentioned, it is well known in the theory of atomic ionization by a point-like charged particle that an important near cancellation may occur in the Lorentz gauge between the contributions of the scalar and vector potentials of the charged particle. A similar situation we may encounter here because of the presence of the term $\left(F_A^0 + \frac{v}{c} F_A^3\right)$. Using (5.56) one can show that

$$F_A^0(\mathbf{q}_A) + \frac{v}{c} F_A^3(\mathbf{q}_A) = \langle u_m \left| \exp(i\mathbf{q}_A \cdot \boldsymbol{\xi}) \right| u_0 \rangle + \frac{v}{c} \langle u_m \left| \alpha_z \exp(i\mathbf{q}_A \cdot \boldsymbol{\xi}) \right| u_0 \rangle$$

$$= \langle u_m \left| \exp(i\mathbf{q}_A \cdot \boldsymbol{\xi}) \right| u_0 \rangle \left(1 + \frac{v}{c^2} \frac{\epsilon_m - \epsilon_0}{q_{min}^A}\right)$$

$$+ \frac{v}{c q_{min}^A} \langle u_m \left| (\alpha_x q_{A,x} + \alpha_y q_{A,y}) \exp(i\mathbf{q}_A \cdot \boldsymbol{\xi}) \right| u_0 \rangle,$$

$$(5.62)$$

where for simplicity we assumed that the atom has just one electron. The right-hand side of (5.62) will be close to $F_A^0(\mathbf{q}_A)$ if one has simultaneously that (i)

$$v(\epsilon_m - \epsilon_0)/\left(c^2 \left| q_{min}^A \right|\right) \ll 1$$

and
(ii)

$$\left| F_A^0(\mathbf{q}_A) \right| \gg \frac{v}{c \left| q_{min}^A \right|} \left| \langle u_m \left| (\alpha_x q_{A,x} + \alpha_y q_{A,y}) \exp(i\mathbf{q}_A \cdot \boldsymbol{\xi}) \right| u_0 \rangle \right|.$$

The inequality (i) means that the condition $\gamma \ll \frac{c^2}{v^2} \frac{\varepsilon_n - \varepsilon_0}{\epsilon_m - \epsilon_0}$ must be fulfilled. Estimating the transition energies $\varepsilon_n - \varepsilon$ and $\epsilon_m - \epsilon_0$ as roughly given by $\sim Z_I^2$ and $\sim Z_A^2$, respectively, the above condition reads $\gamma \ll \frac{c^2}{v^2} \frac{Z_I^2}{Z_A^2}$. Taking into account that $c\boldsymbol{\alpha}$ represents the velocity operator, assuming that a typical 'transition' velocity of the electron in the atom can be approximated as $\sim Z_A$ and keeping in mind the restriction set by the inequality (i), one can show that the inequality (ii) reduces to $\gamma \ll \frac{c^2}{v^2} \frac{Z_I^2}{Z_A^2}$ which is just the same condition as that obtained from the inequality (i). Thus, $F_A^0(\mathbf{q}_A) + \frac{v}{c} F_A^3(\mathbf{q}_A)$ can be approximated by $F_A^0(\mathbf{q}_A)$ provided one has

$$\gamma \ll \frac{c^2}{v^2} \frac{Z_I^2}{Z_A^2}.$$

By combining results of the previous two paragraphs, we obtain the following estimate for the range of the validity of the NRA approximation for the inelastic mode

$$\frac{Z_I^2}{vc} \ll \gamma \ll \frac{Z_I^2}{vc} \frac{c^3}{vZ_A^2}, \tag{5.63}$$

where for collisions with many-electron atoms the atomic number Z_A should be replaced by some 'averaged' nuclear charge $\langle Z_A \rangle$ of the atom which is 'seen' by the majority of the atomic electrons. For light atoms, where $\langle Z_A \rangle$ does not substantially exceed 1, the condition (5.63) is not very restrictive. For collisions with heavy atoms the limitations imposed by (5.63) become rather formal since the inelastic target mode (antiscreening) is of minor importance for such collisions.

In Fig. 5.3 we illustrate the validity of the NRA approximation for the inelastic mode of $100 \, \text{GeV} \, \text{u}^{-1} \, \text{As}^{32+} + \text{H}(1s)$ collisions.

5.9.1 The Effective Atomic Charge

Within the nonrelativistic atom approximation, in which the description of the ion–atom collisions focuses mainly on the electron transitions in the ion, the form-factor coupling reduces according to

$$F_\mu^I \gamma^{-1} \Lambda_\nu^\mu F_A^\nu \rightarrow F_A^0 \left(F_0^I + \frac{v}{c} F_3^I \right). \tag{5.64}$$

The quantity

$$F_A^0 = Z_{A,\text{eff}}^{0 \rightarrow m} = Z_A \delta_{m0} - \langle u_m | \sum_{j=1}^{Z_A} \exp(i\mathbf{q}_A \cdot \boldsymbol{\xi}_j) | u_0 \rangle \tag{5.65}$$

represents in essence the effective charge of the atom which is 'seen' by the electron of the ion and which depends on the momentum transfer in the collision.

In the elastic mode of the collision this charge can be cast into a form which is very convenient for practical calculations. This can be done by noting that for $m = 0$ this charge is simply the elastic form-factor of the atom and can be presented as (see e.g. [9])

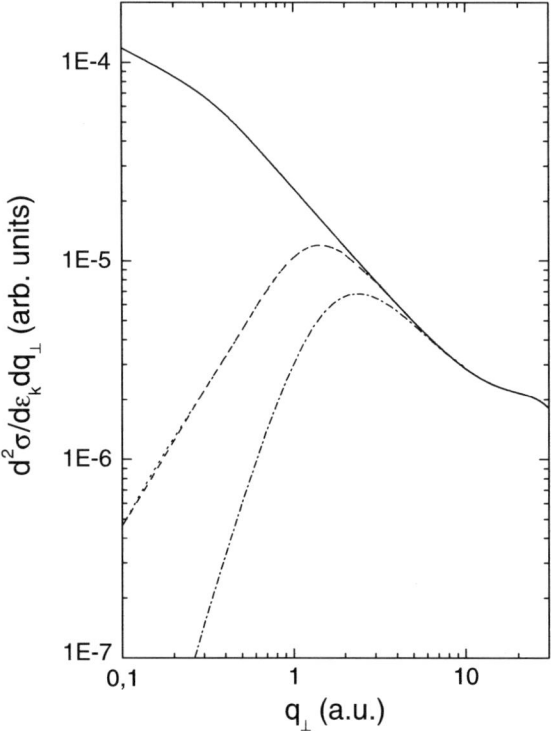

Fig. 5.3. Doubly differential cross section $\mathrm{d}^2\sigma\mathrm{d}q_\perp\,\mathrm{d}\varepsilon_k$ for the electron loss from the projectile-ion in $100\,\mathrm{GeV}\,\mathrm{u}^{-1}$ As^{32+} $+\mathrm{H}(1s)$ collisions. The cross section is given in the projectile frame as a function of the perpendicular part of the momentum transfer for a fixed energy of $\varepsilon_k = 10.8\,\mathrm{keV}$ for the electron emitted from the ion. *Dash and dot curves*: the cross section for collisions, where the hydrogen is ionized, calculated by using the full form-factor coupling (5.36) and the NRA approximation, respectively. In these two calculations the integration over all hydrogen continuum states has been performed. *Dash–dot curve*: the cross section for the elastic mode. For a comparison, the cross section in collisions with bare hydrogen nuclei is shown by a solid curve.

$$Z_{A,\mathrm{eff}}(\mathbf{q}_{A0}) = Z_A - \int \mathrm{d}^3\boldsymbol{\xi}\,\rho_{\mathrm{el}}(\boldsymbol{\xi})\,\exp(\mathrm{i}\mathbf{q}_{A0}\cdot\boldsymbol{\xi}), \qquad (5.66)$$

where $\mathbf{q}_{A0} = (-\mathbf{q}_\perp; -(\varepsilon_n - \varepsilon_0)/v\gamma)$ and $\rho_{\mathrm{el}}(\boldsymbol{\xi})$ is the charge density of the electrons in the ground state of the atom (as viewed in the frame K_A). In [101, 102] the charge density was approximated according to

$$\rho_{\mathrm{el}}(\boldsymbol{\xi}) = \frac{Z_A}{4\pi\xi} \sum_{j=1}^{3} A_j \kappa_j^2 \exp(-\kappa_j\xi), \qquad (5.67)$$

where the parameters A_j satisfy $\sum_{j=1}^{3} A_j = 1$.

The screening parameters A_j and κ_j were found in [101] by using the Thomas–Fermi model whereas in [102] they were obtained for all neutral atoms ranging between hydrogen and uranium by fitting the Dirac–Hartree–Fock–Slater data. Compared to the results of [102] the Thomas–Fermi–Moliere parametrization is somewhat less accurate and should not be applied for light atoms. However, its advantage is that the values of A_j are fixed and are same for all atoms,

$$A_1 = 0.10, \; A_2 = 0.55, \; A_3 = 0.35, \tag{5.68}$$

and the inverse screening radii κ_j have a very simple analytical dependence on the atomic number Z_A,

$$\kappa_1 = 6.7771 \, Z_A^{1/3}, \; \kappa_2 = 1.3554 \, Z_A^{1/3}, \; \kappa_3 = 0.3389 \, Z_A^{1/3}, \tag{5.69}$$

and can be easily used also for atoms heavier than uranium (for which the information is absent in [102]).

With the electron charge density in the form of (5.67) the integral in (5.66) is easily performed resulting in

$$
\begin{aligned}
Z_{A,\mathrm{eff}}(\mathbf{q}_{A0}) &= Z_A q_{A0}^2 \sum_{j=1}^{3} \frac{A_j}{\kappa_j^2 + q_{A0}^2} \\
&= Z_A \left(q_\perp^2 + \frac{(\varepsilon_n - \varepsilon_0)^2}{v^2 \gamma^2} \right) \sum_{j=1}^{3} \frac{A_j}{\kappa_j^2 + q_\perp^2 + \frac{(\varepsilon_n - \varepsilon_0)^2}{v^2 \gamma^2}}. \quad (5.70)
\end{aligned}
$$

With the effective charge in the form (5.70) expressions for the elastic amplitude and cross sections are substantially simplified. For instance, expression (5.47) for the elastic transition amplitude goes over into

$$S_{fi}(\mathbf{q}_\perp) = \frac{2i Z_A}{v} \sum_{j=1}^{3} \frac{A_j \left\langle \psi_n \left| \left(1 - \frac{v}{c}\alpha_z\right) \exp(i \mathbf{q}_{I0} \cdot \mathbf{r}) \right| \psi_0 \right\rangle}{\kappa_j^2 + q_\perp^2 + \frac{(\varepsilon_n - \varepsilon_0)^2}{v^2 \gamma^2}}, \tag{5.71}$$

where $\mathbf{q}_{I0} = (\mathbf{q}_\perp; (\varepsilon_n - \varepsilon_0)/v)$.

A few points can be mentioned here. (a) If we neglect the presence of the atomic electrons by setting all κ_j to be equal to zero, we recover the first order transition amplitude for collisions with a bare nucleus having a charge Z_A (see (5.83) of Sect. 5.11.1). (b) In the limit $c = \infty$ the elastic transition amplitude reduces to its nonrelativistic counterpart. (c) In contrast to the case where only a bare nucleus Z_A would be involved, in the limit $\gamma \to \infty$ the amplitude (5.71) becomes independent of the impact energy and does not have any singularity. This means that in the limit $\gamma \to \infty$ the elastic cross section in collisions with a neutral atomic system becomes a (finite) constant.

There exists an important general feature of the screening effect of the atomic electrons in relativistic collisions. This feature is caused by the presence of the Lorentz factor in the denominator of the right-hand side of (5.70).

In nonrelativistic collisions ($\gamma = 1$), if the ion is a highly charged ion and the atom is a light atom with all screening constants κ_i not exceeding substantially unity, the screening effect is not important because the term $\frac{(\varepsilon_n - \varepsilon_0)^2}{v^2} \sim \frac{Z_I^4}{v^2}$ would dominate over all κ_i^2 in the denominator of the right-hand side of (5.70). This reflects the fact that, according to the nonrelativistic consideration, transitions of a tightly bound electron caused by collisions with light atoms at $\gamma \simeq 1$ occur at so small impact parameters where the electrons of the atom are not able to screen the field of its nucleus.

However, the situation qualitatively changes for ultrarelativistic collisions: it is obvious that for any projectile–target pair the terms κ_i^2 will be larger than $\frac{\omega_{n0}^2}{v^2\gamma^2} \sim \frac{Z_I^4}{v^2\gamma^2}$ provided γ reaches sufficiently high values. The physics behind this formal observation is rather simple. Due to the Lorentz contraction, the range of the field produced by a high-energy point-like charge is effectively reduced by a factor of γ in the direction along its velocity and is increased by the same factor in the plane perpendicular to the velocity. Because of this the upper boundary of the range of impact parameters, at which the point-like charge is effective in inducing the electron transitions, grows linearly with γ and in principle may exceed any given value. However, due to the screening effect of atomic electrons, in collisions with neutral atoms this upper boundary cannot exceed the atomic size.

Thus, in a sharp contrast to nonrelativistic collisions, in ultrarelativistic collisions the shielding of the atomic nucleus by atomic electrons becomes of great importance for any ion–atom pair and, in particular, substantially reduces the excitation and loss cross sections compared to collisions with the unscreened nuclei.

5.10 Manipulations with the Transition Matrix Elements as a Change of Gauge

In the nonrelativistic atom approximation the cross section (5.35), (5.48) reduces to

$$\sigma_{0 \to n}^{0 \to m} = \frac{4}{v^2} \int d^2 \mathbf{q}_\perp \left| Z_{A,\text{eff}}^{0 \to m}(\mathbf{q}_A) \right|^2 \frac{\left| \langle \psi_n | \left(1 - \frac{v}{c}\alpha_z\right) \exp(i\mathbf{q}_I \cdot \mathbf{r}) | \psi_0 \rangle \right|^2}{\left(\mathbf{q}_I^2 - \frac{\omega_{n0}^2}{c^2} \right)^2}, \quad (5.72)$$

where $\omega_{n0} = \varepsilon_n - \varepsilon_0$ is the change in the energy of the electron of the ion.

Let us consider the term

$$\frac{\langle \psi_n | \left(1 - \frac{v}{c}\alpha_z\right) \exp(i\mathbf{q}_I \cdot \mathbf{r}) | \psi_0 \rangle}{\mathbf{q}_I^2 - \frac{\omega_{n0}^2}{c^2}}. \quad (5.73)$$

By applying the continuity equation (5.56) to the ion, which yields

$$\langle \psi_n | \boldsymbol{\alpha} \cdot \mathbf{q}_{\mathrm{I}} \exp(i\mathbf{q}_{\mathrm{I}} \cdot \mathbf{r}) | \psi_0 \rangle = \frac{\omega_{n0}}{c} \langle \psi_n | \exp(i\mathbf{q}_{\mathrm{I}} \cdot \mathbf{r}) | \psi_0 \rangle, \tag{5.74}$$

this term can be cast into different forms which may be more convenient for further calculations. Here we will briefly discuss three of them.

1. One of the forms is obtained as follows. We first rewrite the term in (5.73), which contains the Dirac matrix α_z, according to

$$\frac{v}{c}\alpha_z = \frac{1}{c}\mathbf{v} \cdot \boldsymbol{\alpha} = \boldsymbol{\lambda} \cdot \boldsymbol{\alpha} + \frac{\omega_{n0}}{cq_{\mathrm{I}}^2}\mathbf{q}_{\mathrm{I}} \cdot \boldsymbol{\alpha}, \tag{5.75}$$

where

$$\boldsymbol{\lambda} = \frac{1}{c}\left(\mathbf{v} - \frac{\omega_{n0}}{q_{\mathrm{I}}^2}\mathbf{q}_{\mathrm{I}}\right). \tag{5.76}$$

Then, with the help of (5.75) and making use of the continuity equation (5.74), we obtain

$$
\begin{aligned}
&\frac{\langle \psi_n | \left(1 - \frac{v}{c}\alpha_z\right) \exp(i\mathbf{q}_{\mathrm{I}} \cdot \mathbf{r}) | \psi_0 \rangle}{q_{\mathrm{I}}^2 - \frac{\omega_{n0}^2}{c^2}} \\
&= \frac{\langle \psi_n | \exp(i\mathbf{q}_{\mathrm{I}} \cdot \mathbf{r}) | \psi_0 \rangle}{q_{\mathrm{I}}^2 - \frac{\omega_{n0}^2}{c^2}} - \frac{\omega_{n0}}{cq_{\mathrm{I}}^2} \frac{\langle \psi_n | \mathbf{q}_{\mathrm{I}} \cdot \boldsymbol{\alpha} \exp(i\mathbf{q}_{\mathrm{I}} \cdot \mathbf{r}) | \psi_0 \rangle}{q_{\mathrm{I}}^2 - \frac{\omega_{n0}^2}{c^2}} \\
&\quad - \frac{\langle \psi_n | \boldsymbol{\lambda} \cdot \boldsymbol{\alpha} \exp(i\mathbf{q}_{\mathrm{I}} \cdot \mathbf{r}) | \psi_0 \rangle}{q_{\mathrm{I}}^2 - \frac{\omega_{n0}^2}{c^2}} \\
&= \frac{\langle \psi_n | \exp(i\mathbf{q}_{\mathrm{I}} \cdot \mathbf{r}) | \psi_0 \rangle}{q_{\mathrm{I}}^2 - \frac{\omega_{n0}^2}{c^2}} \left(1 - \frac{\omega_{n0}^2}{c^2 q_{\mathrm{I}}^2}\right) - \frac{\langle \psi_n | \boldsymbol{\lambda} \cdot \boldsymbol{\alpha} \exp(i\mathbf{q}_{\mathrm{I}} \cdot \mathbf{r}) | \psi_0 \rangle}{q_{\mathrm{I}}^2 - \frac{\omega_{n0}^2}{c^2}} \\
&= \frac{\langle \psi_n | \exp(i\mathbf{q}_{\mathrm{I}} \cdot \mathbf{r}) | \psi_0 \rangle}{q_{\mathrm{I}}^2} - \frac{\langle \psi_n | \boldsymbol{\lambda} \cdot \boldsymbol{\alpha} \exp(i\mathbf{q}_{\mathrm{I}} \cdot \mathbf{r}) | \psi_0 \rangle}{q_{\mathrm{I}}^2 - \frac{\omega_{n0}^2}{c^2}}.
\end{aligned}
\tag{5.77}
$$

Inserting the right-hand side of (5.77) into (5.72) we arrive at

$$
\sigma_{0 \to n}^{0 \to m} = \frac{4}{v^2} \int d^2 \mathbf{q}_\perp \left| Z_{\mathrm{A,eff}}^{0 \to m}(\mathbf{q}_{\mathrm{A}}) \right|^2
$$

$$
\times \left| \frac{\langle \psi_n | \exp(i\mathbf{q}_{\mathrm{I}} \cdot \mathbf{r}) | \psi_0 \rangle}{q_{\mathrm{I}}^2} - \frac{\langle \psi_n | \boldsymbol{\lambda} \cdot \boldsymbol{\alpha} \exp(i\mathbf{q}_{\mathrm{I}} \cdot \mathbf{r}) | \psi_0 \rangle}{q_{\mathrm{I}}^2 - \frac{\omega_{n0}^2}{c^2}} \right|^2. \tag{5.78}
$$

It is important to note that the above form of the cross section directly follows from the consideration in the Coulomb gauge. Indeed, by using the nonrelativistic atom approximation for the cross section (5.61), which was derived using explicitly the Coulomb gauge, the latter reduces to exactly the same cross section given by (5.78). Thus, simple manipulations (5.74)–(5.77) with the transition matrix elements for the electron of the ion turn out to be effectively equivalent to the transformation of the four-potentials of the incident atom from (a particular case of) the Lorentz gauge to the Coulomb one.

Since within the nonrelativistic atom approximation only the zero component of the atomic current is kept, the condition (5.60) will in general not be fulfilled. Instead, we obtain that $\boldsymbol{\lambda} \cdot \mathbf{q_I} = (\epsilon_m - \epsilon_0)/\gamma$. Thus, in collisions in the inelastic mode the 'polarization' vector $\boldsymbol{\lambda}$ of the 'transverse' virtual photon is not strictly perpendicular to the photon three-momentum $\mathbf{q_I}$. However, for high-energy ion–atom collisions involving highly charged ions the angle ϑ characterizing the deviation of the polarization vector $\boldsymbol{\lambda}$ from the transverse direction is estimated to be given by $\vartheta \sim \frac{\epsilon_n - \epsilon_0}{\gamma(\varepsilon_n - \varepsilon_0)}$. Therefore, in collisions at high values of γ the deviation is very small and one may assume that one has $\boldsymbol{\lambda} \cdot \mathbf{q_I} = 0$ also for the inelastic target mode.

2. One more form of the term (5.73) can be obtained by removing, with the help of the continuity equation (5.74), that part of the term (5.73) which does not contain the Dirac matrix:

$$\left\langle \psi_n \left| \left(1 - \frac{v}{c}\alpha_z\right) \exp(i\mathbf{q_I} \cdot \mathbf{r}) \right| \psi_0 \right\rangle$$
$$= \frac{c}{\omega_{no}} \left\langle \psi_n \left| \exp(i\mathbf{q_I} \cdot \mathbf{r}) \left(q_{I,x}\alpha_x + q_{I,y}\alpha_y + \frac{1}{\gamma} q_{min}^A \alpha_z \right) \right| \psi_0 \right\rangle . \quad (5.79)$$

Here q_{min}^A is the minimum momentum transferred to the atom (as viewed in its rest frame) which is given by the second line of (5.32).[8] Taking into account (5.79) the cross section is given by

$$\sigma_{0 \to n}^{0 \to m} = \frac{4\,c^2}{v^2 \omega_{n0}^2} \int d^2\mathbf{q_\perp} \frac{\left| Z_{A,\text{eff}}^{0 \to m}(\mathbf{q_A}) \right|^2}{\left(\mathbf{q_I}^2 - \frac{\omega_{n0}^2}{c^2} \right)^2}$$
$$\times \left| \left\langle \psi_n \left| \exp(i\mathbf{q_I} \cdot \mathbf{r}) \left(q_{I,x}\alpha_x + q_{I,y}\alpha_y + \frac{1}{\gamma} q_{min}^A \alpha_z \right) \right| \psi_0 \right\rangle \right|^2 . \quad (5.80)$$

As we have seen, the application of the continuity equation for the ion current in the form given by (5.77) is effectively equivalent to the transformation of the four-potentials of the incident atom to the Coulomb gauge. In turn the transformation (5.79) effectively corresponds to the transition to the electromagnetic gauge where the scalar potential created by the atom in the rest frame of the ion is chosen to be zero. We shall return to a more detailed

---

[8] Note that with the help of the first line of (5.32) the term q_{min}^A/γ can be rewritten as $\frac{1}{\gamma^2} q_{min}^I + \frac{v}{c^2\gamma}(\epsilon_m - \epsilon_0)$ where q_{min}^I is the minimum momentum transfer to the ion (in the rest frame of the ion) and ϵ_0 and ϵ_m are the initial and final energies of the atom (in the rest frame of the atom).

discussion of the correspondence between the manipulations with the transition matrix elements and gauge transformations in Sect. 5.11.1.

3. The third form can be obtained by using for the ion a transformation similar to that given by (5.62) for the atom

$$
\left\langle \psi_n \left| \left(1 - \frac{v}{c}\alpha_z\right) \exp(i\mathbf{q}_\mathrm{I} \cdot \mathbf{r}) \right| \psi_0 \right\rangle
$$
$$
= \frac{v}{cq^\mathrm{I}_\mathrm{min}} \left(\langle \psi_n(\mathbf{r}) \left| \exp(i\mathbf{q}_\mathrm{I} \cdot \mathbf{r}) \right| \psi_0 \rangle \frac{c}{v} \frac{q^\mathrm{A}_\mathrm{min}}{\gamma} \right.
$$
$$
\left. + \langle \psi_n(\mathbf{r}) \left| (q_{\mathrm{I},x}\alpha_x + q_{\mathrm{I},y}\alpha_y) \exp(i\mathbf{q}_\mathrm{I} \cdot \mathbf{r}) \right| \psi_0 \rangle \right). \tag{5.81}
$$

As a result, the cross section (5.72) can be rewritten as

$$
\sigma^{0 \to m}_{0 \to n} = \frac{4}{c^2 (q^\mathrm{I}_\mathrm{min})^2} \int d^2\mathbf{q}_\perp \frac{\left| Z^{0 \to m}_\mathrm{A,eff}(\mathbf{q}_\mathrm{A}) \right|^2}{\left(q_\mathrm{I}^2 - \frac{\omega_{n0}^2}{c^2} \right)^2}
$$
$$
\times \left| \langle \psi_n | \exp(i\mathbf{q}_\mathrm{I} \cdot \mathbf{r}) | \psi_0 \rangle \frac{c}{v} \frac{q^\mathrm{A}_\mathrm{min}}{\gamma} \right.
$$
$$
\left. + \langle \psi_n | (q_x\alpha_x + q_x\alpha_y) \exp(i\mathbf{q}_\mathrm{I} \cdot \mathbf{r}) | \psi_0 \rangle \right|^2. \tag{5.82}
$$

One can show that the use of the continuity equation for the ion current in the form given by (5.81) actually corresponds to the transformation of the four-potentials of the atom to the gauge in which the z component of the vector potential is zero and which is related to the gauge used in (5.72) by the transformation $A'^\mu = A^\mu - \partial^\mu f$ where the gauge function f is defined by $\frac{\partial f}{\partial t} = \frac{v^2}{c}A^0$.

5.10.1 Calculations with Approximate States for the Projectile Electron

In order to derive (5.78) and (5.82) from (5.72) we used the continuity equation for the transition current of the ion. Therefore, the cross sections (5.72), (5.78), (5.80) and (5.82) will yield identical results only if exact electronic states of the ion are used. Approximate electronic states in general cannot provide the near cancellation occurring in the Lorentz gauge between the contributions of the scalar and vector potentials to the term (5.73). Consequently, if in calculations one needs to apply approximations for the states ψ_n and ψ_0, then (5.78), (5.80) or (5.82), where the 'near-cancellation problem' is already not present, should be used as a starting point. Of course, the cross sections (5.78), (5.80) and (5.82) in general are not expected to yield identical results if any approximations are employed for the electron states of the ion. In such a case a careful analysis is necessary in order to find out which of these cross sections leads to a better result.

5.11 Projectile-Electron Transitions as a Three-Body Problem

Here we consider a further simplification in the transition four-current of the atom. This simplification may become possible if the excitation or loss of a very tightly bound electron is considered.

Let us consider a collision between an incident heavy hydrogen-like ion and a many-electron atom. We assume for the moment that we have no interest in what happens with the target-atom in such a collision and concentrate solely on the description of the electron transitions in the projectile-ion. Then, as will be seen below, the consideration of the projectile-electron excitation and loss can, under certain conditions, be reduced to a three-body problem.

As follows from (5.31) the removal (or excitation) of an electron initially occupying a very deeply bound state in a highly charged ion, which is caused by ion–atom collisions, becomes only possible if a sufficiently large momentum q_I is transferred to the electron in the rest frame of the ion. In turn, the atom in its rest frame gets a recoil momentum q_A. Although the relativistic effects caused by the ion–atom motion make q_I and q_A unequal and for a given value of q_I tend to weaken the atomic recoil momentum, it may well be, especially in collisions at low and moderate values of the impact energies, that the momentum q_A is much larger than the typical momenta of the electrons in the ground state of the atom.

If the latter is the case, then the elastic form-factor of the atom can be approximated with a good accuracy by just its nuclear part, Z_A, and the projectile-electron excitation and loss in collisions with a neutral atom which are elastic for the atom can be evaluated as occurring in collisions with a bare nucleus of the atom. If the projectile is initially a hydrogen-like ion, then the elastic mode of the projectile-electron excitation and loss reduces in essence to the relativistic three-body problem involving the electron and two nuclei of the projectile and target.

Of course, in order to really establish the range of the collision parameters, where in a theoretical analysis the elastic collisions with a neutral atom can be substituted by collisions with a nucleus of the atom, one has to evaluate the size of the shielding effect of the atomic electrons.

In Fig. 5.4 results are presented for the elastic contributions to the loss cross section in $U^{91+}(1s)$ – Au and $U^{91+}(1s)$–C collisions for a very broad interval of the impact energies ranging from 1 to $500\,\text{GeV}\,\text{u}^{-1}$. Note, that according to the consideration given in Sect. 6.5 for collisions with Au the first order approximation may noticeably overestimate the loss cross sections at impact energies $\simeq 1\,\text{GeV}\,\text{u}^{-1}$. Nevertheless, the application of this approximation can still yield valuable information about the role of the atomic electrons in the loss process.

It is seen in Fig. 5.4 that the shielding effect of the atomic electrons reduces the loss cross section at all the impact energies shown. As expected, this effect becomes especially important at very high impact energies where it

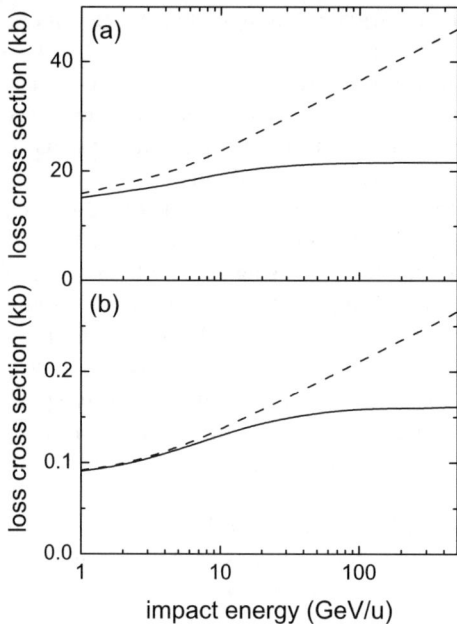

Fig. 5.4. The electron loss from $U^{91+}(1s)$ occurring in collisions with gold ($Z_A = 79$) and carbon ($Z_A = 6$) atoms. The cross section is given as a function of the collision energy. *Dash curves* display the results of the calculation for collisions with bare Au^{79+} (**a**) and C^{6+} (**b**) nuclei. *Solid curves* in (**a**) and (**b**) show the contributions to the loss cross section given by the elastic target mode in collisions with neutral gold and carbon atoms, respectively. The cross sections have been calculated within the first order approximation in the projectile–target interaction. The elastic contribution was evaluated using the nonrelativistic atom approximation and the screening parameters from [102].

is responsible for the saturation of the loss cross section: the cross section, considered as a function of the impact energy, becomes a constant. At the same time, the calculation assuming the unshielded atomic nucleus predicts an unlimited (logarithmic) grows of the cross section with increasing the impact energy.

For our present goal, however, it is more important to note that the shielding effect is rather modest for impact energies close to $1\,\mathrm{GeV\,u}^{-1}$. At these energies the effect is negligible in collisions with light carbon atoms, moreover, even in collisions with very heavy gold atoms it does not exceed 5%. Since the shielding effect decreases with decreasing the impact energy, the first order theory suggests that at all impact energies $\lesssim 1\,\mathrm{GeV\,u}^{-1}$ the role of the atomic electrons in the elastic mode of the electron loss from $U^{91+}(1s)$ is rather weak and can be neglected in collisions with both light and heavy atoms.

For a given ion at a given impact energy the shielding effect is larger in collisions with heavier atoms. Therefore, for the most heavy ions, like e.g. hydrogen-like uranium in the ground state, the elastic mode of the loss process in collisions with practically all atoms at energies relatively low (below $1\,\mathrm{GeV\,u^{-1}}$) can be described in an approximate way as occurring due to the interaction with the unscreened atomic nucleus.[9]

According to the first order consideration the inelastic mode of the projectile-electron excitation and loss comes into play at collision energies above the effective threshold starting with which the ion can be excited or lose an electron by the impact of a free electron having in the rest frame of the ion a velocity equal to that of the incident atom. In collisions with ions carrying tightly bound electrons the relative contribution of this mode to the excitation and loss cross sections scales, roughly speaking, as $1/Z_A$. This means that the inelastic atomic mode of the collision may yield an important contribution to the excitation and loss cross sections only in collisions with light atoms.

Thus, summarizing the above brief discussion, one can say that, provided one is not interested in what happens to the target atom and concentrates just on the transitions in the projectile ion, the projectile-electron excitation and loss in ion–atom collisions can be treated as a three-body problem in the following two situations. First, for collisions involving very highly charged ions and many-electron atoms at not very large impact energies where the elastic mode effectively reduces to the interaction with the unscreened atomic nucleus whereas the inelastic mode of the excitation and loss of a very tightly bound electron in collisions with such atoms is always relatively weak and can simply be ignored. Second, for collisions of highly charged ions with few-electron atoms occurring at impact energies below the effective threshold for the inelastic mode.

5.11.1 Relativistic, Nonrelativistic and Semi-Relativistic Electron Descriptions

Here we shall briefly consider the different electron descriptions used to treat the projectile-electron excitation and loss processes. The consideration will be carried out taking, as the simplest example, the three-body model of these processes discussed in the previous section.

The collision process will be considered in the rest frame of the nucleus of the ion. We take the position of the nucleus as the origin, denote the coordinates of the electron with respect to the origin by $\mathbf{r}$ and assume that in this frame the nucleus of the atom moves along a classical straight-line trajectory $\mathbf{R}(t) = \mathbf{b} + \mathbf{v}t$, where $\mathbf{b} = (b_x, b_y, 0)$ is the impact parameter, $\mathbf{v} = (0, 0, v)$ is the collision velocity and t is the time.

---

[9] Of course, much more care should be taken when considering the electron excitation and loss from not very heavy projectile-ions. Indeed, if such ions collide with heavy atoms the shielding will in general be important at any impact energy.

The expression for the first order transition amplitude $a_{fi}(\mathbf{b})$, written in the impact parameter space, follows from (5.46) after implementing there the necessary simplifications corresponding to the reduction to the three-body problem. Below, however, we shall discuss not the amplitude $a_{fi}(\mathbf{b})$ but its Fourier transform $S_{fi}(\mathbf{q}_\perp)$ written in the momentum space.

Relativistic Electron Description

Within the three-body model the general first-order amplitude (5.47) takes on the much simpler form:

$$S_{fi}(\mathbf{q}_\perp) = \frac{2iZ_A}{v} \frac{\langle \psi_{\mathrm{f}} \mid \exp(i\mathbf{q}_{\mathrm{I}} \cdot \mathbf{r})\left(1 - \frac{v}{c}\alpha_z\right) \mid \psi_{\mathrm{i}}\rangle}{q_{\mathrm{I}}^2 - (\varepsilon_{\mathrm{f}} - \varepsilon_{\mathrm{I}})^2/c^2}, \tag{5.83}$$

where ψ_{i} and ψ_{f} are the initial and final undistorted states of the electron in the ion with corresponding energies ε_{i} and ε_{f} and the momentum $\mathbf{q}_{\mathrm{I}}$, transferred to the ion in its rest frame, is now given by

$$\mathbf{q}_{\mathrm{I}} = (q_{\mathrm{I},x}, q_{\mathrm{I},y}, q_{\mathrm{I},z}) = \left(\mathbf{q}_\perp; q_{\min}^{\mathrm{I}}\right),$$
$$q_{\min}^{\mathrm{I}} = \frac{\varepsilon_{\mathrm{f}} - \varepsilon_{\mathrm{i}}}{v}. \tag{5.84}$$

The form of the amplitude (5.83) corresponds to the case when the scalar, Φ, and vector, $\mathbf{A}$, potentials of the electromagnetic field generated by the incident atomic nucleus, which enter the interaction between the electron of the ion and the atomic nucleus

$$\hat{W}_{\mathrm{d}}(t) = -\Phi(\mathbf{r}, t) + \boldsymbol{\alpha} \cdot \mathbf{A}(\mathbf{r}, t), \tag{5.85}$$

are taken in the Lienard–Wiechert form in which they read (see e.g. [7])

$$\Phi(\mathbf{r}, t) = \frac{\gamma Z_A}{s}$$
$$\mathbf{A}(\mathbf{r}, t) = \frac{\mathbf{v}}{c}\Phi(\mathbf{r}, t). \tag{5.86}$$

In (5.86) Z_A is the charge of the atomic nucleus,

$$\mathbf{s} = (s_x, s_y, s_z) = (x - b_x, y - b_y, \gamma(z - vt)) \tag{5.87}$$

are the electron coordinates with respect to the nucleus of the atom (given in the rest frame of the atom) and $\gamma = 1/\sqrt{1 - v^2/c^2}$ is the collisional Lorentz factor.

By using the conservation of the electron charge expressed as the continuity equation given by (5.74), one can get other forms of the first order transition amplitude. Here we restrict ourselves to quoting three of them[10]:

---

[10] Which directly correspond to the forms of the transition amplitude considered in Sect. 5.10.

$$S_{fi}(\mathbf{q}_\perp) = \frac{2iZ_A c}{v^2 q_{I,z}} \frac{\langle \psi_f \mid \exp(i\mathbf{q}_I \cdot \mathbf{r})\,(q_{I,x}\alpha_x + q_{I,y}\alpha_y + q_{\min}^I \alpha_z/\gamma^2) \mid \psi_i \rangle}{q_I^2 - (\varepsilon_f - \varepsilon_i)^2/c^2}, \quad (5.88)$$

$$S_{fi}(\mathbf{q}_\perp) = \frac{2iZ_A}{v} \frac{\langle \psi_f \mid \exp(i\mathbf{q}_I \cdot \mathbf{r})\left(\frac{1}{\gamma^2} + \frac{v}{cq_{\min}^I}(q_{I,x}\alpha_x + q_{I,y}\alpha_y)\right) \mid \psi_i \rangle}{q_I^2 - (\varepsilon_f - \varepsilon_I)^2/c^2} \quad (5.89)$$

and

$$S_{fi}(\mathbf{q}_\perp) = \frac{2iZ_A}{v} \left(\frac{\langle \psi_f \mid \exp(i\mathbf{q}_I \cdot \mathbf{r}) \mid \psi_i \rangle}{q_I^2} - \frac{\langle \psi_f \mid \exp(i\mathbf{q}_I \cdot \mathbf{r})\,\boldsymbol{\lambda} \cdot \boldsymbol{\alpha} \mid \psi_i \rangle}{q_I^2 - (\varepsilon_f - \varepsilon_i)^2/c^2} \right), \quad (5.90)$$

where the vector $\boldsymbol{\lambda}$ is given by (5.76).[11]

The changes in the form of the first order transition amplitude (compare (5.88)–(5.90) with (5.83)) correspond to gauge transformations of the potentials of the electromagnetic field of the nucleus of the atom:

$$\Phi(\mathbf{r},t) \to \Phi'(\mathbf{r},t) = \Phi(\mathbf{r},t) - \frac{1}{c}\frac{\partial f}{\partial t}$$
$$\mathbf{A}(\mathbf{r},t) \to \mathbf{A}'(\mathbf{r},t) = \mathbf{A}(\mathbf{r},t) + \boldsymbol{\nabla} f, \quad (5.91)$$

where $\Phi(\mathbf{r},t)$ and $\mathbf{A}(\mathbf{r},t)$ are given by (5.86).

Indeed, the amplitude (5.88) can be arrived at directly if the field potentials entering (5.85) are obtained from (5.86) using the transformation (5.91) in which the gauge function f is taken as

$$f = -\frac{c}{v} Z_A \ln(vs + vs_z). \quad (5.92)$$

The transformation (5.91) with the function (5.92) leads to the gauge in which the scalar potential of the atomic nucleus is zero and the electromagnetic field generated by this nucleus is described solely by the vector potential:

$$\Phi'(\mathbf{r},t) = 0$$
$$\mathbf{A}'(\mathbf{r},t) = -\frac{cZ_A}{vs}\left(\frac{s_x}{s+s_z}; \frac{s_x}{s+s_z}; \frac{1}{\gamma}\right). \quad (5.93)$$

On the other hand, the amplitude (5.89) is obtained directly if the field potentials entering (5.85) are derived from (5.86) using the transformation (5.91) with the gauge function f given by

$$f = \frac{v\gamma}{c} \int_0^{vt-z} d\xi \frac{Z_A}{\sqrt{(x-b_x)^2 + (y-b_y)^2 + \gamma^2\xi^2}}. \quad (5.94)$$

---

[11] And is now strictly perpendicular to the momentum transfer $\mathbf{q}_I$. The term in (5.90) containing the vector $\boldsymbol{\lambda}$ describes the contribution to the transition amplitude given by that part of the electromagnetic field of the atomic nucleus which is transversely polarized.

The above choice of a gauge function leads to the gauge in which the scalar potential, compared to its Lienard–Wiechert form, is reduced by a factor of $1/\gamma^2$ and the z-component of the vector potential is simply equal to zero:

$$\Phi'(\mathbf{r}, t) = \frac{Z_A}{\gamma s}$$

$$A'_z(\mathbf{r}, t) = 0$$

$$A'_{x,y}(\mathbf{r}, t) = -\frac{v}{c} \frac{Z_A}{\gamma s} \frac{(vt - z)s_{x,y}}{s_x^2 + s_y^2}. \tag{5.95}$$

Finally, in the case of (5.90) one can prove that this form of the amplitude can be obtained directly provided the field potentials entering (5.85) are chosen to be in the coulomb gauge in which they read

$$\Phi'(\mathbf{r}, t) = -\frac{Z_A}{\xi},$$

$$A'_x(\mathbf{r}, t) = -\frac{Z_A c}{v} \frac{\xi_x \xi_z}{\xi_x^2 + \xi_y^2} \left(\frac{1}{\xi} - \frac{\gamma}{s} \right),$$

$$A'_y(\mathbf{r}, t) = -\frac{Z_A c}{v} \frac{\xi_y \xi_z}{\xi_x^2 + \xi_y^2} \left(\frac{1}{\xi} - \frac{\gamma}{s} \right),$$

$$A'_z(\mathbf{r}, t) = \frac{Z_A c}{v} \left(\frac{1}{\xi} - \frac{1}{\gamma s} \right), \tag{5.96}$$

where

$$\xi_x = x - b_x, \xi_y = y - b_y, \xi_z = z - vt, \xi = \sqrt{\xi_x^2 + \xi_y^2 + \xi_z^2}. \tag{5.97}$$

The continuity equation holds provided ψ_i and ψ_f are exact eigenstates of the Hamiltonian $\hat{H}_d^0$. Therefore, if exact Coulomb–Dirac wavefunctions are employed to describe the initial and final undistorted states of the atom, the first order electron current is conserved, the first order transition amplitude is gauge-independent and expressions (5.83) and (5.88)–(5.90) for the amplitude are fully equivalent. The same, of course, holds true for any other expressions for this amplitude which can be obtained by manipulations with the continuity equation.

Nonrelativistic Electron Description

If the ion is relatively light, one can attempt to describe bound states of the electron using the nonrelativistic Schrödinger–Pauli equation. Then the description of the process of the excitation of the electron of the ion will contain two parts: relativistic and nonrelativistic. The electromagnetic field of the atomic nucleus will be described as before by relativistic potentials whereas the response of the electron to this relativistic field will be treated purely nonrelativistically.

Moreover, even for the electron loss process the overwhelming majority of the emission from relatively light ions is represented by electrons having non-relativistic velocities in the ionic rest frame. Since the nonrelativistic electron description in most cases is much simpler, it is tempting to treat not only the excitation but also the loss processes by considering the electron dynamics purely nonrelativistically.

One has to note, however, that too a straightforward application of the Schrödinger–Pauli equation for calculating the cross sections in relativistic collisions can easily lead to a certain confusion. Indeed, the first applications of the Schrödinger–Pauli equation for relativistic collisions had encountered substantial difficulties of principal character and had been for long time a subject of certain controversy.[12] All this is discussed in detail in Sect. A.2 of the Appendix and here we restrict ourselves just to briefly quoting correct results for the first order transition amplitude obtained with the nonrelativistic description of the electron.

Choosing the potentials of the projectile field in the Lienard–Wiechert form one obtains that the amplitude for electron transitions without spin-flip is given by

$$S_{fi}^{\text{no-flip}}(\mathbf{q}_\perp) = \frac{2iZ_A}{v} \frac{1}{\mathbf{q}_I^2 - (\varepsilon_f - \varepsilon_i)^2/c^2} \Big(\langle \varphi_f \mid \exp(i\mathbf{q}_I \cdot \mathbf{r}) \mid \varphi_i \rangle $$
$$- \frac{v}{2c^2} \langle \varphi_f \mid (\exp(i\mathbf{q}_I \cdot \mathbf{r})\hat{p}_z + \hat{p}_z \exp(i\mathbf{q}_I \cdot \mathbf{r})) \mid \varphi_i \rangle \Big). \quad (5.98)$$

Here φ_i and φ_f are (the space parts of) the initial and final electron states and $\mathbf{q}_I$ is defined by (5.84) where the relativistic energies ε_i and ε_f have to be replaced by their nonrelativistic counterparts.

It is also useful to have the expression for this amplitude obtained in the coulomb gauge:

$$S_{fi}^{\text{no-flip}} = \frac{2iZ_A}{v} \left(\frac{\langle \varphi_f \mid \exp(i\mathbf{q}_I \cdot \mathbf{r}) \mid \varphi_i \rangle}{\mathbf{q}_I^2} + \frac{\langle \psi_f \mid \exp(i\mathbf{q}_I \cdot \mathbf{r}) \, \boldsymbol{\lambda} \cdot \hat{\mathbf{p}} \mid \varphi_i \rangle}{\mathbf{q}_I^2 - (\varepsilon_f - \varepsilon_i)^2/c^2} \right),$$
$$(5.99)$$

where $\boldsymbol{\lambda}$ is defined similarly as in (5.76). Further, using the gauge with zero scalar potential $\Phi = 0$ one can show that this amplitude reads

$$S_{fi}^{\text{no-flip}} = \frac{iZ_A}{v^2} \frac{1}{q_{I,z}} \frac{\langle \varphi_f \mid \hat{G} \exp(i\mathbf{q}_I \cdot \mathbf{r}) + \exp(i\mathbf{q}_I \cdot \mathbf{r}) \, \hat{G} \mid \varphi_i \rangle}{\mathbf{q}_I^2 - (\varepsilon_f - \varepsilon_i)^2/c^2}, \quad (5.100)$$

where

$$\hat{G} = q_{I,x}\hat{p}_x + q_{I,y}\hat{p}_y + \frac{q_{I,z}}{\gamma^2}\hat{p}_z. \quad (5.101)$$

---

[12] For instance, there had been attempts to claim that even the very form of the nonrelativistic Schrödinger equation for the electron in such collisions has to be corrected.

Using the charge conservation condition for spin-no-flip transitions of the non-relativistic electron the following continuity equation is obtained

$$(\epsilon_f - \epsilon_i) \langle \varphi_f \mid e^{i\mathbf{q_I} \cdot \mathbf{r}} \mid \varphi_i \rangle = \frac{1}{2} \langle \varphi_f \mid \left(\mathbf{q_I} \cdot \hat{\mathbf{p}} \, e^{i\mathbf{q_I} \cdot \mathbf{r}} + e^{i\mathbf{q_I} \cdot \mathbf{r}} \, \mathbf{q_I} \cdot \hat{\mathbf{p}} \right) \mid \varphi_i \rangle. \quad (5.102)$$

(5.102), like its relativistic counterpart (5.74), enables to manipulate easily a form of the first order transition amplitude obtained by treating the electron nonrelativistically. In particular, one can show that, provided φ_i and φ_f are exact solutions of the Schrödinger equation, the above expressions for the transition amplitude without spin-flip are equivalent.

The amplitude for electron spin-flip transitions is obtained to be

$$S_{fi}^{\text{flip}}(\mathbf{q}_\perp) = -\frac{iZ_A}{c^2} \frac{q_{I,x} + iq_{I,y}}{\mathbf{q}_I^2 - (\varepsilon_f - \varepsilon_i)^2/c^2} \langle \varphi_f \mid \exp(i\mathbf{q_I} \cdot \mathbf{r}) \mid \varphi_i \rangle. \quad (5.103)$$

The initial expression for this amplitude directly involves the magnetic field strength rather than the field potentials. Therefore, (5.103) is obviously gauge independent.

Semi-Relativistic Electron Description

The description of the motion of the electron on the base of the Dirac equation very often involves the application of the Darwin [105] and Furry (or Furry–Sommerfeld–Maue) [106,107]) approximate wave functions. In particular, these functions have been extensively used in considerations of the various aspects of relativistic ion–atom collisions (see, for instance, [4,76] and references therein).

The main reason for the 'popularity' of the Darwin and Furry wave functions is that, compared to the exact Coulomb–Dirac wave functions, their form is much simpler and, as a result, calculations which employ these wave functions are much easier to perform. Both the Darwin and Furry wave functions are termed as semi-relativistic as well as the calculations performed using these wave functions.

Assuming that the spin of the electron in the initial state is quantized along the collision velocity $\mathbf{v}$, within the semi-relativistic Darwin approximation [105], which is accurate to first order in Z_I/c, the wavefunction is approximated by[13]

$$\psi(\mathbf{r}) = \left(1 + \frac{1}{2mc} \boldsymbol{\alpha} \cdot \hat{\mathbf{p}} \right) u^{(\pm)} \varphi(\mathbf{r}). \quad (5.104)$$

---

[13] The wavefunctions (5.104) are often multiplied by normalization factors of the form $\left(1 + Z_I^2/4c^2\right)^{-1/2}$ and $\left(1 + k^2/4c^2\right)^{-1/2}$, where k ($k \sim Z_I$) is the electron momentum in the continuum. The introduction of such factors, however, is clearly not compatible with the accuracy of the Darwin approximation.

Here $\varphi(\mathbf{r})$ is the solution of the Schrödinger equation for the undistorted atom with the Hamiltonian:

$$\hat{H}_s^0 = \frac{\hat{\mathbf{p}}^2}{2m} - \frac{Z_I}{r}, \tag{5.105}$$

$(u^{(+)})^\dagger = (1,0,0,0)$ and $(u^{(-)})^\dagger = (0,1,0,0)$.

The Furry wave function describes the electron motion in a Coulomb continuum. In the case of an electron state with the incoming boundary conditions this function reads

$$\psi_{\mathbf{k}}^{(-)}(\mathbf{r}) = \frac{1}{(2\pi)^{3/2}} \exp\left(\pi\eta/2\right) \Gamma(1+i\eta) \exp(i\mathbf{k}\cdot\mathbf{r})$$
$$\times \left(1 - \frac{c}{2E_k}\boldsymbol{\alpha}\cdot\nabla\right) {}_1F_1\left(-i\eta,1;-i(kr+\mathbf{k}\cdot\mathbf{r})\right) u^{(\lambda)}(\mathbf{k}). \tag{5.106}$$

Here $\mathbf{k} = \mathbf{v}_e/\sqrt{1-v_e^2/c^2}$ is the asymptotic momentum of the electron, $\mathbf{v}_e$ is the asymptotic electron velocity, $E_k = c\sqrt{k^2+c^2}$ is the total electron energy, and $\eta = Z_I E_k/(kc^2) = Z_I/v_e$ is the Sommerfeld parameter. Further,

$$u^{(\lambda)}(\mathbf{k}) = \sqrt{\frac{E_k+c^2}{2c^2}} \begin{pmatrix} \xi^{(\lambda)} \\ \frac{c\boldsymbol{\sigma}\cdot\mathbf{k}}{E_k+c^2}\xi^{(\lambda)} \end{pmatrix} \tag{5.107}$$

is the relativistic four-component spinor for a free electron moving with a momentum $\mathbf{k}$. In (5.107) $\boldsymbol{\sigma} = (\sigma_x, \sigma_y, \sigma_z)$ are the spin matrices given by (4.39), $\xi^{(\lambda)}$ $(\lambda = \pm)$ are the Pauli spinors,

$$\xi^{(+)} = \begin{pmatrix} 1 \\ 0 \end{pmatrix}, \quad \xi^{(-)} = \begin{pmatrix} 0 \\ 1 \end{pmatrix}, \tag{5.108}$$

and ${}_1F_1$ is the confluent hypergeometric function (see e.g. [108], [109]). In contrast to the Darwin approximation, the Furry approximation is not just an expansion in Z_I/c and can be used even for $Z_I/c \sim 1$ provided the motion of the electron with respect to the nucleus Z_I is ultrarelativistic [110].

The semi-relativistic amplitudes for the loss process obtained with using the Darwin and Furry wave functions in the three gauges discussed above are (formally) given by the expressions (5.83), (5.88) and (5.90). Important to keep in mind, however, that since these wave functions are not exact solutions of the same Hamiltonian, now all these forms of the transition amplitude are no longer equivalent.

Although (5.104) is normally considered as a bound state approximation, it is quite often used to write down electron continuum states as well and then applied to calculate the ionization/loss cross sections. Such an approach, in which both the initial and final electron states are approximated by Darwin wave functions, is also termed 'semirelativistic' and its results are widely regarded in the literature as being superior to those obtained for the excitation and loss/ionization cross sections using the purely nonrelativistic electron description. However, the latter in fact is not true [111], as will be seen below.

In the expansion over Z_I/c the Darwin states only account for the zero and first order terms and, thus, the transition matrix elements, amplitudes and cross sections evaluated with these states are accurate only to order of Z_I/c. Taking this into account and using the Lienard–Wiechert form for the projectile potentials one can show [111] that the amplitudes for electron transitions without and with spin flip are given by

$$S_{fi}^{\text{no-flip}}(\mathbf{q}_\perp) = \frac{2iZ_A}{v}\frac{1}{\mathbf{q}_I^2 - (\varepsilon_f - \varepsilon_i)^2/c^2}\left(\langle\varphi_f \mid \exp(i\mathbf{q}_I \cdot \mathbf{r}) \mid \varphi_i\rangle\right.$$
$$\left. - \frac{v}{2c^2}\langle\varphi_f \mid (\exp(i\mathbf{q}_I \cdot \mathbf{r})\hat{p}_z + \hat{p}_z \exp(i\mathbf{q}_I \cdot \mathbf{r})) \mid \varphi_i\rangle\right) \quad (5.109)$$

and

$$S_{fi}^{\text{flip}}(\mathbf{q}_\perp) = -\frac{iZ_A}{c^2}\frac{q_{I,x} + iq_{I,y}}{\mathbf{q}_I^2 - (\varepsilon_f - \varepsilon_i)^2/c^2}\langle\varphi_f \mid \exp(i\mathbf{q}_I \cdot \mathbf{r}) \mid \varphi_i\rangle. \quad (5.110)$$

Comparing (5.109)–(5.110) with (5.98) and (5.103) we see that if the initial and final states in the relativistic transition amplitude are approximated by the Darwin wave functions then the corresponding expressions for the transition amplitudes simply coincide with those obtained with the Schrödinger–Pauli equation [111]. Thus, within the first order consideration of the excitation and loss/ionization processes the 'semi-relativistic' electron description employing the Darwin approximation for both the initial and final electron states in fact does not represent any improvement over the nonrelativistic electron description based on the Schrödinger–Pauli equation.

Comparison of the Relativistic and Nonrelativistic Electron Descriptions

In Fig. 5.5 we show cross sections for the electron excitation from the ground state into states with the principal quantum number $n = 2$ and the total angular momentum $j = 1/2$ and $j = 3/2$ in collisions with protons. These cross sections were calculated by using the relativistic (Dirac) and nonrelativistic (Schrödinger–Pauli) descriptions of the electron and are given in the figure as a function of the collision energy for different hydrogen-like ions whose atomic number ranges from 20 to 92. One interesting observation which can be drawn from the figure is that the difference between the results of both electron descriptions for the excitation into the states with $j = 1/2$ remains rather small even for $Z_I = 92$ whereas the corresponding difference in the case of the excitation into the states with $j = 3/2$ becomes noticeable already at $Z_I \simeq 30$ and reaches about a factor of 2 at $Z_I = 92$.

For the excitation of light and intermediately heavy ions both the relativistic and nonrelativistic descriptions of the electron predict that the total cross section for the excitation into the states with $n = 2, j = 3/2$ is larger than that for the excitation into the states with $n = 2, j = 1/2$. For the excitation of very heavy ions (like Bi^{82+} and U^{91+} in the figure) the nonrelativistic

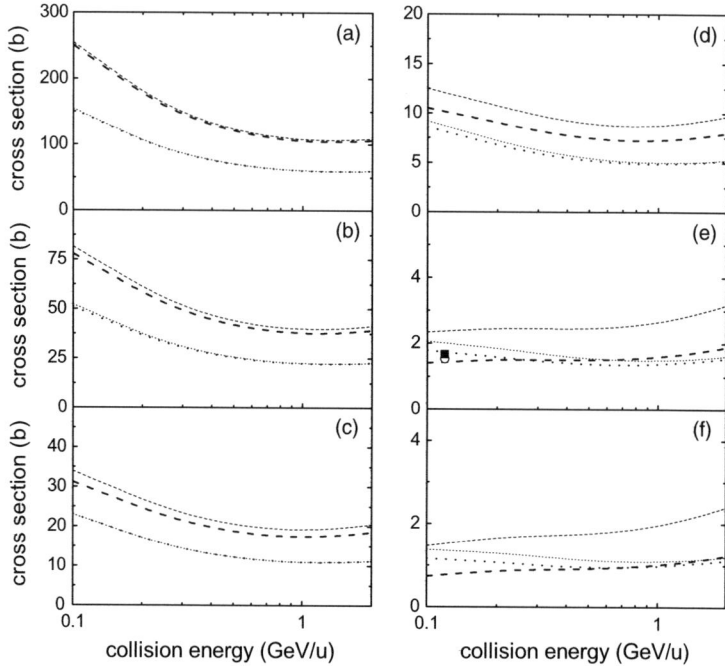

Fig. 5.5. Cross sections for the proton-impact excitation of hydrogen-like ions ((a) Ca^{19+}, (b) Zn^{29+}, (c) Zr^{39+}, (d) Xe^{53+}, (e) Bi^{82+}, (f) U^{91+}) from the ground state into the states with $n = 2, j = 1/2$ and $n = 2, j = 3/2$. Results of the calculation with the Dirac equation are displayed by *thick curves* (*dashed curve*: $n = 2, j = 3/2$; *dot curve*: $n = 2, j = 1/2$). Results obtained with the Schrödinger–Pauli equation are shown by *thin curves* (*dashed curve*: $n = 2, j = 3/2$; *dot curve*: $n = 2, j = 1/2$). In figure (**e**) there are also experimental results from [95, 96] for the excitation of $119\,\mathrm{MeV\,u^{-1}}$ $Bi^{82+}(1s)$ into the states with $n = 2, j = 1/2$ (*solid square*) and $n = 2, j = 3/2$ (*open circle*). These results were measured in collisions with carbon ($Z_A = 6$) and have been scaled in the figure by a factor of $1/36$. From [111].

description suggests that the above correspondence between the cross section still holds for the whole impact energy interval shown in the figure. However, according to the relativistic electron description, at the lower impact energies the excitation into the states with $n = 2, j = 1/2$ turns out to be larger compared to that into the states with $n = 2, j = 3/2$.

In Fig. 5.6 we display cross sections for the electron loss from $Au^{79+}(1s)$ and $Pb^{81+}(1s)$ in collisions with carbon calculated by using the relativistic and nonrelativistic descriptions for the electron of the ion. In these calculations (except that depicted in Fig. 5.6a) it was assumed that the loss of the electron of the ion in collisions with neutral carbon atoms is caused only by the interaction between this electron and the unscreened nucleus of the atom. For such very asymmetric collision systems this assumption is very well

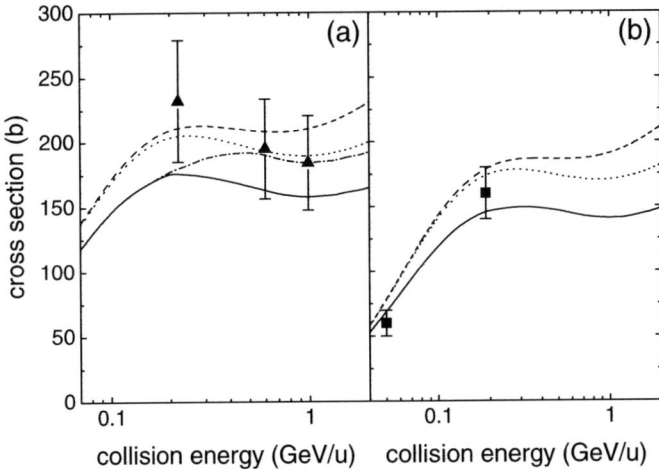

Fig. 5.6. Cross sections for the electron loss from $Au^{78+}(1s)$ (a) and $Pb^{81+}(1s)$ (b) in collisions with carbon given as a function of the collision energy. *Dash curve*: the nonrelativistic electron description. *Dot curve*: the nonrelativistic electron description, only no-spin-flip transitions are taken into account. *Solid curve*: the relativistic electron description. Experimental data are from [112] (*triangles*) and [113] (*squares*). In panel (a) *dash–dot curve* displays results which include also the contribution to the electron loss due to the interaction with the atomic electrons (the inelastic target mode). From [111].

fulfilled below the threshold energy for the antiscreening target mode of the loss process which in the case of collisions with $Au^{78}(1s)$ and $Pb^{81}(1s)$ is $\simeq 170$ and $\simeq 185\,MeV\,u^{-1}$, respectively. Above this impact energy the antiscreening mode gives a noticeable contribution increasing the total loss cross section (see dash–dot curve in Fig. 5.6a). Compared to the relativistic electron description, the description with the Schrödinger–Pauli equation yields larger loss cross sections and this difference tends to increase with the impact energy. Except the lowest energy experimental point in Fig. 5.6a the comparison with experimental data is in favor of the relativistic electron description. Note, however, that although there is no question of which kind of the description is better suited for the electron in a very heavy ion, the results of this comparison *per se* should be taken with certain reservation since the experimental data shown in the figure were obtained in collisions with solid state carbon target.

In Fig. 5.7 the comparison of the relativistic and nonrelativistic electron descriptions is presented for the case of the loss from $Nd^{59+}(1s)$. It is seen that in collisions with moderate γ the difference between the results of these descriptions does not exceed 25–30% which is close to typical accuracy of collisional experiments with heavy ions at these γ.

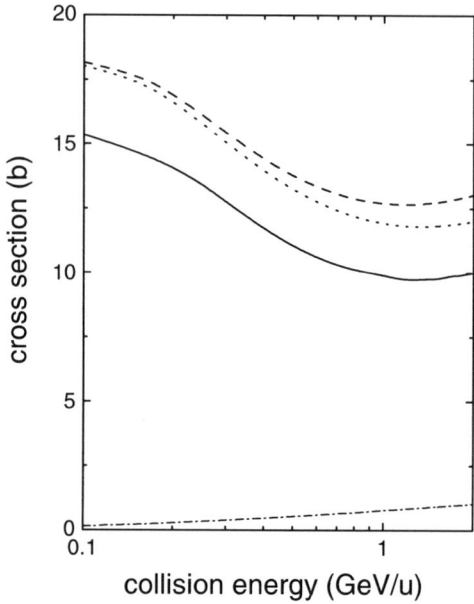

Fig. 5.7. The total cross section for the electron removal from $Nd^{59+}(1s)$ by the proton impact given as a function of the collision energy. *Dot curve*: the nonrelativistic electron description, the spin-flip transitions are not included. *Dash–dot curve*: the nonrelativistic electron description, the contribution of the spin-flip transitions. *Dash curve*: the nonrelativistic electron description, the total contribution. *Solid curve*: the relativistic electron description. From [111].

5.12 Relativistic Ion–Atom Collisions and Nonrelativistic Form-Factors

If in high-velocity collisions both the ion and atom are light atomic particles, the three-body model is clearly not applicable. However, for such collisions one can still considerably simplify the treatment by taking into account that: (a) in the rest frame of the ion the motion of the electron(s) of the ion is nonrelativistic before the collision and remains nonrelativistic also during and after the collision; (b) the same holds true for the motion of the atomic electrons considered in the rest frame of the atom.

The form-factors of the ion and atom, which enter the expressions (5.35), (5.47), (5.48) and (5.36) for the transition amplitudes and cross sections, are given in the corresponding rest frames where, in the case of light ions and atoms, the behavior of the electrons is practically nonrelativistic. Therefore, the form-factors in these formulas, which were obtained by assuming that the electrons should be described relativistically, can be replaced by their nonrelativistic counterparts which follow from the Schrödinger equation. For the most important part of the electron phase-space this step introduces no noticeable inaccuracy into the description while making practical calculations much simpler to perform.

Neglecting the electron spin degrees of freedom it is not difficult to show that in the Lorentz gauge the nonrelativistic form-factors are given by

$$F_0^{\mathrm{I}}(n0; \mathbf{q}) = -\int d^3\mathbf{r}\, \varphi_n^*(\mathbf{r})\, \exp(i\mathbf{q}\cdot\mathbf{r})\, \varphi_0(\mathbf{r}) = -\langle \varphi_n \mid \exp(i\mathbf{q}\cdot\mathbf{r}) \mid \varphi_0 \rangle,$$

$$F_l^{\mathrm{I}}(n0; \mathbf{q}) = \frac{1}{2c}\langle \varphi_n \mid (\exp(i\mathbf{q}\cdot\mathbf{r})\hat{p}_z + \hat{p}_z \exp(i\mathbf{q}\cdot\mathbf{r})) \mid \varphi_0 \rangle,$$

$$F_0^{\mathrm{A}}(m0; \mathbf{q}) = Z_{\mathrm{A}}\delta_{n0} - \left\langle \phi_m \mid \sum_j^{Z_{\mathrm{A}}} \exp(i\mathbf{q}\cdot\mathbf{r}_j) \mid \phi_0 \right\rangle,$$

$$F_l^{\mathrm{A}}(m0; \mathbf{q}) = \frac{1}{2c}\langle \phi_m \mid (\sum_j^{Z_{\mathrm{A}}} \exp(i\mathbf{q}\cdot\mathbf{r}_j)\hat{p}_{z,j} + \sum_j^{Z_{\mathrm{A}}} \hat{p}_{z,j} \exp(i\mathbf{q}\cdot\mathbf{r})) \mid \phi_0 \rangle.$$

$$(5.111)$$

5.13 Electron–Positron Pair Production in Collisions of Bare Ions with Neutral Atoms

The general expressions (5.35), (5.47), (5.48) and (5.36), which were derived within the first order theory in the projectile–target interaction for the cross sections of the projectile-electron excitation and loss and their consequent simplifications can be applied, with minimum changes, to the process of electron–positron pair production occurring in relativistic collisions between a bare ionic nucleus and a neutral atom.

Indeed, within the famous Dirac sea picture the pair production is viewed as a transition between electronic states with negative and positive total energy. If we (a) make the assumption that these states are strongly influenced only by the field of the ionic nucleus while the field of the atom acts merely as a collisional perturbation (which couples these states leading to the pair production) and (b) neglect the interaction between the created electron and positron, then the very close analogy with the projectile-electron excitation and loss processes becomes obvious.

In the case of the free pair production, in which both created particles move in the continuum, the above assumption looks quite natural for asymmetric collisions where the charge of the ionic nucleus is larger than that of the neutral atom. This assumption is also valid for the free-bound pair production where the electron is created in a bound state of the ion. Besides, since the difference in the velocities of the electron and positron is typically much larger than 1 a.u., the neglect of the electron–positron interaction clearly represents a good approximation.

Although in the case of the pair production the simplification of the corresponding general expressions can be done in a way very similar to that discussed for the projectile-electron excitation and loss, a few particular points should be mentioned. First, the momenta transfers necessary to produce an electron–positron pair are on overall substantially larger compared to those typical for the process of the projectile-electron excitation and loss. This means that the shielding effect of the atomic electrons will become of importance at larger impact energies and the effective three-body model for the pair production in collisions between a bare ion and a neutral atom will be valid for a broader range of the impact energies. Second, since a substantial part of the positron emission (and, in the case of free–free pair production, of the electron emission) occurs into states with high energies, the nonrelativistic description of the electron and positron is not expected to yield reasonable results for the pair production, even if the process takes place in collisions of relatively light ions and atoms.

On the other hand, since in collisions with very high impact energies most of the particles created in the continuum have ultrarelativistic energies, the pair production at such impact energies can be successfully calculated using the Furry wave function [75, 114–116].

5.14 Two-Center Dielectronic Transitions

In high-velocity collisions between sufficiently light projectile ion and target atom ($Z_I \ll v$, $Z_A \ll v$) the interaction between the projectile and the target normally proceeds only via the exchange of a single virtual photon. However, already just the single photon exchange between two electrons, one of which is initially bound in the projectile and the second one belongs to the target, may result in simultaneous electron transitions in the projectile and the target. Therefore, in relative terms the two-center dielectronic interaction (TCDI) becomes of especial importance in high-energy collisions.

5.14.1 Mutual Projectile–Target Ionization

Here, following [118,119], we shall consider the two-center dielectronic interaction in the case of mutual projectile–target ionization occurring at relativistic collision velocities.

The most basic situation for studying the TCDI is to explore collisions of hydrogen-like ions with hydrogen atoms. From the theoretical point of view it is also the simplest situation since the internal states of both colliding particles are exactly known.

In the target reference frame K_A the incident projectile-ion is assumed to have a velocity $\mathbf{v} = (0, 0, v)$ whose magnitude can be comparable to the speed of light $c \simeq 137$ a.u. The corresponding Lorentz factor of the collision is given by $\gamma = 1/\sqrt{1 - v^2/c^2}$.

According to (5.35) and (5.36) the fully differential cross section for the mutual ionization of the target and projectile in a single collision event can be written as

$$\frac{d\sigma}{d^3\mathbf{k}_a d^3\mathbf{k}_i d^2\mathbf{q}_\perp} = \frac{4}{v^2} \frac{1}{\left(q_A^2 - (\epsilon_{k_a} - \epsilon_0)^2/c^2\right)^2}$$

$$\times \left| \left(F_0^I(\mathbf{k}_i, \mathbf{q}_I) - \frac{v}{c} F_3^I(\mathbf{k}_i, \mathbf{q}_I) \right) \left(F_A^0(\mathbf{k}_a, \mathbf{q}_A) - \frac{v}{c} F_A^3(\mathbf{k}_a, \mathbf{q}_A) \right) \right.$$

$$+ \frac{F_3^I(\mathbf{k}_i, \mathbf{q}_I) F_A^3(\mathbf{k}_a, \mathbf{q}_A)}{\gamma^2}$$

$$\left. + \frac{F_1^I(\mathbf{k}_i, \mathbf{q}_I) F_A^1(\mathbf{k}_a, \mathbf{q}_A) + F_2^I(\mathbf{k}_i, \mathbf{q}_I) F_A^2(\mathbf{k}_a, \mathbf{q}_A)}{\gamma} \right|^2 . \qquad (5.112)$$

Here, ϵ_0 and ϵ_{k_a} are initial and final internal energies of the atom, respectively, $\mathbf{k}_a$ is the three-momentum of the electron emitted from the atom and $\mathbf{q}_A$ is the three-momentum transferred to the atom; all these quantities are given in the atomic rest frame. The quantities ε_0, ε_{k_i}, $\mathbf{k}_i$ and $\mathbf{q}_I$ have similar meanings but are for the ion and given in the rest frame of the ion. Since we consider light ions and atoms the inelastic form-factors for these particles can be evaluated with the nonrelativistic wave functions (see (5.111)).

Below we shall concentrate on the electron transitions in the target and it is more convenient to redefine the momentum transfers to the atom and ion according to $\mathbf{q}_A = (\mathbf{q}_\perp, q_{min}^A)$ and $\mathbf{q}_I = (-\mathbf{q}_\perp, -q_{min}^I)$, where q_{min}^I and q_{min}^A are given by (5.31) and (5.32), respectively.[14]

Fivefold Differential Cross Section

Let us begin with the cross section

$$\frac{d\sigma}{d^3\mathbf{k}_a d^2\mathbf{q}_\perp} = \int d^3\mathbf{k}_i \frac{d\sigma}{d^3\mathbf{k}_a d^3\mathbf{k}_i d^2\mathbf{q}_\perp}, \qquad (5.113)$$

where the integration is performed over all possible final continuum states of the electron emitted from the projectile. Compared to the fully differential cross section, given by (5.112), the cross section (5.113) yields less detailed but more 'compressed' and 'compact' information about the collision dynamics.[15]

---

[14] Note that this definition of the momentum transfers differs from that in (5.33) by the change in sign. This change as well as the minus signs in the form-factor coupling in (5.112) are caused by the fact that now $\mathbf{v}$ denotes the velocity of the ion in the frame K_A.

[15] Besides, one should also note that theoretical results for the cross section (5.112) would be hard to verify in experiment since it is very difficult to analyze fragments of a very fast projectile simultaneously with those of a target. At the same time the cross section of the type given by (5.113) is expected to be more accessible for observations with state-of-the-art experimental techniques which allow one to detect relatively easy target fragments in coincidence with a charge state of the residual projectile-ion.

The very substantial part of the electron emission from the atom occurs in collisions where the momentum of the emitted electron lies in the plane defined by the vectors of the incident projectile velocity $\mathbf{v}$ and the momentum transfer $\mathbf{q}_A$. In this plane the cross section (5.113) reaches its largest values. Such a collision geometry is commonly referred to as the coplanar geometry. For this collision geometry the cross section (5.113) is displayed in Fig. 5.8a–c for $O^{7+}(1s) + H(1s) \rightarrow O^{8+} + e^- + H^+ + e^-$ collisions. The energy of the electron emitted from hydrogen is taken to be 5 eV and the magnitude of the transverse momentum transfer is $q_\perp = 0.1$ a.u.

Let us first consider the cross section (5.113) in the nonrelativistic limit $c \rightarrow \infty$. According to the nonrelativistic consideration the cross section (5.113)

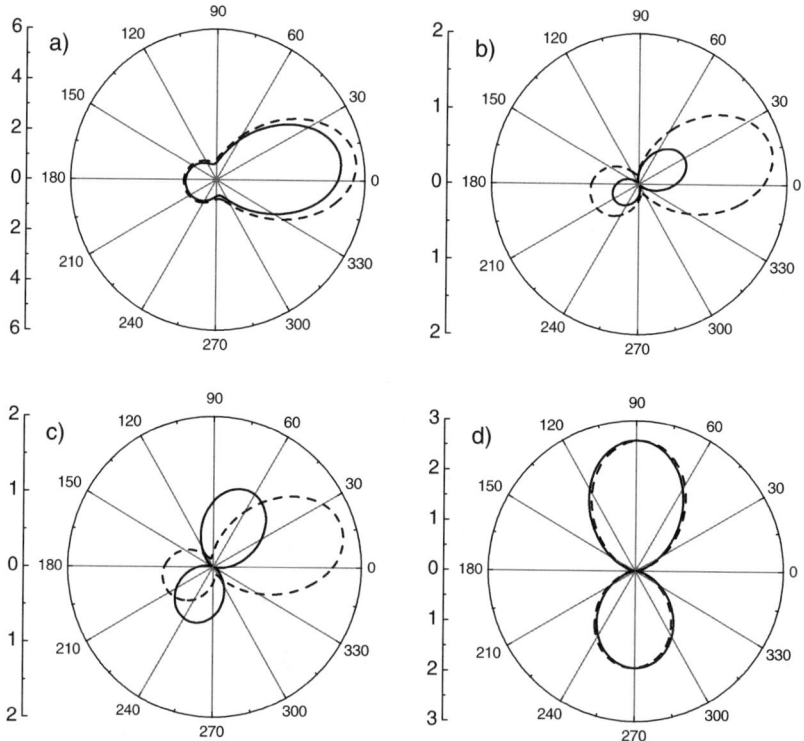

Fig. 5.8. (a)–(c) 'Fully' differential cross section (in arb. units), as a function of the polar electron emission angle, $\vartheta_a = \arccos(\mathbf{k}_a \cdot \mathbf{v}/(k_a v))$, in the coplanar geometry for ionization of hydrogen in $O^{7+}(1s) + H(1s) \rightarrow O^{8+} + H^+ + 2e^-$ collisions, $\varepsilon_{k_a} = 5$ eV, $\mathbf{q}_\perp = 0.1$ a.u. (a) 100 MeV u^{-1} ($v \simeq 60$ a.u., $\gamma \simeq 1.11$); (b) 1 GeV u^{-1} ($v \simeq 120$ a.u., $\gamma \simeq 2.07$); (c) 5 GeV u^{-1} ($\gamma \simeq 6.37$). (d) Fully differential cross section (in arb. units) in the coplanar geometry for ionization of hydrogen by a free electron impact, $\varepsilon_{k_a} = 5$ eV, $q_\perp = 0.1$ a.u. and $\gamma = 1,000$. Results of relativistic ($c \simeq 137$ a.u.) and nonrelativistic ($c = \infty$) calculations are depicted by *solid* and *dashed curves*, respectively.

reaches its largest value along the nonrelativistic mean momentum transfer $\langle \mathbf{q}_A \rangle$ which is obtained by averaging the momentum $\mathbf{q}_A$ over the final states of the electron emitted from the projectile-ion. This maximum in the cross section can be called the binary peak (see Fig. 5.8a–c). The second and smaller maximum appears in the direction opposite to $\langle \mathbf{q} \rangle$ (the recoil peak). The binary peak is to a large extend determined only by a (single) collision between the electron of the atom and that of the ion: in collisions leading to the binary peak the role of the atomic nucleus in the final atomic state is rather modest. In contrast, the 'post-collision' interaction between the atomic electron and the atomic nucleus is crucial for the very existence of the recoil peak since it is the only mechanism capable of producing the back-scattering of the atomic electron after it has collided with the electron of the projectile.

Let us now turn to the relativistic consideration. Figure 5.8a–c show that relativistic effects in ionization of hydrogen via the TCDI can be noticeable already at $\gamma \simeq 1.1$ and may become very substantial at $\gamma \simeq 2$ and higher. They can change both the shape and absolute values of the calculated cross section. For instance, we observe that the positions of the binary and recoil peaks are shifted compared to predictions of the nonrelativistic treatment. At the collisions parameters considered in Fig. 5.8a–c, the main effect arises due to the presence of the Lorentz factor in the momentum $q_{\min}^A$. For rather asymmetric collision systems, this can result in a substantial reduction of the magnitude of $\langle q_{\min}^A \rangle$ (which is obtained by averaging $q_{\min}^A$ over the final states of the electron emitted from the projectile) compared to its value in the nonrelativistic theory.[16] This may have a very pronounced impact on ionization of the lighter partner.

The relativistic reduction of $\langle q_{\min}^A \rangle$ can substantially affect both the shape and absolute values of the cross section (5.113). In particular, it changes positions of the binary and recoil peaks with respect to predictions of the nonrelativistic treatment. However, in Fig. 5.8a–c the positions of relativistic peaks still remain to be determined by the direction of the momentum transfer $\langle \mathbf{q}_A \rangle$ and, in this respect, the predictions of the relativistic consideration are similar to those of the nonrelativistic treatment according to which the positions of the peaks are also determined by the direction of the (nonrelativistic) momentum transfer.

When the collision Lorentz factor increases, the virtual photon transmitting the TCDI interaction may closely resemble a real photon. In particular, under certain conditions polarization of the virtual photon can be nearly perpendicular to its three-momentum vector and the photon becomes almost transversal. In the target rest frame the overwhelming majority of electrons

---

[16] The reduction in the momentum transfer is directly related to the relativistic compression of the atomic electron cloud which occurs in the projectile frame along the target velocity. In that frame this compression changes for the atomic electron the scale of the momentum transfer along the velocity (which in this frame is given by $q_{\min}^I$).

emitted from a light target has energies which are very far from being relativistic (see Fig. 5.11). The dynamics of the interaction between a nonrelativistic electron and a photon are governed mainly by photon polarization properties rather than by the photon three-momentum. Therefore, if the atomic electron absorbs a transversal photon the maxima in the electron emission pattern will be not along the vector of the momentum transfer but perpendicular to it. For ionization of light targets by point-like projectiles such an effect has been discussed in detail in [117]. A similar effect may occur in the emission pattern produced via the TCDI at high γ provided $q_\perp \ll \langle q_{min}^A \rangle$. We, however, will not discuss such an effect here because, as we shall show below, for the target ionization via the TCDI the range of so small $q_\perp$ is of minor importance for the total ionization cross section.

The profound modifications in the electron emission pattern, observed in Fig. 5.8b–c for the case of the target ionization by a 'bound' electron, can be contrasted with the limited changes which are introduced by the relativity in the target ionization by a free electron with the same transverse momentum transfer $q_\perp = 0.1$ a.u. In the latter process, even at $\gamma = 1,000$ the dynamics of the emission remain basically nonrelativistic (see Fig. 5.8d). Moreover, at such transverse momentum transfers like $q_\perp = 0.1$ a.u., which are large compared to the minimum momentum transfer $(\epsilon_{k_a} - \epsilon_0)/v \sim 0.01$ a.u. to the target in collisions with a free point-like charge, no noticeable signatures of the relativity appear with a further increase in the impact energy and they remain very weak in the limit $\gamma \to \infty$.

Doubly Differential Cross Section. The Saturation Effect

Important information about the collision dynamics is obtained by considering the cross section differential in energy and solid angle of the emitted electron,

$$\frac{d^2\sigma}{d\varepsilon_{k_a} d\Omega_{k_a}} = \int d^2\mathbf{q}_\perp \int d^3\mathbf{k}_i \frac{d\sigma}{d^3\mathbf{k}_a d^3\mathbf{k}_i d^2\mathbf{q}_\perp}. \tag{5.114}$$

In Fig. 5.9a–c this doubly differential cross section is shown for the reaction $O^{7+}(1s) + H(1s) \to O^{8+} + H^+ + 2e^-$. It is plotted as a function of the polar emission angle, $\vartheta_a = \arccos(\mathbf{k}_a \cdot \mathbf{v}/(k_a v))$ of the electron for a fixed emission energy of 5 eV. The relativistic effects in the TCDI tend to make the angular distribution of the emitted electrons more symmetric with respect to the direction perpendicular to the collision velocity ($\vartheta_a = 90°$).

At relatively low values of γ the relativistic effects rapidly increase with increasing collision energy. However, for the collision system under consideration the calculations predict (see Fig. 5.9c) that starting with collision energies $\simeq 5$–10 GeV u^{-1} these effects 'saturate': the shape and absolute values of the cross section do not change with a further increase in the collision energy.

One has to stress that such a 'saturation effect' in the emission spectra is absent in the case of the target ionization by a free electron impact. In the

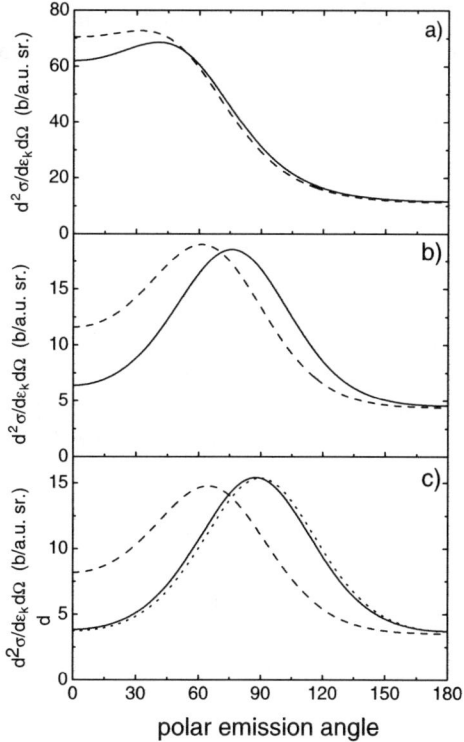

Fig. 5.9. Angular distribution of $5\,\mathrm{eV}$ electrons emitted from hydrogen in $\mathrm{O}^{7+}(1s)$ $+\mathrm{H}(1s)\ \rightarrow\ \mathrm{O}^{8+}\ +\ \mathrm{H}^+\ +\ 2e^-$ collisions. **(a)** $100\,\mathrm{MeV\,u}^{-1}$; **(b)** $1\,\mathrm{GeV\,u}^{-1}$; **(c)** $10\,\mathrm{GeV\,u}^{-1}$ ($\gamma \simeq 11.7$); Results of relativistic and nonrelativistic calculations are depicted by *solid* and *dashed curves*, respectively. In addition, the *dot curve* in figure **(c)** shows results of relativistic calculation for a collision energy of $1\,\mathrm{TeV\,u}^{-1}$ ($\gamma \approx 1,000$).

latter case the emission spectra depend on γ and change when the collision energy increases, whatever the magnitude of the energy is. In collisions with a high-energy free electron the dependence on γ appears due to the contribution given by collisions with very small momentum transfers (or very large impact parameters) and has its origin in the Lorentz contraction of the electromagnetic field generated by relativistic point-like charges: compared to nonrelativistic predictions the range of this field is effectively reduced by a factor of γ in the longitudinal direction and is increased by the same factor in the transverse direction.

The origin of the saturation effect in the ionization by a 'bound' electron becomes apparent if we consider the role played by the distant collisions for this process. It turns out that for ionization via the TCDI such collisions are just of minor importance. The basic physical reason, why very large impact parameters (or very small $q_\perp$) are essentially cut off in the target ionization via the TCDI, is that the projectile is also excited in the collision. In particular, this excitation manifests itself in the following two important points.

Firstly, by using the continuity equation (5.74) for the transition four-current of the projectile, one can easily show that at $\gamma \gg (\epsilon_{k_a} - \epsilon_0)/(\varepsilon_{k_i} - \varepsilon_0)$ one has

$$F_0^I - \frac{v}{c}F_3^I = -\langle \psi_f | \exp(i\mathbf{q}_I \cdot \mathbf{r})|\psi_i\rangle \left(\frac{1}{\gamma^2} + \frac{\epsilon_{k_a} - \epsilon_0}{\gamma(\varepsilon_{k_i} - \varepsilon_0)} \right)$$

$$- \frac{v^2/c}{\varepsilon_{k_i} - \varepsilon_0} \langle \psi_f | \exp(i\mathbf{q}_I \cdot \mathbf{r}) \, \mathbf{q}_\perp \cdot \boldsymbol{\alpha}_{tr}|\psi_i\rangle, \qquad (5.115)$$

where $\boldsymbol{\alpha}_{tr} = (\alpha_1, \alpha_2)$. Thus, the term $\left(F_0^I - \frac{v}{c}F_3^I\right)\left(F_A^0 - \frac{v}{c}F_A^3\right)$ in the second line of (5.112) is small if γ is high and $q_\perp$ is low. Besides, because of the factors γ^{-1} and γ^{-2}, those terms in (5.112) which are proportional to $F_3^I F_A^3$ and $(F_1^I F_A^1 + F_2^I F_A^2)$ are also small at high γ. Remark that (5.112) can be used also for the treatment of target ionization by a point-like charged projectile with a charge Z_p if we replace F_0^I by Z_p and set $F_1^I = F_2^I = F_3^I = 0$ and $\varepsilon_{k_i} - \varepsilon_0 = 0$.[17] Considering the corresponding couplings of the target and projectile form-factors in (5.112) we see that, compared to the ionization by a free point-like charge, at sufficiently high γ the target ionization via the TCDI in collisions with small $q_\perp$ is strongly suppressed.

Secondly, let us compare the photon propagator for target ionization by a point-like charged projectile, which is proportional to

$$\frac{1}{q_\mu q^\mu} = \frac{1}{q_\perp^2 + \frac{(\epsilon_{k_a} - \epsilon_0)^2}{v^2 \gamma^2}}, \qquad (5.116)$$

with the photon propagator in (5.112), which is proportional to

$$\frac{1}{q_\mu q^\mu} = \frac{1}{\mathbf{q}_A^2 - (\varepsilon_{k_a} - \varepsilon_0)^2/c^2}$$

$$= \frac{1}{q_\perp^2 + \frac{(\epsilon_{k_a} - \epsilon_0 + \varepsilon_{k_i} - \varepsilon_0)^2}{v^2 \gamma^2} + 2(\gamma - 1)\frac{(\epsilon_{k_a} - \epsilon_0)(\varepsilon_{k_i} - \varepsilon_0)}{v^2 \gamma^2}}. \qquad (5.117)$$

We see that in collisions where both the target and projectile are excited, $\epsilon_{k_a} - \epsilon_0 > 0$ and $\varepsilon_{k_i} - \varepsilon_0 > 0$, the length of the four-momentum of the photon transmitting the TCDI at equal values of $q_\perp$ more strongly differs from the mass shell condition $q_\mu q^\mu = 0$ inherent for a real photon. Thus, compared to the photon exchanged in collisions with a point-like projectile, the photon responsible for the TCDI has a more pronounced virtuality. This fact also reduces the range of this interaction compared to the range of the interaction with a point-like charged projectile.

---

[17] After such replacements one obtains a point-like charged particle of spin zero. However, collisions with very small momentum transfers, $q_A \ll m_p v$ (m_p is the projectile mass), in which the projectile suffers just a very small deflection, are of semi-classical character. Therefore, the actual spin value of a point-like projectile is of minor importance for such collisions.

Singly Differential Cross Sections

Valuable insights into the collision dynamics can be also obtained by calculating cross sections which are singly differential in the longitudinal, $k_{a,lg} = k_a \cos \vartheta_a$, and transverse, $k_{a,tr} = k_a \sin \vartheta_a$, components of the electron momentum and the emission energy, $\varepsilon_{k_a} = k_a^2/2$. We start the discussion of these spectra with consideration of the cross section differential in the longitudinal component $k_{a,lg}$

$$\frac{d\sigma}{dk_{a,lg}} = \int d^2 \mathbf{k}_{a,tr} \int d^2 \mathbf{q}_\perp \int d^3 \mathbf{k}_i \frac{d\sigma}{d^3 \mathbf{k}_a d^3 \mathbf{k}_i d^2 \mathbf{q}_\perp}, \qquad (5.118)$$

which often provides most important information about the collision dynamics. The cross section (5.118) is displayed in Fig. 5.10a–c for $N^{6+}(1s) + H(1s) \rightarrow N^{7+} + H^+ + 2e^-$ collisions. In the calculation the integration over the absolute value of the transverse part $\mathbf{k}_{a,tr}$ of the momentum $\mathbf{k}_a$ of the emitted electron was performed for $0 \leq k_{a,tr} \leq 7$ a.u. According to nonrelativistic

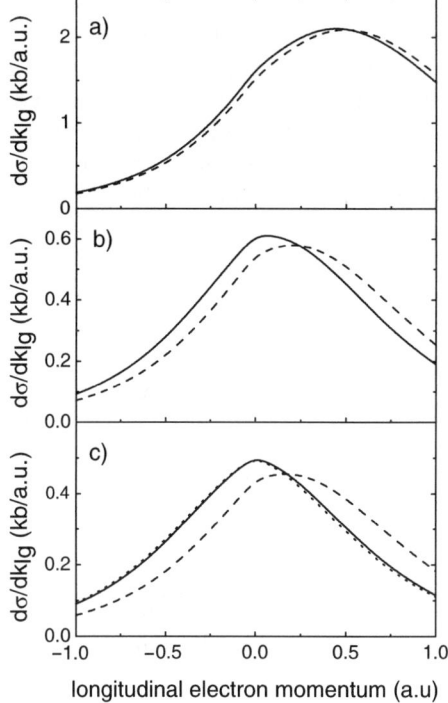

Fig. 5.10. Longitudinal momentum distribution of electrons emitted from hydrogen in $N^{6+}(1s) + H(1s) \rightarrow N^{7+} + H^+ + 2e^-$ collisions. (a) $100 \, \text{MeV} \, u^{-1}$; (b) $1 \, \text{GeV} \, u^{-1}$; (c) $10 \, \text{GeV} \, u^{-1}$. Results of relativistic and nonrelativistic calculations are depicted by *solid* and *dashed curves*, respectively. In addition, the *dotted curve* in (c) shows results of relativistic calculations for $1 \, \text{TeV} \, u^{-1}$.

calculations, even at $v = 137$ a.u. there should be a strong forward–backward asymmetry in the longitudinal electron spectrum. However, relativistic calculations predict that starting with $\gamma \sim 1.5$–2 this asymmetry should be less pronounced and that it can be substantially reduced when γ increases further. Clearly, this 'symmetrization' in the longitudinal spectrum corresponds to the 'symmetrization' effect which was discussed for the angular distribution of the emitted electron (see Fig. 5.10a–c). Similarly to the consideration of the doubly differential cross section (5.114), we observe that for the collision system in question the relativistic effects in the longitudinal emission spectrum saturate at collision energies of $\sim$5–10 GeV u^{-1} and that a further increase in the collision energy actually changes the shape and absolute values of this spectrum very weakly.

To complete this subsection let us briefly discuss the energy spectrum of electrons emitted from hydrogen. This spectrum is represented by the singly differential cross section

$$\frac{d\sigma}{d\varepsilon_{k_a}} = \int d\Omega_{k_a} \int d^2\mathbf{q}_\perp \int d^3\mathbf{k}_i \frac{d\sigma}{d^3\mathbf{k}_a d^3\mathbf{k}_i d^2\mathbf{q}_\perp}. \tag{5.119}$$

The cross section (5.119) is shown in Fig. 5.11 for $O^{7+}(1s)$ +H(1s) $\rightarrow$ O^{8+} + H$^+$ + 2e^- collisions. Hydrogen ionization by relativistic point-like charged projectiles occurs mainly in such collisions where the momentum transfer to the target is smaller (or even much smaller) than the typical momentum of the electron in the ground state of hydrogen. However, hydrogen ionization via the TCDI involves on average substantially larger momentum transfers ($q_\perp$

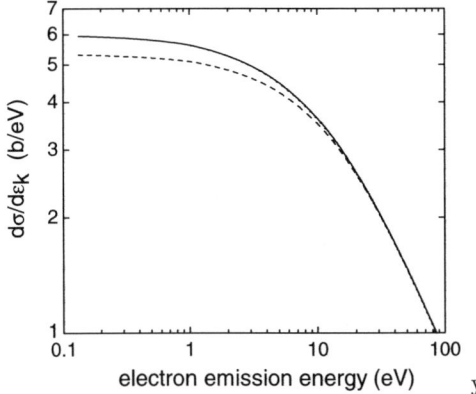

Fig. 5.11. Energy distribution of electrons emitted from hydrogen in 10 GeV u^{-1} $O^{7+}(1s)$ +H(1s) $\rightarrow$ O^{8+} + H$^+$ + 2e^- collisions. Results of relativistic and nonrelativistic calculations are depicted by *solid* and *dashed curves*, respectively. In addition, the *dotted curve* (which is almost indistinguishable from the solid one) shows results of relativistic calculations for 1 TeV u^{-1}.

up to $\sim Z_I \gg 1$ a.u.). This results in the extension of the electron emission spectrum to much larger energies.[18]

It follows from the calculations that the relativistic effects in the TCDI enhance the lower energy part of the spectrum. This enhancement is rather modest already for small emission energies (about 10% at $\epsilon_{k_a} \approx 1\,\mathrm{eV}$) and becomes negligible at $\epsilon_{k_a} \gtrsim 20\,\mathrm{eV}$. Since the energy spectrum produced via the TCDI extends to relatively large energies, the influence of the above enhancement on the total cross section for hydrogen ionization via the TCDI is quite weak and increases the total emission from the target by no more than a few per cent for the collision system under consideration.

As one could expect and as Fig. 5.11 illustrates, the 'saturation' effect holds also for the energy spectrum.

Thus, summarizing the above consideration, we see that for the ionization of a light target by an electron bound in a light projectile the theory predicts noticeable relativistic effects already in collisions with low values of γ and that these effect 'saturate' at not very high γ-s. It is known that both these features do not hold for the ionization of a light target by a free electron.

In the case of collisions between light projectile-ion and target-atom the relativistic effects in the emission spectrum of the target electron are solely caused by the collision velocity approaching the speed of light and are mainly of 'redistributive' character: compared to predictions of nonrelativistic calculations, they affect rather weakly the total amount of emitted electrons but may substantially redistribute the electron emission between different parts of the electron momentum space.

5.14.2 Radiation Field and Resonant Two-Center Dielectronic Transitions

The square of the four-momentum transfer in the collision, $q^2 = q_\mu q^\mu$, is given by

$$
\begin{aligned}
q^2 &= \mathbf{q}_I^2 - \frac{\omega_{n0}^2}{c^2} = \mathbf{q}_A^2 - \frac{\Omega_{m0}^2}{c^2} \\
&= q_\perp^2 + \frac{\omega_{n0}^2 + \Omega_{m0}^2}{\gamma^2 v^2} + 2\frac{\omega_{n0}\Omega_{m0}}{\gamma v^2} \\
&= q_\perp^2 + \frac{(\omega_{n0} + \Omega_{m0})^2}{\gamma^2 v^2} + 2\frac{\omega_{n0}\Omega_{m0}}{v^2}\left(\frac{1}{\gamma} - \frac{1}{\gamma^2}\right),
\end{aligned}
\tag{5.120}
$$

where $\omega_{n0} = \varepsilon_n - \varepsilon_0$, $\Omega_{m0} = \epsilon_m - \epsilon_0$.

As we have seen in the previous subsection, in the ion–atom collisions resulting in the excitation of both these particles (ω_{n0} and Ω_{m0} are positive), the square of the four-momentum $q_\mu q^\mu$ is never zero and the effective range

---

[18] For example, in hydrogen ionization by a fast point-like charge the energy spectrum of electrons ejected from hydrogen also reaches its maximum values at very small emission energies but decreases by almost an order of magnitude when the emission energy increases from, say, ≈ 1 to $\approx 13.6\,\mathrm{eV}$.

of the ion–atom interaction does not exceed a few atomic units becoming a constant at sufficiently high impact energies.

The projectile–target interaction acquires qualitatively new features when $q_\mu q^\mu = 0$. For instance, in such a case the singularity at real $q_\perp$ appears in the integrand of the amplitude (5.46). This occurs when the sum of the last two terms in (5.120) is negative and is only possible if $\omega_{n0}\Omega_{m0} < 0$, i.e. when in the collision one of the colliding centers gets excited and the other one is de-excited. Such collisions were studies in some detail in [120–122] and below we shall follow the considerations given in these papers.

Let us for the moment assume for definiteness that the projectile is excited ($\omega_{n0} > 0$) and the target is de-excited ($\Omega_{m0} < 0$) in the collision. Then we obtain that the equality $q_\mu q^\mu = 0$ holds if the following conditions are fulfilled

$$\omega_{n0}\,\Omega_{m0} < 0$$

$$|\,\omega_{n0}\,|\,\sqrt{\frac{c-v}{c+v}} \le \Omega_{m0} \le |\,\omega_{n0}\,|\,\sqrt{\frac{c+v}{c-v}}. \tag{5.121}$$

In such a case the projectile and target exchange in the collision an on-mass-shell photon. As a result, the denominator in the integrand of the amplitude (5.46) becomes singular and, since the numerator in the integrand remains in general non-zero, we face a particular case of a resonance in atomic collisions.

The physical meaning of the above inequalities becomes obvious if we recall that, according to the relativistic Doppler effect (see (4.15)), the radiation spectrum, which is monochromatic, say, in the projectile frame where it has a frequency ω_P, in the target frame spreads to a continuous spectrum with frequencies ω_T occupying the interval given by

$$\omega_P\,\sqrt{\frac{c-v}{c+v}} \le \omega_T \le \omega_P\,\sqrt{\frac{c+v}{c-v}}. \tag{5.122}$$

Comparing (5.120) and (5.122) we see that the ion–atom interaction can be interpreted as occurring via the emission of a photon with the energy $\omega_{n0} = |\,\varepsilon_n - \varepsilon_0\,|$ by the projectile in its rest frame and the absorption of the same photon, but now having the energy $\Omega_{m0} = \epsilon_m - \epsilon_0$, by the target in the target frame. Thus, it is the Doppler effect which makes the resonance possible in collisions between systems with different energy scales.

One should mention that according to the standard nonrelativistic consideration of the collision the above resonance is not possible. Indeed, within the latter the nonrelativistic limit of the transition amplitude is obtained by setting $c \to \infty$ in (5.46) that yields

$$a_{0\to n}^{0\to m}(\mathbf{b}) = \lim_{\varsigma\to+0}\frac{-i}{\pi v}\int d^2 q_\perp \exp(-i q_\perp \cdot \mathbf{b})$$

$$\times \frac{\langle u_m|\exp(-i\mathbf{q}\cdot\mathbf{r})|u_0\rangle\langle\psi_n|\exp(i\mathbf{q}\cdot\boldsymbol{\xi})|\psi_0\rangle}{\mathbf{q}^2 + i\varsigma}, \tag{5.123}$$

where $\psi_{0,n}$ and $u_{0,m}$ are solutions of the Schrödinger equation for the inner motion of the electrons in the projectile and target, respectively, and $\mathbf{q}$ denotes the three-momentum transfer to the projectile.[19] Since at $c \to \infty$ the 'time' component of the four-momentum transfer has vanished, the singularity in the photon propagator can arise only if the three-momentum transfer $\mathbf{q}$ is equal to zero. The latter, however, means that actually no collision occurs and thus there is no transitions related to the 'collision'. This is seen in the expression (5.123) for the transition amplitude where at $\mathbf{q} = 0$ the transition form-factors become equal to zero that, together with the factor $q_\perp$ from $d^2\mathbf{q}_\perp$, overcompensates the singularity in the photon propagator.

When the exchanged photon is on mass shell, one can expect that the range of the ion–atom interaction is much longer than that in collisions where the atomic systems interact via the transmission of an off-shell photon. Indeed, in the resonance case at large impact parameters the transition amplitude (5.46) turns out to be proportional to $1/\sqrt{b}$. The corresponding transition probability $\mid a_{0 \to n}^{0 \to m} \mid^2 \propto 1/b$ and at any arbitrarily large b the product $b \mid a_{0 \to n}^{0 \to m} \mid^2$ remains a constant. This shows that the physical 'agent' transmitting the ion–atom interaction in the resonance case is the radiation field.

The attempt to calculate the cross section according to the standard procedure, $\sigma_{0 \to n}^{0 \to m} \propto \int_0^\infty db\, b \mid a_{0 \to n}^{0 \to m} \mid^2$, leads to an infinite result. This infinity arises not because the transition probability itself is large (at any b it is, in fact, very small compared to 1) but because at large b it decreases too slowly with increasing the impact parameter. Therefore, it cannot be regularized by considering higher order terms in the projectile–target interaction (two-photon exchanges e.t.c.). Indeed, not only the contributions of such terms to the transition amplitude are very small even at small b but also they decrease very rapidly when b increases and become completely negligible at those large impact parameters which are 'responsible' for the problem. The striking situation with the diverging cross section makes it necessary to look for some physical factors which normally are not taken into account and even not mentioned.

The appearance of infinite cross sections on a certain step of a theoretical consideration, of course, does not necessarily mean that actual cross sections are infinite. As it is known, in real physical systems there always exists a kind of 'damping' factor which becomes of especial importance when a resonance process is considered setting finite limits on cross section values in the vicinity of resonances. In our case such a 'damping' factor can formally be introduced by adding a finite imaginary part iI into the denominator in the integrand of (5.46). The physical reasons leading to the appearance of the imaginary part are discussed below.

The Size Effect

The most obvious way to make the cross section finite is to point out that the integration over the impact parameter must, in fact, be restricted to $b \le b_0$

---

[19] Note that in the nonrelativistic approximation $\mathbf{q}_\mathrm{I} = -\mathbf{q}_\mathrm{A}$.

where b_0 is related to the transverse target size (the size in the direction perpendicular to $\mathbf{v}$) and the projectile beam diameter. Then the cross section can be written as

$$\sigma_{0 \to n}^{0 \to m} = A + B b_0, \qquad (5.124)$$

where A and B are some parameters with A accounting for the contribution to the cross section in collisions with small b. Such a size effect can be incorporated into the consideration in a more convenient way if the integration over b is still performed for $0 \le b < \infty$ but the photon propagator is regularized by adding the imaginary part $i I_1$, where $I_1 \simeq \frac{2|\omega_{n0}|}{c b_0}$, in the denominator of (5.46).

Finite Lifetime of the Excited State

The inspection of (5.120) suggests that an imaginary part in the denominator of the photon propagator can appear if the natural width of the excited state is taken into consideration by the replacement $\omega_{n0} \to \omega_{n0} + i\Gamma/2$. The resulting contribution I_2 to the total 'damping' factor I is then given by $I_2 \simeq 2 \, |\omega_{n0}| \, \Gamma/c^2$. The term I_2 introduces the new characteristic length, $D \simeq c/\Gamma$, into the consideration. If this length is of the order of or smaller than other characteristic lengths in the problem, then the natural width of the excited state will noticeably influence the resonance collisions.

Target Density Effect

One more 'damping' factor and thus yet one more contribution to the imaginary part I appears if we take into account the following. In obtaining the transition amplitude (5.46) it was assumed that the collision between the projectile and the target is not influenced by the presence of other atoms in the target medium and other particles in the projectile beam. Thus, the other target atoms and projectile particles are considered to be completely 'passive' when the 'active' projectile and target collide. This is the standard procedure in the physics of energetic ion–atom collisions and it is normally supposed to be quite suitable for considering collisions of ionic beams with rarefied gaseous targets. However, when the range of the projectile–target interaction becomes 'abnormally' long, as it is for the resonance case considered here, the presence of the 'passive' atoms in the target medium may in general no longer be ignored. Indeed, this presence may be of especial importance in our case because the electromagnetic field transmitting the projectile–target interaction 'works' on a frequency which is resonant to the atomic transition and, therefore, the cross section for the interaction between the field and the 'passive' atoms becomes rather large.

In our case the mean free path of the electromagnetic field in the target medium is given by $\Lambda = 1/(N_{\mathrm{gr}} \sigma_{\mathrm{R}})$, where

$$\sigma_R = \pi \frac{c^2}{\omega^2} \frac{\Gamma^2}{(\omega_0 - \omega)^2 + \Gamma^2/4} \tag{5.125}$$

is the cross section for the scattering of the electromagnetic field by the atom in the vicinity of the resonance atomic transition frequency ω_0 and N_{gr} is the density of the atoms. At distances of order Λ and larger the target medium ceases to be 'transparent' for the projectile–target interaction. Therefore, when the impact parameter b for the collision between the ion and the 'active' atom starts to exceed Λ this interaction essentially vanishes. Thus, the presence of the 'passive' atoms effectively cuts the impact parameters at $b_d \sim \Lambda$. Since Λ depends on the target density, this reduction in the effective range of the ion–atom interaction can be called the density effect.[20] The density effect can be incorporated into the treatment by adding the 'damping' term $iI_3 = i2 \mid \omega_{n0} \mid N_{gr}\sigma_R/c$ into the denominator of the photon propagator.[21]

Equation (5.125) does not account for the broadenings of the spectral line ω_0 caused by the atom–atom interaction/collisions in the medium and by the Doppler effect due to the thermal motion of the atoms. For rarefied gases the collision broadening can normally be neglected. The thermal Doppler broadening should be taken into account if it becomes close or larger than the natural linewidth Γ. In such a case results for cross sections, obtained by using (5.125), should be averaged over the thermal velocity distribution of the target atoms. Note that the thermal motion does not affect the resonance condition (5.120) if $\gamma \ll \sqrt{c/u}$, where $u \sim \sqrt{T}$ is the mean thermal velocity of the target atoms.

In Fig. 5.12 results are shown for the electron loss from $B^{4+}(1s)$ projectiles. It is supposed that (a) the target has a 15% fraction of the excited $(H(2p_1)$ or $H(2p_0)$ atoms and is contained in a cylindric volume; the excited states are quantized along the collision velocity. Further, the target has a density of 5×10^{12} cm^{-3} and is characterized by a Maxwell distribution with temperature of 40 K. A narrow beam of $B^{4+}(1s)$ ions penetrates the target volume along its symmetry axis. Under these target conditions the mean free path Λ of the radiation field at the frequency resonant to $2p \rightarrow 1s$ transition in the hydrogen medium is $\simeq 5$ cm. Assuming that the target transverse size is larger than Λ the resonance loss cross section will be independent of the target size.

Considering the inequalities (5.121) at a fixed value of the ratio $\chi = \omega_{n0}/ \mid \Omega_{m0} \mid$ we obtain that the threshold Lorentz factor for the resonance to

---

[20] A well known example of an effect related to the target density is the response of target medium to the field of a point-like ultra-relativistic projectile. At large impact parameters this response leads to the screening of the projectile field and reduces the projectile energy loss in the medium. This effect was first estimated in [123] and is nicely discussed in [124] (see also [7]).

[21] Concerning the effect of the projectile beam density on the resonance interaction one can note that the density of ions in a projectile beam is normally orders of magnitude smaller than the density of target atoms. Under such conditions the presence of 'passive' particles in the projectile beam can be neglected.

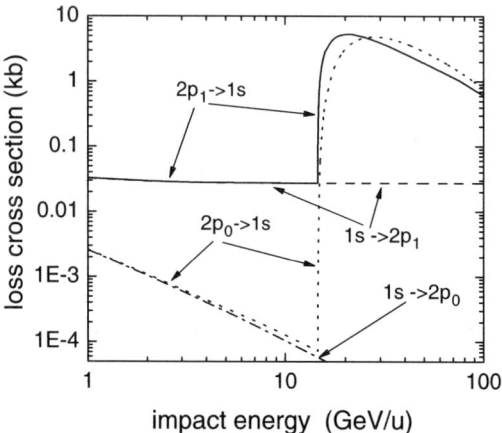

Fig. 5.12. Cross sections for the electron loss from $B^{4+}(1s)$ as a function of the impact energy. *Solid curve*: collisions in which $H(2p_1) \rightarrow H(1s)$. *Dash curve* collisions in which $H(1s) \rightarrow H(2p_1)$. *Dot curve*: collisions in which $H(2p_0) \rightarrow H(1s)$. *Dash–dot curve*: collisions in which $H(1s) \rightarrow H(2p_0)$. The excited states are quantized along the collision velocity. From [120].

occur is given by $\gamma_{th} = (\chi^2 + 1)/(2\chi)$. For the electron loss from $B^{4+}(1s)$, accompanied by the de-excitation of $H(2p)$, we get $\gamma_{th} \simeq 16.7$ that corresponds to the threshold impact energy of $\simeq 14.6\,\text{GeV}\,\text{u}^{-1}$. At this energy the loss cross sections begin to increase tremendously and reach values of about 5.5 kb before starting to decrease slowly with a further increase in the impact energy.

In symmetric collisions, where $\omega_{n0} = -\Omega_{m0}$ and, thus, the Doppler shift would just spoil the resonance effect, the coupling to the radiation field can occur at much smaller impact energies.[22] In Fig. 5.13 results are displayed for the transfer of excitation in $H(1s)+H(2p)$ collisions. It is supposed that (a) the target consisting of 90% atoms in the ground state and 10% atoms in $2p_0$ (or $2p_1$) state is contained in a cylindric volume with the diameter of 10 mm, (b) the target has the total density of $10^{11}\,\text{cm}^{-3}$ (c) and is characterized by a Maxwell distribution with temperature of 40 K, (d) a narrow beam of $H(1s)$ penetrates the target volume along its symmetry axis. For a comparison, in Fig. 5.13 results are shown (a) for $H(1s)+H(2p_1) \rightarrow H(2p_1)+H(1s)$ collisions obtained assuming $c \rightarrow \infty$, (b) for $H(1s)+H(1s) \rightarrow H(2p_1)+H(2p_1)$ and $H(1s)+H(1s) \rightarrow H(2p_1)+H(2p_1)$ collisions.

It is seen that under the condition of Fig. 5.13 the coupling to the radiation field turns out to be much more important mechanism for the transfer

---

[22] It is known that for symmetric atomic systems, which are at rest with respect to each other, the atom–atom interaction can acquire much longer range compared to that of the 'normal' Van der Waals force (see e.g. [125] pp. 522–524). Such kind of an atom–atom coupling, however, rapidly becomes nonresonant when one of the atoms starts to move and is negligible for collision velocities which are considered here.

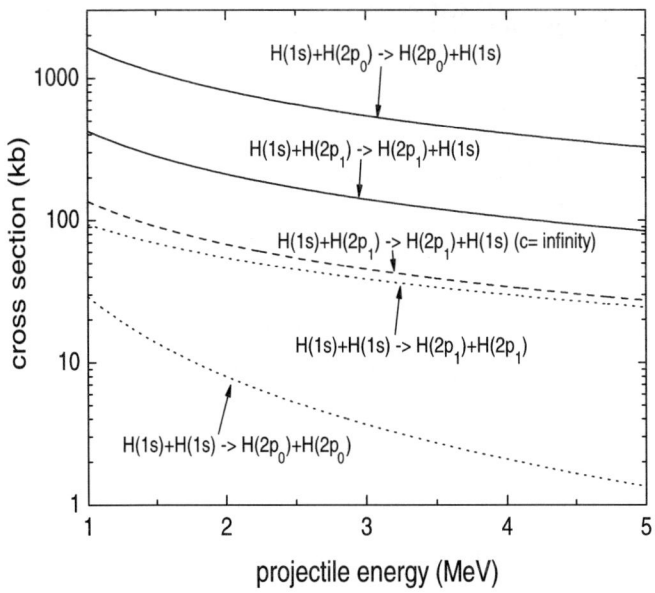

Fig. 5.13. Cross sections for the transfer of excitation in hydrogen–hydrogen collisions: $H(1s)+H(2p_0) \rightarrow H(2p_0)+ H(1s)$ and $H(1s)+H(2p_1) \rightarrow H(2p_1)+ H(1s)$. These cross sections are plotted as a function of the collision energy. For a comparison, results for $H(1s)+H(1s) \rightarrow H(2p_0)+ H(2p_0)$ and $H(1s)+H(1s) \rightarrow H(2p_1)+ H(2p_1)$ collisions are also shown. The excited states are quantized along the collision velocity.

of excitation than one would expect. Note also that at the above collision parameters the cross section for the transfer of excitation is independent of the target density and temperature but rather strongly (linearly) depends on the target diameter.

In contrast to the situation considered in Fig. 5.12, the role of the resonance coupling to the radiation field in the case considered in Fig. 5.13 monotonously diminishes with the increase in the collision energy. This is because the colliding system consisting of atomic particles with equal energy differences, due to the increase of the Doppler broadening of the spectrum of the radiation field, gradually comes out of the resonance that makes the radiation field less effective in the transfer of excitation.

Thus, the excitation and loss cross sections, which are *microscopic* quantities, may in principle become dependent on such *macroscopic* parameters as the target density, temperature and size. Except the well known process of the radiative electron capture,[23] we are not aware about any other processes occurring in fast ion–atom collisions where the coupling to the radiation field might be of importance for the transitions between internal states of the colliding particles.

---

[23] For a review on the topic of the radiative electron capture see [126].

6

Theoretical Methods Extending beyond the First Order Approximation

6.1 Collisions with Light Atoms: Preliminary Remarks

In the previous chapter the relativistic collisions between an ion and an atom both carrying electron(s) were considered within the first order theory in which the interaction between the ion and atom is restricted just to a single photon exchange. The latter implies that for the first order results to be reliable the ion–atom interaction should be sufficiently weak.

The parameters Z_A/v and Z_I/v are often used to characterize the effective strength of the field of the atom acting on the ion and that of the ion acting on the atom during the collision. Although it is not always very clear what the words 'sufficiently weak' mean, normally one assumes that the first order theories should be strictly valid provided both the conditions $Z_A/v \ll 1$ and $Z_I/v \ll 1$ are simultaneously fulfilled.

Of course, in order to really establish the correct limits for the validity of the first order theory one has to have more general theories which would go beyond the first order approximation in the ion–atom interaction. The development of such theories is of basic interest and importance and is also necessary from the point of view of applications since in experiments on the electron excitation and loss in relativistic collisions the net charge of the projectile is usually so high (for instance, Pb^{81+}, Bi^{82+}, U^{91+} projectiles are often used) that the ratio Z_I/v may not be much smaller than 1 since the collision velocity v cannot exceed the speed of light in vacuum.

If such a very highly charged ion collides with a light atom, for which the condition $Z_A/v \ll 1$ is fulfilled, the atom *per se* represents merely a weak perturbation for the electron of the ion. This, however, does not mean that all cross sections for the projectile-electron excitation and loss can be calculated within the first order theory.

Indeed, the field exerted by the highly charged ion during the collision on the electrons of the atom can effectively be quite strong on the atomic scale. The latter may lead to considerable deviations in the behavior of the atomic electrons from predictions of the first order theory. A different behavior of the atomic electrons means that the atomic field acting on the ion will also differ from that predicted by the first order theory. As a result, since the atom and the ion are strongly coupled in the collision, the electron of the ion will in general also be affected differently compared to the case when the collision process proceeds in accord with first order predictions.

In the next section we shall consider various aspects of collisions between highly charged projectiles and light atoms using an eikonal-like model which will be discussed in great detail. Compared to the first order approximation the advantage of this model is based on taking into account the distortion of the atomic transition four-current by the field of the projectile.

6.2 Symmetric Eikonal Model

Let us consider a collision between a highly charged ion and a light atom. We begin with the semiclassical approximation and assume that in the target frame K_A, in which the target nucleus rests at the origin, the projectile nucleus moves along a straight-line classical trajectory $\mathbf{R}(t) = \mathbf{b} + \mathbf{v}t$. For the sake of simplicity we shall give a detailed derivation of results only for targets and projectiles with one electron.

The starting point of our consideration is the semiclassical transition amplitude given by (5.37). Since the atom is assumed to be light and, hence, its field is weak ($Z_A/v \ll 1$) the expressions for the transition current of the ion and its field will formally remain exactly the same as in the first order consideration. Now, however, the transition current of the atom $J_\mu^A(x)$ will be evaluated by using initial and final states of the atom which include additional factors describing the distortion of these states by the field of the ion, i.e. the atom will be described by a distorted transition four-current.

The transition amplitude, which makes use of the distorted transition four-current, can be calculated making steps similar to those used to derive the first order transition amplitude. Namely, we first evaluate the atomic transition four-current $J_\mu^A(x)$ in the target frame K_A, where the atomic nucleus is at rest. Then the transition four-current $J'^\mu_I(x_I)$ and the four-potential $A'^I_\mu(x_I)$ of the ion are calculated in the ion frame K_I in which the ionic nucleus is at rest (x_I are the coordinates in K_I). Finally, the four-potential of the ion is transformed into the frame K_A where the transition amplitude is evaluated.

In the frame K_A the four-current J_μ^A of the atom reads

$$J_0^A(x) = c \int d^3r \phi_f^\dagger(\mathbf{r}, t) \left(Z_A \delta^{(3)}(\mathbf{x}) - \delta^{(3)}(\mathbf{x} - \mathbf{r}) \right) \phi_i(\mathbf{r}, t),$$

$$J_l^A(x) = c \int d^3r \phi_f^\dagger(\mathbf{r}, t) \, \alpha_l \, \delta^{(3)}(\mathbf{x} - \mathbf{r}) \phi_i(\mathbf{r}, t); \quad l = 1, 2, 3. \qquad (6.1)$$

In (6.1) $\mathbf{r}$ is the coordinate of the atomic electron with respect to the atomic nucleus given in the frame K_A and α_l are the Dirac matrices for the electron of the atom. The initial and final internal states of the target, in the spirit of the symmetric eikonal approximation, are now taken as

$$\phi_i(\mathbf{r}, t) = (vs + \mathbf{v} \cdot \mathbf{s})^{-i\nu} u_0(\mathbf{r}) \exp(-i\varepsilon_0 t),$$
$$\phi_f(\mathbf{r}, t) = (vs - \mathbf{v} \cdot \mathbf{s})^{i\nu} u_n(\mathbf{r}) \exp(-i\varepsilon_n t), \tag{6.2}$$

where u_0 and u_n represent the solutions of the Dirac equation for the undistorted atom having energies ε_0 and ε_n, respectively. The three-dimensional space vector $\mathbf{s}$ is the coordinate of the atomic electron with respect to the ionic nucleus given in the *frame* K_I and $\nu = Z_I/v$.

Inserting (6.2) into (6.1) and using the Fourier representation for the eikonal phase-factor $s_\perp^{-2i\nu}$,

$$s_\perp^{-2i\nu} = \int d^2\mathbf{p}_\perp f(\mathbf{p}_\perp, \nu) \exp(i\mathbf{p}_\perp \cdot \mathbf{s}_\perp) \tag{6.3}$$

with

$$f(\mathbf{p}_\perp, \nu) = \lim_{\substack{\alpha \to +0 \\ \lambda \to +0}} \frac{\Gamma(1 - i\nu)\Gamma(1/2 + i\nu)}{2\pi\Gamma(1/2)\Gamma(2i\nu)} p_\perp^{\alpha - 2 + 2i\nu} \exp(-\lambda p_\perp), \tag{6.4}$$

where $\Gamma(x)$ is the gamma-function (see e.g. [108]) and the integration is performed over the two-dimensional transverse vector $\mathbf{p}_\perp$ ($\mathbf{p}_\perp \cdot \mathbf{v} = 0$), we obtain

$$J_0^A(x) = c \exp(i(\varepsilon_n - \varepsilon_0)t) \int d^2\mathbf{p}_\perp f(\mathbf{p}_\perp, \nu) \exp(-i\mathbf{p}_\perp \cdot \mathbf{b})$$
$$\times \int d^3 r u_n^\dagger(\mathbf{r}) \exp(i\mathbf{p}_\perp \cdot \mathbf{r}) \left(Z_A \delta^{(3)}(\mathbf{x}) - \delta^{(3)}(\mathbf{x} - \mathbf{r}) \right) u_0(\mathbf{r}),$$

$$J_l^A(x) = c \exp(i(\varepsilon_n - \varepsilon_0)t) \int d^2\mathbf{p}_\perp f(\mathbf{p}_\perp, \nu) \exp(-i\mathbf{p}_\perp \cdot \mathbf{b})$$
$$\times \int d^3 r u_n^\dagger(\mathbf{r}) \alpha_l \exp(i\mathbf{p}_\perp \cdot \mathbf{r}) \delta^{(3)}(\mathbf{x} - \mathbf{r}) u_0(\mathbf{r}); \quad l = 1, 2, 3. \tag{6.5}$$

With the help of the integral representation (e1) the atomic four-current (6.5) can be cast into

$$J_\mu^A(x) = \frac{c}{(2\pi)^3} \int d^2\mathbf{p}_\perp f(\mathbf{p}_\perp, \nu) \exp(-i\mathbf{p}_\perp \cdot \mathbf{b})$$
$$\times \int d^3 k \exp(i(\varepsilon_n - \varepsilon_0)t - i\mathbf{k} \cdot \mathbf{x}) \Phi_\mu^A(n0; \mathbf{k}; \mathbf{p}_\perp), \tag{6.6}$$

where the components of the eikonal form-factor of the atom $\Phi_\mu^A(n0; \mathbf{k}; \mathbf{p}_\perp)$ are defined according to

$$\Phi_0^A(n0; \mathbf{k}; \mathbf{p}_\perp) = \int d^3 r\, u_n^\dagger(\mathbf{r}) \left(Z_A \exp(i\mathbf{p}_\perp \cdot \mathbf{r}) - \exp(i(\mathbf{k} + \mathbf{p}_\perp) \cdot \mathbf{r}) \right) u_0(\mathbf{r}),$$

$$\Phi_l^A(n0; \mathbf{k}; \mathbf{p}_\perp) = \int d^3 r\, u_n^\dagger(\mathbf{r}) \exp(i(\mathbf{k} + \mathbf{p}_\perp) \cdot \mathbf{r}) \alpha_l u_0(\mathbf{r}). \tag{6.7}$$

Our present consideration assumes that the field of the atom is weak and, therefore, the initial and final states of the electron of the ion are simply taken as undistorted ionic states. As a result, within the present model the four-potential $A_I^\mu(x)$ generated by the projectile-ion in the frame K_A can be found along the lines which are exactly similar to those used to derive such a potential within the first order semiclassical treatment and the final expression for $A_I^\mu(x)$ is given by (5.44).[1]

Inserting this four-potential as well as (6.7) into the general expression (5.37) for the transition amplitude, we obtain that the eikonal transition amplitude in the impact parameter space is given by [41]

$$a_{\text{fi}}(\mathbf{b}) = -\frac{i}{\pi v} \int d^2\mathbf{p}_\perp f(p_\perp, \nu) \exp(-i\mathbf{p}_\perp \cdot \mathbf{b}) \int d^2\mathbf{k}_\perp \exp(-i\mathbf{k}_\perp \cdot \mathbf{b})$$
$$\times \frac{\Phi_\mu^A \left(n0; \mathbf{k}_\perp, q_{\text{min}}^A; \mathbf{p}_\perp\right) \gamma^{-1} \Lambda_\alpha^\mu F_I^\alpha \left(m0; -\mathbf{k}_\perp, -q_{\text{min}}^I\right)}{k_\perp^2 + (q_{\text{min}}^A)^2 - \frac{(\varepsilon_n - \varepsilon_0)^2}{c^2}}, \quad (6.8)$$

where the integration runs over the two-dimensional transverse vectors $\mathbf{k}_\perp$ and $\mathbf{p}_\perp$. In turn, the eikonal transition amplitude converted into the momentum space reads [41]

$$S_{\text{fi}}(\mathbf{q}_\perp) = \frac{1}{2\pi} \int d^2\mathbf{b}\, a_{\text{fi}}(\mathbf{b}) \exp(i\mathbf{q}_\perp \cdot \mathbf{b})$$
$$= -\frac{2i}{v} \int d^2\mathbf{p}_\perp f(p_\perp, \nu)$$
$$\times \frac{\Phi_\mu^A \left(n0; \mathbf{q}_A - \mathbf{p}_\perp; \mathbf{p}_\perp\right) \gamma^{-1} \Lambda_\alpha^\mu F_I^\alpha \left(m0; \mathbf{q}_I + \mathbf{p}_\perp\right)}{(\mathbf{q}_\perp - \mathbf{p}_\perp)^2 + (q_{\text{min}}^A)^2 - \frac{(\varepsilon_n - \varepsilon_0)^2}{c^2}}. \quad (6.9)$$

As before, $\mathbf{q}_I = (-\mathbf{q}_\perp; -q_{\text{min}}^I)$ and $\mathbf{q}_A = (\mathbf{q}_\perp; q_{\text{min}}^A)$ denote the total momenta transferred to the ion (in K_I) and the atom (in K_A), respectively. q_{min}^I and q_{min}^A represent their (absolute) minimum values, which are, of course, the same as in the first order consideration, and $\mathbf{q}_\perp$ is the transverse part of the momentum transfer to the atom. The explicit form of the coupling of the atomic and ionic form-factors in (6.9) reads

$$\Phi_\mu^A \gamma^{-1} \Lambda_\alpha^\mu F_I^\alpha = \left(\Phi_0^A + \frac{v}{c}\Phi_3^A\right)\left(F_I^0 + \frac{v}{c}F_I^3\right) + \frac{\Phi_3^A F_I^3}{\gamma^2} + \frac{\Phi_1^A F_I^1 + \Phi_2^A F_I^2}{\gamma}.$$
$$(6.10)$$

6.2.1 The Nonrelativistic Limit

In the limit $c \to \infty$, when all relativistic effects vanish, the following simplifications occur in the transition amplitude (6.9). In the denominator of the integrand in (6.9) the retardation term $(\varepsilon_n - \varepsilon_0)^2/c^2$ disappears and the minimum

---

[1] Where the obvious changes have to be done.

momentum q_{min}^A takes on its nonrelativistic limit $q_{min}^A = (\varepsilon_n - \varepsilon_0 + \epsilon_m - \epsilon_0)/v$. In the coupling (6.10) only the time components survive, $\Phi_\mu^A \gamma^{-1} \Lambda_\alpha^\mu F_I^\alpha \to \Phi_0^A F_I^0$. Besides, now these components are evaluated using the nonrelativistic states of the atom and ion. As a result, after all these simplifications the amplitude (6.9) reduces to (3.17) which was obtained in Chap. 1 using the nonrelativistic symmetric eikonal approximation.

Let us remind that we arrived at the amplitude (3.17) by assuming that the initial and final internal states of the ion are different, $n \neq 0$, (while there were no such restrictions on the final state of the atom). This important condition was explicitly embodied in the derivation of the distorted wave amplitude (3.11) from which the amplitude (3.17) was obtained by choosing the eikonal distortion factors.

In the derivation of the relativistic eikonal amplitude given in this section the condition $n \neq 0$ was not explicitly used and, in this sense, the situation with the applicability of the amplitude (6.9) is somewhat less clear. However, comparing the amplitude (6.9) with the symmetric eikonal amplitude (6.86), obtained for collisions with point-like charges (see Sect. 6.5), it is not difficult to find out that in the case $n = 0$ the amplitude (6.9) does not reduce to (6.86).

Taking this into account and also remembering that the nonrelativistic amplitude (3.17) may not be used for collisions with $n = 0$ we may infer that the relativistic amplitude (6.9) is only valid to describe collisions in which the final and initial internal states of the projectile are different, $m \neq 0$. At the same time, there is no restrictions on the possible final internal state of the atom: this state may differ from the initial state $(n \neq 0)$ or be the same $(n = 0)$.

6.2.2 The Relationship with the First Order Theory

The transition amplitude in the momentum space, following from the first order theory in the ion–atom interaction, is given by (5.47), in which the form-factor of the ion F_I^α is exactly the same as in the amplitude (6.9). Besides, it is not difficult to convince oneself that the first order (undistorted) form-factor of the atom, F_μ^A, represents the simple limiting case of the eikonal atomic form-factor, Φ_μ^A:

$$F_\mu^A \left(n0; \mathbf{q}_\perp, q_{min}^A \right) = \lim_{\mathbf{p}_\perp \to 0} \Phi_\mu^A \left(n0; \mathbf{q}_\perp - \mathbf{p}_\perp, q_{min}^A; \mathbf{p}_\perp \right). \qquad (6.11)$$

The peculiarity of the first order consideration is that for collisions, in which the internal state of the projectile is changed $(m \neq 0)$, it predicts that the interaction between the projectile nucleus and the atomic target does not have impact on the collision physics no matter how high the charge of the projectile is. Besides, the first order consideration also suggests that the interaction between the target nucleus and the projectile electron in the inelastic target mode, $n \neq 0$, does not contribute to the amplitude (5.47) as well. Thus, according to the first order treatment, the projectile–target interaction in collisions, where both projectile and target change their internal states, reduces to the two-center dielectronic interaction.

The first order transition amplitude is remarkably symmetric with respect to the (mutual) exchange of the quantities of the atom and ion in (5.47). In contrast, such a symmetry is broken in the eikonal transition amplitude (6.9) since within the eikonal model the projectile and the target are not treated on an equal footing due to the high charge state of the projectile ion.

There are two important general differences between the eikonal amplitude (6.9) and the first order one (5.47). Firstly, according to (6.9), the interaction of the projectile nucleus with the target does influence the physics of the projectile–target collision. Secondly, the interaction between the target nucleus and the projectile electron contributes to the projectile-electron excitation and loss also in the case of inelastic (for the target) collisions. Note, that actually these two points cannot be disentangled since the contribution of the latter interaction appears automatically once the eikonal distortion factors due to the field of the projectile nucleus are included into the initial and final states of the target (6.2). Thus, also in the relativistic consideration of ion–atom collisions, the incorporation of the interaction between the highly charged projectile nucleus and the target electron 'wakes up' the interaction between the target nucleus and the projectile electron in the inelastic target mode.

The actual differences between results, obtained with the eikonal and first order transition amplitudes, depend on the magnitude of $\nu = Z_{\mathrm{I}}/v$ and the momentum transfer $\mathbf{q}$ in the collision. The true perturbative limit in our case is given by the condition $\nu = Z_{\mathrm{I}}/v \to 0$. The analysis of the amplitude (6.9) shows that in this limit the contribution to the integral over the virtual transverse momentum transfer $\mathbf{p}_\perp$ is given solely by the infinitesimally small vicinity of the point $p_\perp = 0$. At $p_\perp = 0$ the form-factors and the denominator of the integrand in (6.9) become identical to those of (5.47). The integration over the infinitesimally small range $\mathbf{p}_\perp = 0$ is easily performed yielding 1. As a result, in the limit $\nu = Z_{\mathrm{I}}/v \to 0$ the eikonal transition amplitude (6.9) reduces to the first order amplitude (5.47).

6.2.3 Projectiles with More Than One Electron

Let us now assume that the projectile initially carries more than one electron but that the absolute value of the total charge of these electrons is still much less than the charge of the projectile nucleus. In order to generalize the amplitude (6.9) to such a case one has just to replace there the single-electron ionic form-factor by the many-electron form-factor

$$F_0^I(m0;\mathbf{k}) = -\int \prod_{i=1}^{N_{\mathrm{I}}} \mathrm{d}^3\boldsymbol{\xi}_i\, \chi_{\mathrm{m}}^\dagger(\boldsymbol{\tau}_{N_{\mathrm{I}}}) \sum_{j=1}^{N_{\mathrm{I}}} \exp(i\mathbf{k}\cdot\boldsymbol{\xi}_j)\, \chi_0(\boldsymbol{\tau}_{N_{\mathrm{I}}}),$$

$$F_l^I(m0;\mathbf{k}) = \int \prod_{i=1}^{N_{\mathrm{I}}} \mathrm{d}^3\boldsymbol{\xi}_i\, \chi_{\mathrm{m}}^\dagger(\boldsymbol{\tau}_{N_{\mathrm{I}}}) \sum_{j=1}^{N_{\mathrm{I}}} \alpha_{l(j)}\, \exp(i\mathbf{k}\cdot\boldsymbol{\xi}_j)\, \chi_0(\boldsymbol{\tau}_{N_{\mathrm{I}}}), \quad (6.12)$$

where χ_0 and χ_m are the initial and final internal states of the undistorted ion, $\tau_{N_I} = \{\boldsymbol{\xi}_1, \boldsymbol{\xi}_2, \ldots, \boldsymbol{\xi}_{N_I}\}$ represents the set of coordinates of the $N_I \geq 1$ electrons of the ion with respect to the ionic nucleus given in the frame K_I and $\alpha_{l(j)}$ is the Dirac matrix for the jth electron. The sum and the product in (6.12) run over the projectile electrons.

6.2.4 Inclusion of the Nuclear–Nuclear Interaction

Like in the case of nonrelativistic collisions, the inclusion of the interaction between the nuclei of the projectile ion and the target atom into the treatment becomes necessary when cross sections differential in the projectile deflection angle (in the transverse momentum transfer) are considered. This inclusion is also necessary when cross sections differential in the transverse momentum of the target recoil ions are studied. In fast ion–atom collisions this interaction, however, influences neither spectra of emitted electrons integrated over the transverse momentum transfer nor the distribution of the target recoil ions over the longitudinal momentum.

The inclusion of the nuclear–nuclear interaction can be done within the symmetric eikonal approximation by incorporating additional eikonal phases associated with this interaction into the initial and final distorted states of the target according to

$$\phi_i(\mathbf{r}, t) = (vs + \mathbf{v} \cdot \mathbf{s})^{-i\nu} (vs_0 + \mathbf{v} \cdot \mathbf{s}_0)^{i\eta} \varphi_0(\mathbf{r}) \exp(-i\varepsilon_0 t),$$
$$\phi_f(\mathbf{r}, t) = (vs - \mathbf{v} \cdot \mathbf{s})^{i\nu} (vs_0 - \mathbf{v} \cdot \mathbf{s}_0)^{-i\eta} \varphi_n(\mathbf{r}) \exp(-i\varepsilon_n t). \quad (6.13)$$

In (6.13) the three-dimensional space vector $\mathbf{s}_0$ is the coordinate of the target nucleus relative to the nucleus of the ion given in the *rest frame* of the ion and $\eta = Z_I Z_A / v$.

Using (6.13) one can obtain (see [41]) that the transition amplitude is now given by

$$S_{fi}(\mathbf{q}_\perp) = -\frac{2i}{v} \int d^2 \mathbf{p}_\perp f(p_\perp, \nu) \int d^2 \mathbf{t}_\perp f(t_\perp, -\eta)$$
$$\times \frac{\Phi_\mu^A \left(n0; \mathbf{Q}^A; \mathbf{p}_\perp \right) \gamma^{-1} \Lambda_\alpha^\mu \, F_I^\alpha \left(m0; \mathbf{Q}^I \right)}{(\mathbf{q}_\perp + \mathbf{t}_\perp - \mathbf{p}_\perp)^2 + (q_{\min}^A)^2 - \frac{(\varepsilon_n - \varepsilon_0)^2}{c^2}}, \quad (6.14)$$

where $\mathbf{Q}^A = \left(\mathbf{q}_\perp + \mathbf{t}_\perp - \mathbf{p}_\perp, q_{\min}^A \right)$, $\mathbf{Q}^I = -\left(\mathbf{q}_\perp + \mathbf{t}_\perp - \mathbf{p}_\perp, q_{\min}^I \right)$ and $\mathbf{t}_\perp$ is the two-dimensional transverse momentum vector, $\mathbf{t}_\perp \cdot \mathbf{v} = 0$.

6.2.5 Collisions with Two-Electron Atoms

In the case when the target atom has two electrons the initial and final internal states of the atom are chosen within the symmetric eikonal approximation as

$$\phi_i(\mathbf{r}_1, \mathbf{r}_2, t) = (vs_1 + \mathbf{v} \cdot \mathbf{s}_1)^{-i\nu} (vs_2 + \mathbf{v} \cdot \mathbf{s}_2)^{-i\nu} \varphi_0(\mathbf{r}_1, \mathbf{r}_2) \exp(-i\varepsilon_0 t),$$
$$\phi_f(\mathbf{r}_1, \mathbf{r}_2, t) = (vs_1 - \mathbf{v} \cdot \mathbf{s}_1)^{i\nu} (vs_2 - \mathbf{v} \cdot \mathbf{s}_2)^{i\nu} \varphi_n(\mathbf{r}_1, \mathbf{r}_2) \exp(-i\varepsilon_n t). \quad (6.15)$$

In the above expression φ_0 and φ_n are the initial and final undistorted states of the atomic target with the energies ε_0 and ε_n, respectively. The three-dimensional vectors $\mathbf{s}_1$ and $\mathbf{s}_2$ represent the space coordinates of the first and second target electron with respect to the ionic nucleus given in the *rest frame* of the ion and $\nu = Z_I/v$.

The derivation of the transition amplitude with the atomic two-electron states (6.15) is similar to that given for atoms with one active electron. Therefore, we shall not discuss this derivation and directly quote the final result

$$S_{fi}(\mathbf{q}_\perp) = -\frac{2i}{v} \int d^2\mathbf{p}_\perp \int d^2\boldsymbol{\kappa}_\perp \frac{f(p_\perp,\nu)f(\kappa_\perp,\nu)}{(\mathbf{q}_\perp - \mathbf{p}_\perp - \boldsymbol{\kappa}_\perp)^2 + (q^A_{min})^2 - \frac{(\varepsilon_n-\varepsilon_0)^2}{c^2}}$$
$$\times \Phi^A_\mu\left(n0; \mathbf{K}^A; \mathbf{p}_\perp; \boldsymbol{\kappa}_\perp\right) \gamma^{-1} \Lambda^\mu_\alpha F^\alpha_I\left(m0; \mathbf{K}^I\right), \tag{6.16}$$

where $\mathbf{K}^A = \left(\mathbf{q}_\perp - \mathbf{p}_\perp - \boldsymbol{\kappa}_\perp, q^A_{min}\right)$ and $\mathbf{K}^I = -\left(\mathbf{q}_\perp - \mathbf{p}_\perp - \boldsymbol{\kappa}_\perp, q^I_{min}\right)$. The integration in (6.16) runs over the two-dimensional transverse vectors $\mathbf{p}_\perp$ and $\boldsymbol{\kappa}_\perp$ ($\mathbf{p}_\perp \cdot \mathbf{v} = 0$ and $\boldsymbol{\kappa}_\perp \cdot \mathbf{v} = 0$). Further, the two-electron atomic form-factor in (6.16) reads

$$\Phi^A_0(n0; \mathbf{k}; \mathbf{p}_\perp; \boldsymbol{\kappa}_\perp) = \int d^3\mathbf{r}_1 d^3\mathbf{r}_1 \, \varphi^\dagger_n(\mathbf{r}_1, \mathbf{r}_2) \, \varphi_0(\mathbf{r}_1, \mathbf{r}_2) \, \exp(i\mathbf{p}_\perp \cdot \mathbf{r}_1 + i\boldsymbol{\kappa}_\perp \cdot \mathbf{r}_2)$$
$$\times \{Z_A - \exp(i\mathbf{k}\cdot\mathbf{r}_1) - \exp(i\mathbf{k}\cdot\mathbf{r}_2)\},$$
$$\Phi^A_l(n0; \mathbf{k}; \mathbf{p}_\perp; \boldsymbol{\kappa}_\perp) = \int d^3\mathbf{r}_1 \int d^3\mathbf{r}_2 \, \varphi^\dagger_n(\mathbf{r}_1, \mathbf{r}_2) \, \exp(i\mathbf{p}_\perp \cdot \mathbf{r}_1 + i\boldsymbol{\kappa}_\perp \cdot \mathbf{r}_2)$$
$$\times \{\alpha_{l,1}\exp(i\mathbf{k}\cdot\mathbf{r}_1) + \alpha_{l,2}\exp(i\mathbf{k}\cdot\mathbf{r}_2)\}\varphi_0(\mathbf{r}_1, \mathbf{r}_2), \tag{6.17}$$

where the indices 1 and 2 refer to the first and second atomic electron. In (6.16) the inelastic form-factor of the ion $F^I_\alpha(m0; \mathbf{K}^I)$ is the same as in (6.9).

6.2.6 Some Applications

Below we shall consider a few applications of the eikonal model and compare its results with those given by the first order treatment.

Spectra of Electrons Emitted from the Target

We start with the basic process: the collision between an hydrogen-like highly charged ion and a hydrogen atom. As an example, we consider the excitation of $Kr^{35+}(1s)$ projectile into its states with the principal quantum number $n = 2$. In Fig. 6.1a we show the contributions to the excitation cross section from the elastic target mode: $Kr^{35+}(1s) + H(1s) \rightarrow Kr^{35+}(n = 2) + H(1s)$. The elastic contributions were calculated using the relativistic eikonal and first order transition amplitudes and are depicted in Fig. 6.1a by solid and dash curves, respectively. They are given as a function of the collision energy. For comparison, dot and dash-dot curves in the figure display results of the

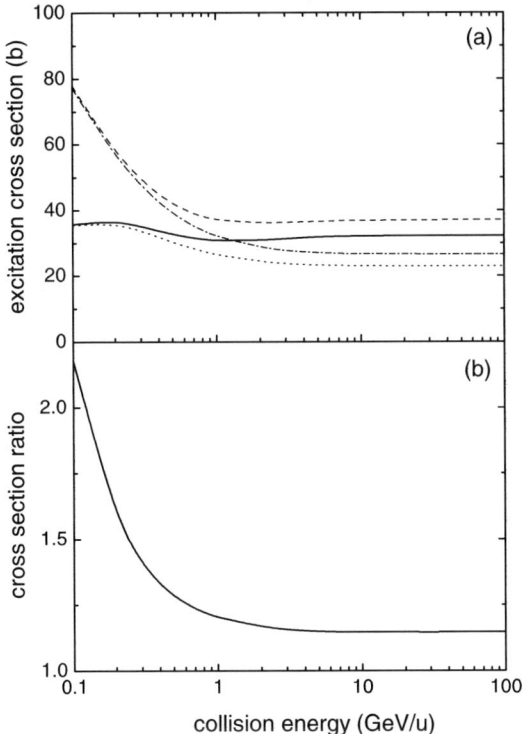

Fig. 6.1. Projectile excitation in $Kr^{35+}(1s) + H(1s) \rightarrow Kr^{35+}(n = 2) + H(1s)$ collisions. (a) *Solid curve*: the relativistic eikonal approximation. *Dash curve*: the relativistic first order approximation. *Dot curve*: the nonrelativistic ($c \rightarrow \infty$) eikonal approximation. *Dash-dot curve*: the nonrelativistic first order approximation. (b) The ratio of the elastic target mode contributions to the projectile excitation cross section calculated in the relativistic first order and relativistic eikonal approximations. From [41].

nonrelativistic ($c \rightarrow \infty$) eikonal and first order calculations, respectively. In Fig. 6.1b shown is the ratio of the relativistic eikonal and first order cross sections for the elastic atomic mode. It is seen in the figure that the ratio decreases from more than 2 at an impact energy of 100 MeV u^{-1} ($v \approx 59$ a.u.) to 1.15 at asymptotically high impact energies where the ratio becomes already energy-independent. Thus, according to the eikonal theory, the interaction between the projectile nucleus and the target electron decreases the contribution of the elastic target mode to the projectile excitation cross section.

The calculation of the contribution to the projectile excitation from the inelastic target mode is much more complicated. Therefore, we performed such a calculation only for a few impact energies ranging between 1 and 100 GeV u^{-1}. According to the calculation, the eikonal model predicts that the contribution of the inelastic target mode is larger than that suggested by the first order consideration.

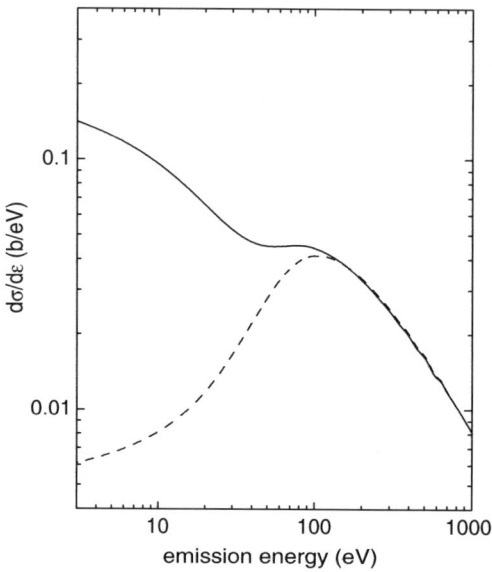

Fig. 6.2. Energy distribution of electrons emitted from the target in 1 Ge u^{-1} MeV Kr^{35+}(1s)+H(1s) $\rightarrow$ Kr^{35+}(n = 2)+p$^+$ + e$^-$ collisions. *Solid curve*: calculation using the eikonal amplitude. *Dash curve*: first order calculation. From [41].

The difference between the inelastic target contributions, calculated with the eikonal and first order theories, is illustrated in Fig. 6.2 where the energy spectrum of the electron emitted from the target is shown (the spectrum is given in the target frame). It is seen in this figure that the eikonal model predicts the very strong enhancement of the low-energy part of the target emission. This enhancement can be understood if we note the following. In order to excite a tightly bound electron in a highly charged projectile the hydrogenic target must transfer to the projectile electron a momentum which is large on the momentum scale of hydrogen. According to the first order consideration, in the inelastic target mode this momentum transfer can only occur via the two-center dielectronic interaction (TCDI) which was discussed in Sect. 5.14 and which directly couples the electrons of the projectile and the target. Because of this direct coupling, the target electron acquires a large re-coil momentum and tends to populate highly lying energy states in the target continuum. This is clearly seen in Fig. 6.2 where the maximum of the emission spectrum, predicted by the first order treatment, is reached at energies of ~100 eV and even at emission energies as high as 1 keV the number of emitted electrons is still larger than at 1–10 eV. Within the eikonal model, in addition to the first order process caused by the TCDI, the more complicated process becomes possible in which the projectile electron is excited by the interaction with the target nucleus and, simultaneously, the target electron is emitted due to its interaction with the projectile nucleus. Within such a two-step process the direct momentum exchange between the electrons does not occur and the momentum transfer to the target electron is smaller, compared to that in the first order process, since the large recoil is now taken by the target nucleus. As a result, the two-step process strongly enhances the low-energy part of the target emission.

In the examples considered above the ratio $\nu = Z_I/v$ ranges between 0.6 and 0.3 when the impact energies run between $100\,\mathrm{MeV\,u^{-1}}$ and $1\,\mathrm{GeV\,u^{-1}}$, and even at the asymptotically high energies, where $v \to c$, this ratio is still not very small (0.26). Therefore, the substantial deviations of the eikonal results from the first order ones are not unexpected. Below we shall see, however, that the deviations can be noticeable even at much smaller values of ν.

Let us consider the cross section, $\mathrm{d}^2\sigma/\mathrm{d}\varepsilon\,\mathrm{d}\Omega$, which is differential in energy and angle of the electron emitted from the target and which is obtained by integrating over all final continuum states of the electron of the projectile. This cross section is shown in Fig. 6.3 for the reaction $O^{7+}(1s)+He(1s^2) \to O^{8+} + He^+(1s)+2e^-$ at collision energies of $100\,\mathrm{MeV\,u^{-1}}$, 1 and $10\,\mathrm{GeV\,u^{-1}}$ corresponding to collision velocities of 59, 120 and 136.5 a.u., respectively. In this figure the cross section is given for a fixed electron energy of $\varepsilon = k^2/2 = 5\,\mathrm{eV}$ as a function of the polar emission angle of the target electron, $\theta = \arccos\,(\mathbf{k}\cdot\mathbf{v}/kv)$, where $\mathbf{k}$ is the electron momentum with respect to the target nucleus and $k = |\mathbf{k}|$.

The collision system under consideration involves five bodies and is difficult to treat without making additional approximations. It follows from experimental data on $O^{7+}(1s)+He(1s^2)$ collisions [64] that at an impact energy of $1.25\,\mathrm{MeV\,u^{-1}}$ ($v = 7.1$ a.u.) the numbers of the electron loss events accompanied by single and double ionization of helium are approximately equal. However, when for the same collision system the collision energy increases to $4.7\,\mathrm{MeV\,u^{-1}}$ ($v = 13.9$ a.u.) the number of the electron loss events accompanied by double ionization of helium becomes already just a quarter of that for the loss events resulting in helium single ionization [64]. Thus, one can expect that for $O^{7+}(1s)+He(1s^2)$ collisions at much higher impact energies only one target electron will be 'active' in the electron loss process. Therefore, in our estimate we assume that there is one active and one passive electron in the helium target and that the vectors $\mathbf{p}_\perp$ and $\boldsymbol{\kappa}_\perp$ in the amplitude (6.16) are related to the active and passive electrons, respectively. Taking into account that the main contribution to the eikonal amplitude (6.16) at $\nu = Z_I/v \ll 1$ is given by the region of small values of $\kappa_\perp$, one can reduce the amplitude (6.16) to the amplitude (6.9) where the charge of the target nucleus Z_A is replaced by the effective charge Z_{eff} of the target core (the nucleus and the passive electron of the target). The latter is given by the (nonrelativistic) form-factor of the target core and, in contrast to the charge of the nucleus, depends on the momentum transfer $\mathbf{q}$ in the collision, $Z_{\mathrm{eff}} = Z_{\mathrm{eff}}(q)$, that reflects the compound nature of the core. We approximate the charge $Z_{\mathrm{eff}}(q)$ by the elastic nonrelativistic form-factor of a hydrogen-like ion with the nucleus charge of 1.69. Further, we assume that the active target electron moves

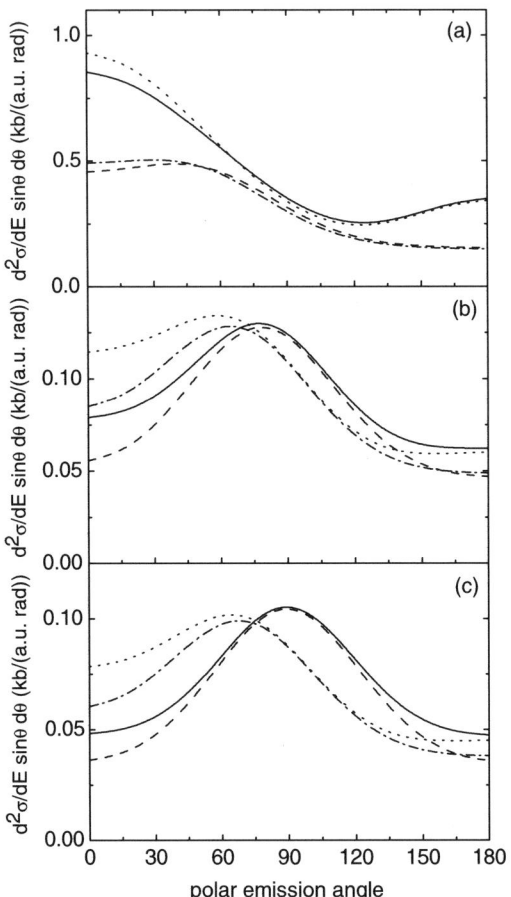

Fig. 6.3. Angular distribution of 5 eV electrons emitted from helium in $O^{7+}(1s)$ $+He(1s^2) \rightarrow O^{8+} + He^+(1s) + 2 e^-$ collisions. **(a)** $100\,MeV\,u^{-1}$; **(b)** $1\,GeV\,u^{-1}$ ($v = 120\,a.u.,\ \gamma \simeq 2.1$); **(c)** $10\,GeV\,u^{-1}$ ($v = 136.5\,a.u.,\ \gamma \simeq 11.7$); *Solid curve*: the relativistic eikonal calculation. *Dash curve*: the first order result. *Dot curve*: the nonrelativistic ($c = \infty$) eikonal calculation. *Dash-dot curve*: the nonrelativistic first order result. From [41].

in the Coulomb field created by the target core which acts on the active electron as a point-like charged object with an effective core charge $Z_c = 1.345$ chosen to fit the ionization potential of helium.

According to Fig. 6.3a there is the very substantial difference between results of the first order and eikonal calculations at an impact energy of $100\,MeV\,u^{-1}$ where the ratio $Z_I/v \approx 0.14$. Moreover, the difference remains quite noticeable even at $1\,GeV\,u^{-1}$ (see Fig. 6.3b) where the ratio is just 0.067. Besides, at this impact energy there is also the substantial difference between results of the relativistic and nonrelativistic calculations. Thus, at $1\,GeV\,u^{-1}$

both the higher order and relativistic effects are clearly visible in the angular distribution of the electron emitted from the target. While at 1 GeV u^{-1} the appearance of the relativistic effects is not surprising the noticeable deviation from the first order predictions at this impact energy is rather unexpected.[2]

Comparing relativistic and nonrelativistic results of the first order theory one can conclude that the relativistic effects make the angular distribution of the emitted electrons more symmetric with respect to the direction perpendicular to the collision velocity ($\theta = 90^0$) [118]. Comparing first order and eikonal results we see that the higher order terms in the projectile–target interaction tend in general to break this symmetry. However, since the virtual momentum transfer $\mathbf{p}_\perp$ is a transverse vector, the higher order effects contained in the eikonal theory are *per se* not able to break the symmetry and can only enhance the asymmetry already present in the first order calculation. At very high impact energies the latter becomes negligible and the eikonal spectrum also becomes symmetric (see Fig. 6.3c).

At comparatively low values of the collision Lorentz factor γ the relativistic effects rapidly increase and the higher order effects rapidly decrease with increasing the collision energy. However, when the collision energy increases further the corresponding increase and decrease become slower and, starting with collision energies $\simeq$10 GeV u^{-1}, the relativistic and higher order effects in the collision under consideration saturate: the shape and absolute values of the cross section do not change substantially with a further increase in the collision energy.

Target Recoil Momentum Spectroscopy in Collisions with Relativistic Highly Charged Ions

In the previous section we have considered, as an example, collisions with not very heavy ions and have seen that the deviations from the first order predictions in general may take place already at rather small values of the parameter $\nu = Z_I/v$ ($\nu < 0.1$).

Now we turn to the consideration of collisions of hydrogen and helium atoms with very highly charged hydrogen-like ions for which the parameter ν is not much smaller than 1 even when the collision velocity approaches the speed of light and, therefore, the deviations from predictions of the first order theory may be especially pronounced. We shall consider only collisions in which the target atom gets ionized.

Important information about the physics of such collision can be obtained by studying the distributions of the emitted electron(s). However, below we shall restrict ourselves to the discussion of the spectra of the target recoil ions concentrating, following to [127], on the spectrum of such ions given as

---

[2] In particular, the latter point was not taken into account in [118] where the mutual projectile–target ionization in O^{7+}–He collisions was considered within the first order theory.

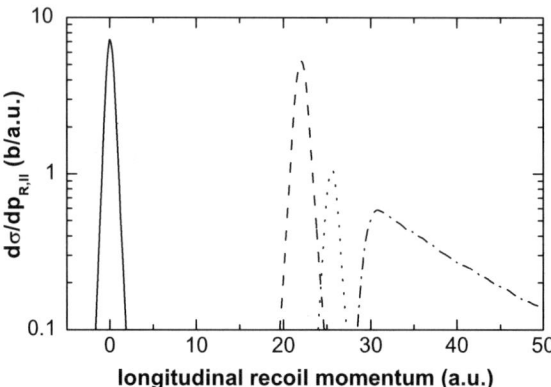

Fig. 6.4. The longitudinal momentum spectrum of the target recoil ions produced in $100\,\mathrm{MeV\,u^{-1}}$ $Nd^{59+}(1s)+H(1s)$ collisions. Dash, dot and dash-dot curves at large $p_{R,\parallel}$ correspond to the $Nd^{59+}(n=2)$, $Nd^{59+}(n=3)$ and $Nd^{60+} + e^-$ states of the projectile, respectively. Besides, solid curve displays the total contributions of the above channels to the recoil spectrum at small $p_{R,\parallel}$. From [127].

a function of the longitudinal component $p_{R,\parallel}$ of the recoil momentum which is parallel to the velocity of the projectile-ion.

In Fig. 6.4 results are shown for the longitudinal momentum spectrum of the target recoil ions, $d\sigma/dp_{R,\parallel}$, generated in the processes of the projectile electron loss and excitation to states with $n = 2$ and $n = 3$, where n is the principal quantum number, in collisions of $100\,\mathrm{MeV\,u^{-1}}$ $Nd^{59+}(1s)$ with atomic hydrogen. The recoil peak centered at $p_{R,\parallel} = 0$ arises due to the two-center dielectronic interaction and thus is basically the effect of the first order in the projectile–target interaction.[3] The peaks at much larger $p_{R,\parallel}$, however, are not predicted by the first order theory and appear in the calculation only if all the interactions between the nuclei and the electrons are taken into account. Hence, these peaks represent clear signatures of the higher order effects in the interaction between the colliding particles.

In Fig. 6.5 we display results for the projectile excitation into states with $n = 2$ accompanied by the single ionization of helium target. We see that the relativistic and nonrelativistic calculations predict rather different positions and shape of the target recoil peak.

The relativistic effects influencing the form of the recoil spectrum can be subdivided into those depending on the collision velocity v and those related to the inner motions of electrons within the colliding centers. As is seen in Fig. 6.5, in the momentum spectrum these two groups of the effects counteract.

---

[3] In collisions with highly charged ions the two-center dielectronic interaction is modulated by multiple interactions between the ionic nucleus and the electron of the target which, under the conditions of Fig. 6.4, decreases the height of the peak at $p_{R,\parallel} \simeq 0$ by about 25% compared to the prediction of the first order theory.

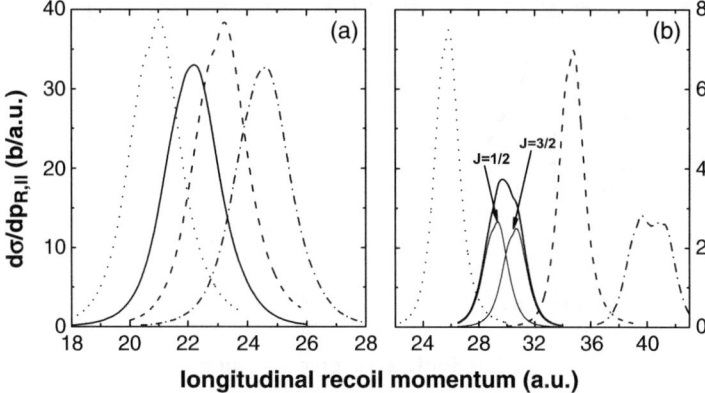

Fig. 6.5. The projectile excitation into states with the principal quantum number $n = 2$ accompanied by target single ionization in collisions of $100\,\mathrm{MeV\,u^{-1}}$ $\mathrm{Nd}^{59+}(1s)$ (a) and $325\,\mathrm{MeV\,u^{-1}}$ $\mathrm{U}^{91+}(1s)$ (b) with $\mathrm{He}(1s^2)$. *Dash* and *solid curves* display results of the purely nonrelativistic ($c = \infty$) and fully relativistic calculations, respectively. *Dot curves* were obtained by assuming $c = \infty$ only in the description of the internal electron states. *Dash-dot curves* were calculated by setting $c = \infty$ only in the treatment of the relative ion–atom motion. From [127].

The relativistic effects in the inner motion of the electron in the ion lead to the energy splitting of the levels having the total angular momentum $j = 1/2$ and $j = 3/2$ and increase the energy differences between them and the ground state. Because of the splitting the peak in the recoil momentum spectrum becomes broader and the splitting is also partly responsible for the decrease in the height of the peak. As a result of the increase in the energy differences the peak position is shifted to larger values of $p_{\mathrm{R},\parallel}$. However, the relativistic effects due to the relative ion–atom motion soften the recoil of the residual atomic core which shifts the peak position to lower values of $p_{\mathrm{R},\parallel}$. In the examples shown in Fig. 6.5 the value of the collision velocity is very close to the value of the typical orbiting velocity of the electron in the ground state of the ion but the shift of the recoil peak due to the relativistic effects in the relative ion–atom motion turns out to be larger by about a factor of 2.

The processes considered above are of course not the only ones which produce target recoil ions. Therefore, the important question is whether the signatures of the projectile-electron excitation and loss in the momentum spectrum of the target recoil ions will not be masked by other processes (especially if the final state of the projectile is not detected). To answer this question one needs to perform a more extensive study of the target recoil momentum spectrum produced in collisions in which the internal state of the projectile may change or may remain the same. Some results of such a study are shown in Fig. 6.6 which actually illuminates the very general features in the formation of the longitudinal recoil spectrum in collisions with very heavy hydrogen-like

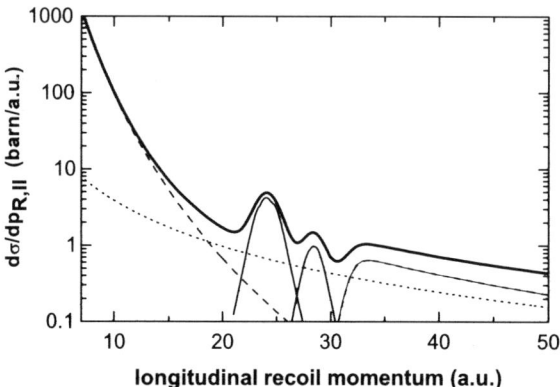

Fig. 6.6. The longitudinal momentum spectrum of He^+ recoils produced in $430\,\mathrm{MeV\,u}^{-1}\,\mathrm{Th}^{89+}(1s)+\mathrm{He}$ collisions. *Dash* and *dot curves* show the contributions to the spectrum from the singly inelastic channel given by the interaction between the ionized electron and the residual atomic ion and by the nucleus–nucleus Rutherford scattering, respectively. *Thin solid curves*: the contribution from the projectile electron loss and excitations into $n = 2, 3$. *Thick solid curve*: the total contribution from the above channels. The contribution from the projectile excitation into bound states with $n \geq 4$ was not calculated. From [127].

ions at moderate values of γ. According to this figure one can separate four different regions of $p_{\mathrm{R},\parallel}$ in which the formation is dominated by qualitatively different mechanisms.

(a) At small values of $p_{\mathrm{R},\parallel}$ the recoil spectrum is generated via the indirect coupling between the projectile, whose internal state *is unchanged*, and the target core: the projectile induces a transition of the target electron and the electron exchanges its momentum with the target core. In relativistic collisions this channel is characterized by very large impact parameters ($\gg 1$ a.u.) and is very efficient in transferring small values of the momentum to the target recoil. At larger $p_{\mathrm{R},\parallel}$, however, it rapidly loses its effectiveness which leads to the very strong decrease in the longitudinal spectrum when $p_{\mathrm{R},\parallel}$ grows.

(b) Eventually with increasing $p_{\mathrm{R},\parallel}$ the direct coupling between the nuclei of the ion and atom (the Rutherford scattering) may start to dominate the formation of the spectrum. This channel 'works' in collisions with very small impact parameters ($\sim Z_{\mathrm{I}} Z_{\mathrm{A}}/v\sqrt{M_{\mathrm{A}} v p_{\mathrm{R},\parallel}}$, where M_{A} is the mass of the atomic nucleus) and, provided $Z_{\mathrm{I}} \sim v$, the nuclear scattering is accompanied by *target ionization* with the probability close to 1 but the inner state of the projectile remains unchanged.[4]

---

[4] To describe this channel of momentum transfer equations (5.32) have to be corrected by adding a term $q_\perp^2/(2M_{\mathrm{A}}v)$ into the equation for $q_{\mathrm{min}}^{\mathrm{A}}$ (a similar change has to be made also in (5.31)). Note that this correction becomes important only at very small impact parameters whose contribution to the doubly inelastic cross sections is negligible and, therefore, can safely be neglected when the latter are calculated.

(c) With the further increase in $p_{R,\parallel}$ ($3Z_I^2/8v \lesssim p_{R,\parallel} \lesssim Z_I^2/v$) the channel involving the excitation and loss of the tightly bound electron of the projectile comes into play. Compared to the nucleus-nucleus collisions this channel is characterized by much larger values of the typical impact parameters, which at $Z_I \sim v$ are of the order of the size of the projectile ground state, and is much more effective. Since the absolute charge of the electron constitutes just about 1% of the net ion charge and its mass is negligibly small compared to the mass of the nuclei, such an effectiveness is quite surprising and may have interesting consequences.

Consider, for instance, two very heavy projectiles whose atomic numbers differ just by 1 and whose masses are very close (e.g. $^{209}_{83}\text{Bi}$ ($A = 208.98$) and $^{209}_{84}\text{Po}$ ($A = 208.98$), A is the atomic mass). Let one of them be represented by a bare nucleus with a charge $Z_I^{(1)}$ and the second be a hydrogen-like ion with a net charge $Z_I^{(2)} - 1 = Z_I^{(1)}$. Thus, the projectiles possess practically equal charge-to-mass ratios and seem to behave almost identically in external electric fields. However, if these projectiles collide with atoms, the atoms can easily 'recognize' whether the projectile is a bare nucleus or an hydrogen-like ion since in the latter case the spectrum of the target recoil ions possesses a prominent resonance-like structure, reflecting the excitation and loss of the projectile's electron (superimposed on the smooth background from collisions elastic for the projectile).

(d) At even higher $p_{R,\parallel}$ the recoil spectrum is formed almost solely by the nucleus-nucleus scattering [5].

In very asymmetric collisions ($M_I \gg \gamma M_A$, where M_I is the mass of the ion nucleus) the contributions to the cross section $d\sigma/dp_{R,\parallel}$ due to the nuclear Rutherford scattering and due to the excitation of the electron of the ion scale as Z_A^2/M_A and Z_A^2, respectively. Therefore, the relative strength of the latter channel in the spectrum formation is weakest for collisions with atomic hydrogen.

In collisions of very highly charged ions with atoms the doubly inelastic cross sections are by several orders of magnitude smaller than the cross section for the pure atomic ionization. However, since at moderate γ the regions of $p_{R,\parallel}$ relevant to the singly and doubly inelastic processes are well separated and because of the 'prominent' behavior of the projectile electron in the transfer of large $p_{R,\parallel}$, the theoretical consideration predicts that a great

---

[5] So far we have not mentioned the capture of a target electron by the projectile. This process occurs via the radiative and coulomb capture channels. At the impact energies under consideration the last channel is quite weak and is characterized by very small recoil peaks located at large negative $p_{R,\parallel}$ ($p_{R,\parallel} = -(I_f/\gamma - I_i)/v - m_e c^2(1 - 1/\gamma)/v$, where I_i and I_f are the initial and final binding energies of the electron). The relatively strong radiative capture channel results only in quite low values of the target recoil momenta where its effect is fully masked by the direct target ionization.

deal of information about the doubly inelastic collisions with relativistic heavy ions could be obtained in an experiment just by measuring the longitudinal momentum spectrum of the target recoil ions.

6.3 Collisions with Heavy Atoms: Preliminary Remarks

The first order treatment of the ion–atom collisions is in principle valid only provided the following two conditions are simultaneously fulfilled: (a) $Z_I/v \ll 1$ and (b) $Z_A/v \ll 1$, where Z_I and Z_A are the atomic numbers of the projectile ion and target atom. Therefore, strictly speaking, the first order theory should be applied neither to collisions of very highly charged ions (for which the ratio Z_I/v can never be much smaller than 1) with arbitrary atoms nor to collisions of arbitrary ions with very heavy atoms (for which the ratio Z_A/v is not much smaller than 1).

In collisions between highly charged projectiles and light atoms the first order approach faces difficulties because the target atom is exposed in the collision to the field of the projectile nucleus which is very strong on the atomic scale. Therefore, the first order perturbation theory is not capable of a correct description of electronic transitions in the target. As a result, *for a fixed final state of the target* the first order theory does not yield proper results also for transitions occurring in the projectile.

Fortunately, however, for collisions with light atoms the situation actually is much more favorable with respect to the 'practical' applications of the first order theories to calculate the electron loss and excitation cross sections. Indeed, results of calculations, performed in the first order approximation and using the eikonal model of the previous section, suggest that the difference between these two descriptions becomes not very essential if one is not interested in what exactly happens to the target in such collisions and just sum over all possible final target states. It turns out that, after such a summation, results for the cross sections the projectile-electron transitions obtained in the eikonal approach and the first order theory become quite similar.

Such a close coincidence of the eikonal and first order results is not fortuitous. In fact, such a coincidence even looks quite expected as long as (a) the atomic field *per se* represent just a weak perturbation for the electron of the ion and (b) the space distribution of the atomic electrons can be regarded as 'frozen' during the short effective collision time when the electron of the ion is exposed to the short-range field of the neutral atom. The point (a) is fulfilled for $Z_A/v \ll 1$ and the assumption (b) is a very reasonable one at the high impact velocities $v \gg Z_A$ where the effective collision time is much shorter than typical orbiting times of the electrons in a light atomic system These are the basic physical reasons why in collisions with light atoms the first order theory can yield reasonable results for the projectile cross sections summed over the final target states even in the case when $Z_I/v \sim 1$.

The situation becomes quite different when collisions with heavy atoms are considered. In collisions with very heavy atoms the atomic field acting on the electron of the ion can effectively reach quite large values which makes the first order approximations invalid. However, even if we would be ready to sacrifice our knowledge about electronic transitions in the target we still want to get a reliable description of the electron transitions occurring in the projectile. The problem now lies not only in the fact that we cannot take a proper account of the action of the highly charged projectile on the atom containing very many electrons but also in the fact that, independent of whether or not the atomic response to the projectile field is treated properly, already the very strength of the atomic field *per se* may make the first order considerations questionable.

Therefore, the question which will be addressed in the next sections of this chapter is how one can describe electron transitions in the projectile when it collides with so heavy atomic target that one has $Z_A/c \sim 1$. The general approach for attacking this problem will be to sacrifice the description of the behavior of the atom in the collision and instead fully concentrate on the treatment of the projectile-electron transitions by making some plausible assumptions about the character of the action of the atom on the electron of the projectile during the collision. Such an approach would be sufficient if we are interested only in the study of electron transitions in the projectile-ion but are not concerned by what will happen with the atomic target.

In what follows we shall consider the projectile-electron excitation and loss which occur either at the extremely high or rather low relativistic impact energies.

6.4 Extreme Relativistic Collisions with Heavy Atoms

Except the resonance transitions, which were discussed in Sect. 5.14.2 of Chap. 5, the projectile-electron excitation and loss caused by collisions with neutral atoms occur at impact parameters typically not exceeding the size of the atom. Because of the effectively short-range character of the interaction between the atom and ion, at any impact energy (including the asymptotic limit $\gamma \to \infty$) the first order treatments may not always be appropriate.

In the case of the extreme relativistic collisions, where $\gamma \gg 1$, the analysis of the atomic action on the electron(s) of the projectile is greatly simplifies by the ultrashort duration of the ion–atom collision. Indeed, although it is highly unlikely that the initial quantum state of the atomic target will not be changed by the collision with a highly charged projectile penetrating the target electron cloud, it is, nevertheless, still reasonable to assume that the spatial distribution of the target electrons will not be considerably altered during the very short effective collision time when the field of a neutral target acts on the projectile electron.[6] Besides, at the extreme relativistic impact energies

---

[6] In collisions at the extreme relativistic energies this time will be very short even on the time scale of the most tightly bound electrons in a heavy atomic target.

the minimum momenta transferred to the ion and atom reduce respectively, to $q^I_{min} = (\varepsilon_n - \varepsilon_0)/c$ and $q^A_{min} = (\epsilon_m - \epsilon_0)/c$ which, in particular, means that the momentum transfer to the ion does not depend on the final internal state of the atom.[7]

Therefore, the projectile-electron excitation and loss cross sections can be evaluated using the assumption that the electrons of the atomic target are 'frozen' during the effective collision time and, hence, the spatial distribution of these electrons during this time can be represented by the wave function of the initial atomic state.

One should note that the assumption that the spatial distribution of the atomic electrons is 'frozen' during the collision does not imply that these electrons will finally remain in the atomic initial state. Therefore, in contrast to the first order theory, the *screening* mode, as it will be regarded in this section, is no longer equivalent to the *elastic* mode.[8]

The assumption about the 'frozen' atomic electrons enables one to treat the field of the atom during the collision as an external field and, thus, to single out the degrees of freedom of the electron of the ion as the only quantum degrees of freedom in the collision. As a result, the many-body problem of the two colliding atomic particles reduces to the much simpler problem of the motion of the electron of the ion in two fields, the field of the nucleus of the ion and the field of the 'frozen' atom.

It is obvious that within such an approach the contribution to the transitions of the electron of the projectile, which is given by the two-center electron–electron correlations, will be lost. Remark, however, that we are now going to consider the higher-order effects in projectile-electron excitation and loss and, therefore, shall concentrate on collisions with very heavy atoms where these effects should be more pronounced.

According to the first order consideration, provided the dimension of the electron orbit in the projectile is much less than the dimension of the neutral target, the interaction of the projectile-electron with the target nucleus screened by the 'frozen' atomic electrons is by far the dominant one in collisions with atoms containing very many electrons. In addition, the deviation from results of the first order consideration is first of all expected for very small impact parameters where, as it will be discussed in detail in Chap. 7, the relative contribution of the two-center electron correlations is even weaker compared to that in the total excitation and loss cross sections. Therefore, the neglect of the contributions yielded by the two-center electronic correlations does not seem to be too a big shortcoming.

---

[7] One could say that the collision becomes so extremely short that the electron is not able to 'accumulate' all the information about the collision.

[8] In this sense the situation which we consider now is very similar to that encountered when discussing the sudden and Glauber approximations in Sects. 3.4 and 3.5 of Chap. 3.

In the rest frame of the nucleus of the ion the motion of its electron is described by the Dirac equation

$$i\frac{\partial\Psi(\mathbf{r},t)}{\partial t} = \left(c\boldsymbol{\alpha}\cdot\left(\mathbf{p}+\frac{1}{c}\mathbf{A}(\mathbf{r},t)\right)+\beta c^2 - \frac{Z_\mathrm{I}}{r} - \Phi(\mathbf{r},t)\right)\Psi(\mathbf{r},t), \quad (6.18)$$

where $\boldsymbol{\alpha}$ and β are the Dirac's matrices, $\mathbf{r}$ is the coordinate of the electron with respect to the nucleus of the ion, Φ and $\mathbf{A}$ are the scalar and vector potentials describing the field of the incident neutral atom. We assume that in the rest frame of the atom its scalar potential is well approximated by a short-range interaction

$$\Phi' = \frac{Z_\mathrm{A}\phi(r')}{r'}, \quad (6.19)$$

where

$$\phi(r') = \sum_j A_j \exp(-\kappa_j r') \quad (6.20)$$

with the screening parameters A_j ($\sum_j A_j = 1$) and κ_j which have already been introduced in Sect. 5.2. An interaction of the type (6.19)–(6.20) can be regarded as originating from the exchange of 'massive photons' with masses $M_j = \kappa_j$: a photon with mass M_j has the source characterized by a charge $Z_j = Z_\mathrm{A}A_j$ ($\sum_j Z_j = Z_\mathrm{A}$).

The scalar and vector potentials of a source Z_j of massive photons with mass M_j, which moves with relativistic velocity v, are described by the Proca equation [7]

$$\Delta\Phi_j - \frac{1}{c^2}\frac{\partial^2\Phi_j}{\partial t^2} - M_j^2\Phi_j = -4\pi Z_j\delta(\mathbf{r}-\mathbf{R}(t))$$

$$\mathbf{A}_j = \frac{\mathbf{v}}{c}\Phi_j. \quad (6.21)$$

We will assume that in the projectile frame the atom moves along a straight-line trajectory $\mathbf{R} = \mathbf{b} + \mathbf{v}t$, where $\mathbf{b} = (b_x, b_y)$ is the impact parameter and $\mathbf{v} = (0,0,v)$ is the velocity of the atom.

With the help of the Fourier transformation the solution of (6.21) can be written as

$$\Phi_j(\mathbf{r},t) = \frac{Z_j}{2\pi^2}\int d^2k_\perp \exp(i\mathbf{k}_\perp(\mathbf{r}_\perp-\mathbf{b}))\int_{-\infty}^{+\infty} dk_z \frac{\exp(ik_z(z-vt))}{k_\perp^2 + \frac{k_z^2}{\gamma^2} + M_j^2}, \quad (6.22)$$

where $\mathbf{r} = (\mathbf{r}_\perp, z)$ with $\mathbf{r}_\perp \cdot \mathbf{v} = 0$.

The straightforward integration of (6.22) results in

$$\Phi_j(\mathbf{r},t) = \frac{\gamma Z_j}{\sqrt{\gamma^2(z-vt)^2 + (\mathbf{r}_\perp-\mathbf{b})^2}}\exp\left(-M_j\sqrt{\gamma^2(z-vt)^2 + (\mathbf{r}_\perp-\mathbf{b})^2}\right).$$

$$(6.23)$$

The potential (6.23) could be easily derived directly from (6.19) and (6.20) by using the Lorentz transformation. However, the advantage of the Fourier representation (6.22) is that, for collision velocities v which very closely approach the speed of light, it allows one to obtain straightforwardly the substantial simplification for the form of the scalar and vector potentials, whereas the limit $v \to c$ of (6.23) is rather delicate.

6.4.1 Light-Cone Potentials

For infinite γ one can drop the term $\frac{k_z^2}{\gamma^2}$ in the integrand of (6.22) and obtain [132]

$$
\Phi_j(\mathbf{r}, t) = \frac{Z_j}{2\pi^2} \int d^2k_\perp \exp(i\mathbf{k}_\perp(\mathbf{r}_\perp - \mathbf{b})) \int_{-\infty}^{+\infty} dk_z \frac{\exp(ik_z(z - ct))}{k_\perp^2 + M_j^2}
$$
$$
= 2Z_j \delta(ct - z) K_0(M_j \mid \mathbf{r}_\perp - \mathbf{b} \mid), \tag{6.24}
$$

where δ is the delta-function and K_0 is Macdonald's function [108]. Then the scalar and vector potentials describing the field of the incident atom in the rest frame of the ion are given by

$$
\Phi(\mathbf{r}, t) = 2Z_A \delta(ct - z) \sum_j A_j K_0(M_j \mid \mathbf{r}_\perp - \mathbf{b} \mid)
$$

$$
A_z(\mathbf{r}, t) = \Phi(\mathbf{r}, t),
$$
$$
A_x(\mathbf{r}, t) = A_y(\mathbf{r}, t) = 0. \tag{6.25}
$$

Using a gauge transformation for the potentials (6.25) with the gauge function f chosen as

$$
f = 2Z_A \theta(ct - z) \sum_j A_j K_0(M_j \mid \mathbf{r}_\perp - \mathbf{b} \mid) \tag{6.26}
$$

we obtain another convenient set of the potentials

$$
A'_x(\mathbf{r}, t) = 2Z_A \theta(ct - z) \sum_j A_j \partial_x K_0(M_j \mid \mathbf{r}_\perp - \mathbf{b} \mid)
$$

$$
A'_y(\mathbf{r}, t) = 2Z_A \theta(ct - z) \sum_j A_j \partial_y K_0(M_j \mid \mathbf{r}_\perp - \mathbf{b} \mid),
$$

$$
\Phi'(\mathbf{r}, t) = A'_z(\mathbf{r}, t) = 0, \tag{6.27}
$$

where $\theta(a)$ is the Heaviside step function: $\theta(a) = 0$ for $a < 0$ and $\theta(a) = 1$ otherwise. Note that the electric and magnetic fields described by the potentials (6.25) (or (6.27)) have nonzero components only in the (x, y)-plane. The strengths of the electric and magnetic field are mutually perpendicular.

The case when the field is produced by a point-like charged particle can be described by (6.25) (or (6.27)) by taking the limit $M_j \to 0$.

The potentials (6.25) and (6.27) possess some interesting properties. In particular, their dependence on the coordinates (x, y, z) and time t is given by the product(s) of two factors. One of them is a function of the transverse coordinates (x, y) only. The other one depends on the coordinate z and time t and this dependence appears only through the light-cone combination $ct - z$. If, when addressing a real problem, the interaction can be approximated as being transmitted by the light-cone potentials, the resulting model problem can often be solved exactly.

Before we proceed with the application of the light-cone potentials to the extreme relativistic ion–atom collisions we shall briefly consider the motion of a free electron in the field generated by a point-like particle moving at the speed of light.

6.4.2 Classical Electron in the Field of a Particle Moving with the Speed of Light

We start with the classical consideration of the electron dynamics. As before we assume that the external electromagnetic field acting on the electron is produced by a particle which move at the speed along a given straight-line classical trajectory $\mathbf{R} = (X, Y, ct) = (\mathbf{b}, ct)$.

For the sake of definiteness we shall choose the light-cone potentials, which describe this field at a time t at a point with the coordinates $(x, y, z) = (\mathbf{r}_\perp, z)$, in the form similar to (6.27)

$$
\begin{aligned}
A_1 &= -2Z\, \partial_x \Lambda(|\mathbf{r}_\perp - \mathbf{b}|)\, \theta(\eta), \\
A_2 &= -2Z\, \partial_y \Lambda(|\mathbf{r}_\perp - \mathbf{b}|)\, \theta(\eta), \\
A_0 &= A_3 = 0,
\end{aligned}
\tag{6.28}
$$

where Z is the charge of the particle and the explicit form of the function Λ depends on whether the light-cone field is of the Coulomb or Yukawa type. The phase of the field $\eta = ct - z$ can be rewritten as $\eta = n_\mu x^\mu$, where $n_\mu = (1, 0, 0, -1)$ is a null-vector ($n_\mu n^\mu = n^2 = 0$). Note that n_μ and A_μ are orthogonal: $n^\mu A_\mu = 0$.

Neglecting the radiation force the motion of a classical electron is described by the Lorentz equations. In our case it is convenient to use these equations in the four-dimensional form, in which they read (see e.g. [8])

$$
\frac{dp_\mu}{d\tau} = \frac{e}{mc} F_{\mu\nu} p^\nu.
\tag{6.29}
$$

Here, e and m are the charge[9] and mass of the electron, respectively, p_μ is the electron four-momentum, $F_{\mu\nu} = \partial_\mu A_\nu - \partial_\nu A_\mu$ is the electromagnetic field tensor and τ is the proper time.

---

[9] In this and the next subsections we keep the explicit notation e for the electron charge.

We first multiply both sides of (6.29) by n^μ. Taking into account that $n^\mu A_\mu = 0$ and $(\partial_0 + \partial_3) A_j = 0$ $(j = 1, 2)$ we obtain from (6.29) that

$$\frac{d n^\mu p_\mu}{d\tau} = 0. \tag{6.30}$$

Hence, the quantity

$$n^\mu p_\mu = p_0 + p_3 = \frac{E}{c} - p_z, \tag{6.31}$$

where E is the total electron energy and p_z is the z-component of its momentum, is conserved in this process. Further, since $p_\mu = m \frac{d x_\mu}{d\tau}$, we see that $n^\mu p_\mu = m \frac{d\eta}{d\tau}$ and have

$$\frac{d\eta}{d\tau} = \frac{E - c p_z}{mc}. \tag{6.32}$$

Taking into account (6.28)–(6.29) and (6.31)–(6.32) we obtain

$$\frac{d p_0}{d\eta} = \frac{2Ze}{E - c p_z} \delta(\eta) \, (p_1 \, \partial_x \Lambda + p_2 \, \partial_y \Lambda) \, \delta(\eta)$$

$$\frac{d p_1}{d\eta} = -\frac{2Ze}{c} \delta(\eta) \, \partial_x \Lambda$$

$$\frac{d p_2}{d\eta} = -\frac{2Ze}{c} \delta(\eta) \, \partial_y \Lambda$$

$$\frac{d p_3}{d\eta} = -\frac{d p_0}{d\eta}. \tag{6.33}$$

The transverse components of the electron momentum are easily found from (6.33). The electron energy and the longitudinal component of its momentum can be conveniently derived from the set of equations, which are obtained from (6.33) if we, instead of the first equation there, employ the energy-momentum relation $(E/c)^2 - p^2 = m^2 c^2$ and also make use of the fact that the quantity $E/c - p_z$ represents the integral of motion (see (6.31)). Then we have

$$p_x = p_x^{in} + \frac{2Ze}{c} \theta(ct - z) \, \partial_x \Lambda,$$

$$p_y = p_y^{in} + \frac{2Ze}{c} \theta(ct - z) \, \partial_y \Lambda,$$

$$E - c p_z = E^{in} - c p_z^{in},$$

$$E + c p_z = c^2 \frac{m^2 c^2 + p_x^2 + p_y^2}{E^{in} - c p_z^{in}}, \tag{6.34}$$

where the derivatives $\partial_x \Lambda$ and $\partial_y \Lambda$ are taken at the position (x, y) of the electron at $ct = z$ and the superscript 'in' denotes the initial values of the

electron energy and momentum. According to (6.34) the electron, as a result
of its collision with the light-cone field, suddenly changes its momentum and
energy. The magnitude of these changes depends on the the value of the charge
Z of the moving particle and the relative position of the electron with respect
to this particle in the plane perpendicular to the velocity of the latter.

The expressions for the electron energy and momentum after the collision
have the most simple form if the electron is initially at rest. In such a case
from (6.34) we obtain

$$p_x = \frac{2Ze}{c}\partial_x \Lambda,$$

$$p_y = \frac{2Ze}{c}\partial_y \Lambda,$$

$$p_z = \frac{p_x^2 + p_y^2}{2mc},$$

$$E = mc^2 + \frac{p_x^2 + p_y^2}{2m}. \tag{6.35}$$

Note that while in (6.35) the transverse components of the electron momentum
are of the first order in Ze, the longitudinal momentum is proportional to
$(Ze)^2$. This difference reflects the fact that, as was already mentioned, the
electric and magnetic strengths of the light-cone field are nonzero only in the
transverse plane. Therefore, the electron, which is initially at rest, first starts
to move in the transverse direction due to the interaction with the electric
part of the Lorentz force and only then the magnetic component of this force
becoming nonzero can drive the electron in the z-direction. Note also that
along the z-axis the electron will be always pushed by the Lorentz force in
the direction of the motion of the light-cone particle, no matter what is the
sign of the charge of the particle.

In the case, when the light-cone field is created by a point-like charged
particle, its potentials can be obtained from (6.27) by taking the limit $M_j \to 0$.
The latter yields $\partial_x \Lambda(|\mathbf{r}_\perp - \mathbf{b}|)) = (x - b_x)/|\mathbf{r}_\perp - \mathbf{b}|^2$, $\partial_y \Lambda(|\mathbf{r}_\perp - \mathbf{b}|)) = (y - b_y)/|\mathbf{r}_\perp - \mathbf{b}|^2$ and from (6.35) we obtain

$$p_x = \frac{2Ze}{c}\frac{x - b_x}{|\mathbf{r}_\perp - \mathbf{b}|^2}$$

$$p_y = \frac{2Ze}{c}\frac{y - b_y}{|\mathbf{r}_\perp - \mathbf{b}|^2}$$

$$p_z = \frac{2Z^2e^2}{mc^3}\frac{1}{|\mathbf{r}_\perp - \mathbf{b}|^2}$$

$$E = mc^2 + \frac{2Z^2e^2}{mc^2}\frac{1}{|\mathbf{r}_\perp - \mathbf{b}|^2}. \tag{6.36}$$

6.4.3 Quantum Electron in the Field of a Particle Moving with the Speed of Light

In order to consider the behavior of a quantum electron in the field (6.28) we start with the Dirac equation written in the covariant form (see (4.47))

$$\left[\gamma^\mu \left(\hat{p}_\mu - \frac{e}{c} A_\mu \right) - mc \right] \psi = 0. \tag{6.37}$$

By acting on (6.37) from the left by the operator $\left[\gamma^\mu \left(\hat{p}_\mu - \frac{e}{c} A_\mu \right) + mc \right]$ and making use of the anticommutation relations (4.48) for the matrices γ^μ we obtain

$$\left(\hat{p}^2 - 2\frac{e}{c} A \cdot \hat{p} - \mathrm{i}\frac{e}{c} (\partial \cdot A) + \frac{e^2}{c^2} A^2 - m^2 c^2 - \mathrm{i}\frac{e}{c} \sigma^{\mu\nu} F_{\mu\nu} \right) \psi = 0, \tag{6.38}$$

where $F_{\mu\nu}$ is the electromagnetic field tensor and

$$\sigma^{\mu\nu} = \frac{1}{4} \left(\gamma^\mu \gamma^\nu - \gamma^\mu \gamma^\nu \right). \tag{6.39}$$

Similarly to the classical consideration we shall assume that the light-cone potentials A_μ are taken in the form (6.28).

It is convenient to introduce the so called light-cone coordinates

$$\eta = ct - z,$$
$$\xi = ct + z. \tag{6.40}$$

Using these coordinates the operator $\hat{p}^2$ can be rewritten according to

$$\hat{p}^2 \equiv \hat{p}_0^2 - \hat{p}_z^2 - \hat{p}_x^2 - \hat{p}_y^2$$
$$= -4\frac{\partial^2}{\partial\eta\partial\xi} - \hat{p}_x^2 - \hat{p}_y^2. \tag{6.41}$$

It is also not difficult to show that the result of the double summation in the last interaction term in parentheses of (6.38) is

$$\Pi \equiv \sigma^{\mu\nu} F_{\mu\nu} = -(1 + \alpha_z) \frac{\partial}{\partial\eta} (\alpha_x A_x + \alpha_y A_y)$$
$$= -2Z\delta(\eta)(1 + \alpha_z) \left(\alpha_x \frac{\partial}{\partial x} + \alpha_y \frac{\partial}{\partial y} \right) \Lambda. \tag{6.42}$$

Taking into account (6.41) and (6.42) and also the fact that $\partial \cdot A = 0$, we arrive at

$$\left(-4\frac{\partial^2}{\partial\eta\partial\xi} - \hat{p}_x^2 - \hat{p}_y^2 - 2\frac{e}{c} (A_x \hat{p}_x + A_y \hat{p}_y) - m^2 c^2 - \mathrm{i}\frac{e}{c} \Pi \right) \psi = 0. \tag{6.43}$$

The interaction terms in (6.43) do not depend on ξ. This means that the quantity $K_\xi = \frac{1}{2}(p_0 + p_3)$, which is the momentum component conjugate to ξ, is conserved in the collision.

Making in (6.43) the ansatz

$$\psi = \varphi \exp\left(-\mathrm{i}\, K_\xi \xi - \mathrm{i}\, \frac{m^2 c^2}{4K_\xi} \eta\right) \tag{6.44}$$

and denoting $2K_\xi = M$ we obtain for φ the following equation:

$$\mathrm{i}\frac{\partial \varphi}{\partial \eta} = \frac{\left(\hat{\mathbf{p}}_\perp - \frac{e}{c}\mathbf{A}_\perp\right)^2}{2M}\varphi + \mathrm{i}\frac{e}{2Mc}\Pi\varphi, \tag{6.45}$$

where $\hat{\mathbf{p}}_\perp = (\hat{p}_x, \hat{p}_y)$ and $\mathbf{A}_\perp = (A_x, A_y)$. The last term in (6.45) appears because the electron has spin $1/2$ and it would be absent for a zero-spin 'electron'. If, for the moment, we disregard this term, then the resulting equation formally coincides with the nonrelativistic Schrödinger equation for a spinless particle with charge e and mass M which moves in the 'time' η and two-dimensional space (x, y) interacting with the 'time'-dependent electromagnetic field described by the vector potentials A_x and A_y.

In (6.45) the interaction of the electron with the light-cone field consists of the terms, which contain the field potentials and are proportional to $\theta(\eta)$, and the terms which contain the field strengths and are proportional to $\delta(\eta)$. In the gauge, which we have employed, the field potentials suddenly 'jump' from zero at $\eta < 0$ to nonzero values at $\eta > 0$. If the change in the interaction at $\eta = 0$ would be limited merely to this step-wise behavior of the potentials, the solution of (6.45) would be continuous at all η, including the point $\eta = 0$. However, because the spin terms contain the factor $\delta(\eta)$, the function φ is discontinuous at $\eta = 0$.[10] Since these terms, which couple the electron spin to the electric and magnetic field strengths, are gauge invariant, this discontinuity cannot be removed by a gauge transformation.

At $\eta < 0$ the solution of (6.45), which describes a free electron with a well defined transverse momentum $\mathbf{p}_\perp$ and polarization ρ, reads

$$\varphi = \varphi_{\mathbf{p}_\perp,\rho} = \exp(\mathrm{i}\mathbf{p}_\perp \cdot \mathbf{r}_\perp)\exp\left(-\mathrm{i}\frac{p_\perp^2}{2M}\eta\right)u^{(\rho)}(\mathbf{p}), \tag{6.46}$$

where $u^{(\rho)}(\mathbf{p})$ is a coordinate-independent four-spinor.

At $\eta > 0$ an electron having a transverse kinetic momentum $\boldsymbol{\kappa}_\perp$ and polarization λ is described, according to (6.45), by

$$\varphi = \varphi_{\boldsymbol{\kappa}_\perp,\lambda} = \exp(\mathrm{i}\boldsymbol{\kappa}_\perp \cdot \mathbf{r}_\perp)\exp\left(-\mathrm{i}\frac{2Ze}{c}\Lambda\right)\exp\left(-\mathrm{i}\frac{\kappa_\perp^2}{2M}\eta\right)u^{(\lambda)}(\boldsymbol{\kappa}). \tag{6.47}$$

---

[10] Since $\alpha_z^2 = 1$ and, hence, one has $(1 - \alpha_z)\Pi \equiv 0$, the four-spinor $(1 - \alpha_z)\varphi$ represents a solution of the 'Schrödinger' equation. In the gauge, which we use, this solution remains continuous at $\eta = 0$. Note also that the latter would not be the case if the field potentials would be chosen in the form (6.25).

Using the partial solutions (6.46) and (6.47) one can construct a solution of the (6.45) describing the behavior of the electron under the action of the light-cone field.

Let us, as an example, consider the solution of (6.45) corresponding to an electron which has before the collision (at $\eta \to -\infty$) a well defined kinetic momentum $\mathbf{p}_0 = (\mathbf{p}_{0,\perp}; p_{0,z})$ and polarization λ. This solution can be written in the following form:

$$\varphi = \exp(\mathrm{i}\mathbf{p}_{0,\perp} \cdot \mathbf{r}_\perp) \exp\left(-\mathrm{i}\frac{p_{0,\perp}^2}{2M}\eta\right) u^{(\rho)}(\mathbf{p}); \quad \eta < 0 \tag{6.48}$$

and

$$\varphi = \int \mathrm{d}^2\boldsymbol{\kappa}_\perp a(\boldsymbol{\kappa}_\perp, \lambda) \exp(\mathrm{i}\boldsymbol{\kappa}_\perp \cdot \mathbf{r}_\perp) \exp\left(-\mathrm{i}\frac{2Ze}{c}\Lambda\right)$$

$$\times \exp\left(-\mathrm{i}\frac{\kappa_\perp^2}{2M}\eta\right) u^{(\lambda)}(\boldsymbol{\kappa}); \quad \eta > 0, \tag{6.49}$$

where $a(\boldsymbol{\kappa}_\perp, \lambda)$ are the unknown expansion coefficients to be determined.

To find these coefficients one cannot simply equate the right-hand-side parts of (6.48) and (6.49) at $\eta = 0$ because, as was already mentioned, φ is not continuous in this point. Therefore, in order to determine $a(\boldsymbol{\kappa}_\perp, \lambda)$ we first note that at $\eta = 0$ the last term on the right-hand-side of (6.45) becomes infinite. Hence only this interaction term and the derivative over η should be kept when we consider (6.45) at $\eta = 0$ and in the infinitely small vicinity of this point. This enable us to find the solution of (6.45) which is valid not only at $\eta < 0$ but also in the interval $[0 - \epsilon < \eta < 0 + \epsilon]$, where $\epsilon \to +0$. This solution reads

$$\varphi = \exp(\mathrm{i}\mathbf{p}_{0,\perp} \cdot \mathbf{r}_\perp) \exp\left(-\mathrm{i}\frac{p_{0,\perp}^2}{2M}\eta\right)$$

$$\times \exp\left(-\frac{Ze\theta(\eta)}{Mc}(1 + \alpha_z)\left(\alpha_x\frac{\partial\Lambda}{\partial x} + \alpha_y\frac{\partial\Lambda}{\partial y}\right)\right) u^{(\rho)}(\mathbf{p}); \quad \eta \le \epsilon. \tag{6.50}$$

Since $\alpha_z\alpha_x = -\alpha_x\alpha_z$, $\alpha_z\alpha_y = -\alpha_y\alpha_z$ and $1 - \alpha_z^2 = 0$, only the first two terms are nonzero in the expansion of that exponent in (6.50), which contains the Dirac matrices. Therefore, the above solution can also be rewritten as

$$\varphi = \exp(\mathrm{i}\mathbf{p}_{0,\perp} \cdot \mathbf{r}_\perp) \exp\left(-\mathrm{i}\frac{p_{0,\perp}^2}{2M}\eta\right)$$

$$\times \left(1 - \frac{Ze\theta(\eta)}{2Mc}(1 + \alpha_z)\left(\alpha_x\frac{\partial\Lambda}{\partial x} + \alpha_y\frac{\partial\Lambda}{\partial y}\right)\right) u^{(\rho)}(\mathbf{p}); \quad \eta \le \epsilon. \tag{6.51}$$

Now, by matching the solutions (6.49) and (6.51) at $\eta = \epsilon$ one can determine the coefficients $a(\boldsymbol{\kappa}_\perp, \lambda)$. The wave function ψ is then obtained by using (6.48), (6.49) and (6.44).

A somewhat more complicated problem of the motion of an electron in the field of two charged particles, which move in the observer's frame with the speed of light in the opposite directions, was considered in [128]– [129] focusing on the pair production.

6.4.4 Light-Cone Approximation for Ion–Atom Collisions

Let us now return to the ion–atom collisions and consider transitions of an electron which is initially bound by the field of the ionic nucleus and interacts with so fast atom that the field of the latter can be described by the light-cone potentials. The wave function of the electron is a solution of the Dirac equation (6.18).

This problem can be solved exactly using a method proposed in [131] to calculate electron transitions caused by collisions with a point-like charge moving at the speed of light. This method relies on the light-cone potentials in the form similar to (6.25) and the gauge, in which $A_0 = -A_3 \sim \delta(ct - z)$ and $A_1 = A_2 = 0$, shall be used below.

Following this method we first expand the electron wave function according to

$$\Psi(\mathbf{r}, t) = \sum_n a_n(t) \exp(-i\varepsilon_n t)\psi_n(\mathbf{r}), \tag{6.52}$$

where $\psi_n(\mathbf{r})$ are the states of the electron in the undistorted ion with energies ε_n and we assume that the sum includes also the continuum states [11]. Inserting the expansion (6.52) into the left hand side of (6.18), multiplying both sides of the resulting equation by an arbitrary state $\psi_f^\dagger(\mathbf{r}) \exp(i\varepsilon_f t)$ and integrating over the spatial coordinates of the electron we obtain

$$i\frac{a_f(t)}{dt} = \exp(i\varepsilon_f t)\langle\psi_f \mid cG\,\delta(ct - z)\,(1 - \alpha_z) \mid \Psi(t)\rangle, \tag{6.53}$$

where

$$G = G(\mathbf{r}_\perp - \mathbf{b}) = \frac{2Z_A}{c}\sum_j A_j K_0\left(M_j \mid \mathbf{r}_\perp - \mathbf{b}\mid\right). \tag{6.54}$$

Assuming that before the collision the electron is in a state ψ_0, the initial conditions are given by

$$a_n(t \to -\infty) = \delta_{n0},$$
$$\Psi(t \to -\infty) = \exp(-i\varepsilon_0 t)\psi_0(\mathbf{r}). \tag{6.55}$$

Due to the presence of the delta-function in expression (6.53) all what one needs to know, in order to obtain the transition amplitude, is the product

---

[11] with both positive and negative energies.

$(1 - \alpha_z)\Psi$ at $z = ct$. For this purpose we rewrite the Dirac equation (6.18) using the light-cone coordinates (see Eqs.(6.40)) which yields

$$i(1 - \alpha_z)\frac{\partial\Psi}{\partial\eta} - G\,\delta(\eta)\,\Psi$$
$$= -i(1 + \alpha_z)\frac{\partial\Psi}{\partial\xi} + \left(\boldsymbol{\alpha}_\perp \cdot \mathbf{p}_\perp + \beta c - \frac{Z_I}{cr}\right)\Psi. \qquad (6.56)$$

We shall now integrate the above equation over the infinitesimally small vicinity of the point $\eta = 0$ (which includes both $\eta \leq 0$ and $\eta > 0$). Taking into account that the integration of the right hand side part of (6.56) obviously yields an infinitely small quantity, which can be neglected, and making use of the initial conditions one obtains that in the vicinity of $ct = z$

$$(1 - \alpha_z)\Psi(t) = (1 - \alpha_z)\exp(-i\theta(ct - z)\,G)\psi_i\exp(-i\varepsilon_i t), \qquad (6.57)$$

where θ is the theta-function. With the help of (6.57) one has

$$i\frac{a_f(t)}{dt} = \exp(i\omega_{fi} t)\langle\psi_f \mid G\,\delta(ct - z)\,(1 - \alpha_z)\exp(-iG\,\theta(ct - z)) \mid \psi_i\rangle, \qquad (6.58)$$

where ω_{fi} is the transition frequency.

Taking into account that $d\theta(ct - z)/dt = c\delta(ct - z)$ and using (6.55), one can easily integrate (6.58) over time from $-\infty$ to $+\infty$ and obtain that the exact amplitude $a_{0\to n}$ for the electron of the ion to undergo a transition between states ψ_0 and ψ_n is given by

$$a_{0\to n}(\mathbf{b}) = \delta_{0n} + \left\langle\psi_n \left|(1 - \alpha_z)\exp\left(i\frac{\omega_{n0}z}{c}\right)(\exp(-iG) - 1)\right| \psi_0\right\rangle. \qquad (6.59)$$

By making use of the identity

$$\langle\psi_n \mid \alpha_z\exp\left(i\frac{\omega_{n0}z}{v}\right)\mid \psi_0\rangle \equiv \frac{v}{c}\langle\psi_n \mid \exp\left(i\frac{\omega_{n0}z}{v}\right)\mid \psi_0\rangle, \qquad (6.60)$$

which holds for $n \neq 0$ [12], and taking into account that $\langle\psi_0 \mid 1 - \alpha_z \mid \psi_0\rangle = 1$ the expression (6.59) for the transition amplitude can be reduced to

$$a_{0\to n}(\mathbf{b}) = \left\langle\psi_n \left|(1 - \alpha_z)\exp\left(i\frac{\omega_{n0}z}{c}\right)\exp(-iG)\right| \psi_0\right\rangle, \qquad (6.61)$$

which is fully equivalent to (6.59) both at $n \neq 0$ and $n = 0$.

Similarly as in the case of the exchange of gravitons or zero mass photons (see [133–135]), the expression (6.61) can be seen as the 'eikonalization' of the transition amplitude in the case of the exchange of massive photons in collisions at $\gamma \to \infty$. Below, the transition amplitude (6.61) is referred to as the light-cone amplitude.

---

[12] Equation (6.60) is a particular case of the continuity equation (5.56) for the electron transition current. More simply it can be obtained from (5.74) by setting there $\mathbf{q}_\perp = 0$ and $q^I_{min} = \omega_{n0}/v$.

The Transition Amplitude in the Momentum Space

In general, in addition to the transition amplitude given in the impact parameter space, it is useful to have also the corresponding amplitude written in
the momentum space, $S_{0\to n}(\mathbf{q}_\perp)$, which is the two-dimensional Fourier transform of the amplitude $a_{0\to n}(\mathbf{b})$. It is not difficult to show that, within the
light-cone approximation, the amplitude $S_{0\to n}(\mathbf{q}_\perp)$ is given by

$$S_{0\to n}(\mathbf{q}_\perp) = \frac{1}{2\pi} \int d^2\mathbf{b} \exp(i\mathbf{q}_\perp \cdot \mathbf{b})\, a_{0\to n}(\mathbf{b})$$
$$= F(\mathbf{q}_\perp)\, \langle \psi_n \,|(1-\alpha_z)\exp(i\mathbf{q}\cdot\mathbf{r})|\, \psi_0 \rangle. \qquad (6.62)$$

In the above expression $\mathbf{q} = \left(\mathbf{q}_\perp; \frac{\omega_{n0}}{c}\right)$ is the momentum transfer and

$$F(\mathbf{q}_\perp) = \frac{1}{2\pi} \int d^2\mathbf{s}\, \exp\left(-i\,\mathbf{q}_\perp \cdot \mathbf{s}\right)\left(\exp(-iG(\mathbf{s})) - 1\right), \qquad (6.63)$$

where the integration runs over the two-dimensional transverse vector $\mathbf{s}$ ($\mathbf{s}\cdot\mathbf{v} =$
0, $0 \le s < \infty$, $0 \le \phi_s \le 2\pi$). The expressions (6.63)–(6.63) were obtained
using (6.59) as the starting point.

In the light-cone amplitude (6.63) the term $\langle \psi_n \,|(1-\alpha_z)\exp(i\mathbf{q}\cdot\mathbf{r})|\, \psi_0 \rangle$ is
exactly the same as in the first order consideration (compare e.g. with (5.72)
or (5.83)). It is the function F given by (6.63) which accounts for the higher
order effects in the interaction between the electron of the ion and the field of
the fast moving atom.

Unitarity of the Light-Cone Approximation

The total probability for the electron to occupy, after the collision with a given
impact parameter $\mathbf{b}$, any of the states of the ion (including bound states and
states with positive and negative energies) reads

$$P(\mathbf{b}) = \sum_n |\, a_{0\to n}(\mathbf{b})\,|^2. \qquad (6.64)$$

Since the amplitude (6.61) represents the exact solution (although formally
only for $\gamma = \infty$) the total probability $P(\mathbf{b})$ is expected to be equal to 1.
In what follows we shall briefly discuss how the unitarity of the light-cone
approximation can be proved.

Following [136] we define the operator,

$$\hat{D} = \sum_k \frac{i^k}{c^k k!} \hat{H}_0^k z^k, \qquad (6.65)$$

where $\hat{H}_0$ is the electronic hamiltonian of the ion. Its hermitian conjugate is
given by

$$\hat{D}^\dagger = \sum_k \frac{(-i)^k}{c^k k!} z^k \hat{H}_0^k. \qquad (6.66)$$

Taking into account that (1) $\sum_n |\psi_n\rangle\langle\psi_n| = 1$, (2) $\hat{D}^\dagger |\psi_n\rangle = \exp(-iz\varepsilon_n/c) |\psi_n\rangle$, (3) $\langle\psi_n| \hat{D} = \langle\psi_n| \exp(i\varepsilon_n z/c)$ and using (6.61) we obtain

$$\sum_n |a_{0\to n}|^2 = \sum_n \left\langle \psi_0 \left| \exp\left(i\frac{\varepsilon_0 z}{c}\right) \exp(iG)\,(1-\alpha_z) \exp\left(-i\frac{\varepsilon_n z}{c}\right) \right| \psi_n \right\rangle$$
$$\times \left\langle \psi_n \left| \exp\left(i\frac{\varepsilon_n z}{c}\right) (1-\alpha_z) \exp(-iG) \exp\left(-i\frac{\varepsilon_0 z}{c}\right) \right| \psi_0 \right\rangle$$
$$= \left\langle \psi_0 \left| \exp\left(i\frac{\varepsilon_0 z}{c}\right) \exp(iG)\,(1-\alpha_z)\,\hat{D}^\dagger \right.\right.$$
$$\left.\left. \times \hat{D}\,(1-\alpha_z) \exp(-iG) \exp\left(-i\frac{\varepsilon_0 z}{c}\right) \right| \psi_0 \right\rangle. \qquad (6.67)$$

Using the identity

$$(1-\alpha_z)\hat{D}^\dagger \hat{D} = 1, \qquad (6.68)$$

which follows from the transformation

$$(1-\alpha_z)\hat{D}^\dagger = \hat{D}^{-1}\hat{D}(1-\alpha_z)\hat{D}^\dagger$$
$$= \hat{D}^{-1} \sum_{n,m} |\psi_m\rangle\langle\psi_m| \hat{D}(1-\alpha_z)\hat{D}^\dagger |\psi_n\rangle\langle\psi_n|$$
$$= \hat{D}^{-1} \sum_{n,m} |\psi_m\rangle\langle\psi_m| \exp(i\omega_{mn}z/c)(1-\alpha_z) |\psi_n\rangle\langle\psi_n|$$
$$= \hat{D}^{-1} \sum_{n,m} |\psi_m\rangle\delta_{nm}\langle\psi_n| = \hat{D}^{-1}, \qquad (6.69)$$

we finally arrive at

$$\sum_n |a_{0\to n}|^2 = \langle\psi_0| \exp(i\varepsilon_0 z/c) \exp(iG)$$
$$\times (1-\alpha_z) \exp(-iG) \exp(-i\varepsilon_0 z/c) |\psi_0\rangle$$
$$= \langle\psi_0 | 1 - \alpha_z | \psi_0\rangle = 1. \qquad (6.70)$$

Thus, the light-cone amplitude (6.61) does preserve the unitarity, as it should be for an exact solution.

The Limit of the Vanishing Screening

In the limit of vanishing screening ($M_j \to 0$) one can use the relation $K_0(x) \approx -\ln\left(\frac{x}{2}\right) - \Gamma$ for $|x| \ll 1$ (see e.g. [108]), where Γ is Euler's constant. Then the transition amplitude (6.61) can be transformed into

$$a^C_{0\to n}(\mathbf{b}) = \left\langle \psi_n \left| (1-\alpha_z) \exp\left(i\frac{\omega_{n0} z}{c}\right) \exp\left(\frac{2iZ_A}{c} \ln\left(\frac{|\mathbf{r}_\perp - \mathbf{b}|}{b}\right)\right) \right| \psi_0 \right\rangle. \qquad (6.71)$$

The transition amplitude (6.71) coincides with that derived in [131] for the electron transition in collisions with a point-like charge.

It is well known that in the case, when electron transitions are caused by collisions with particles, whose net electric charge is not zero, calculations with the first order amplitude in the limit $\gamma \to \infty$ may yield infinite cross sections. In particular, this occurs when the electron loss (or ionization) cross section is evaluated. The loss cross section behaves like $\ln \gamma$ and the mild logarithmic divergence arises because at $\gamma \to \infty$ for any impact parameter the collision becomes sudden for the electron and, for the dipole allowed transitions, effectively all the impact parameters contribute to the loss cross section. In the literature on the ion–atom collisions this divergence is sometimes considered as a peculiarity of the first order consideration which should disappear when the higher order contributions in the projectile–target interaction are properly addressed. It is important to keep in mind, however, that such a divergence is also inherent for the light-cone amplitude (6.71).[13]

Effective Collision Time and Collisions at Finite γ

The light-cone approximation if formally exact only when $\gamma = \infty$. In practical terms such a situation never happens. However, the valuable property of this approximation is that it can, under certain conditions, be used for collisions with finite values of γ [132, 137]. Based on the general grounds one can expect that the light-cone transition amplitude (6.61) will yield good results provided the effective duration time of the interaction between the electron and the field of the incident particle, $T(b) \sim b/(v\gamma)$, is small compared to the characteristic electron transition time in the ion $\tau \sim w_{n0}^{-1}$. The latter condition holds for impact parameters $b \ll b_0 = v\gamma/w_{n0}$ and at these impact parameters the collision is sudden for the electron.

For collisions with neutral atoms there is another characteristic distance, the dimension of the neutral atom a_0. If $b_0 \gg a_0$ then the amplitude (6.61) can be used for any impact parameter because for larger impact parameters $b \gtrsim b_0$, where collisions are no longer sudden for the electron, the electron–atom interaction is already negligible.

---

[13] One should add that although from the point of view of theory the above divergence is certainly of interest, this unlimited growth in the excitation/ionization cross sections appearing in the atomic physics calculations cannot be observed in experiment. The reasons for this are similar to those discussed in Sect. 5.14.2 of Chap. 5: (a) the impact parameter cannot exceed the size of an experimental camera where the collisions are studied and (b) even in a very dilute gas target the field of a point-like charge moving with a velocity very closely approaching the speed of light will be completely screened at sufficiently large impact parameters because of the collective response of the target medium [7, 123, 124].

Comparison with the First Order Amplitude

The light-cone amplitude (6.61) is to be compared with the transition amplitude for the screening mode, which is obtained in the first order semiclassical perturbation theory (see (7.7) of Chap. 7),

$$a^1_{0 \to n}(\mathbf{b}) = \frac{2\mathrm{i}Z_A}{v} \sum_j A_j$$

$$\times \langle \psi_n \left| \left(1 - \frac{v}{c}\alpha_z \right) \exp\left(\mathrm{i}\frac{\omega_{n0} z}{v} \right) K_0 \left(B_j \mid \mathbf{r}_\perp - \mathbf{b} \mid \right) \right| \psi_0 \rangle, \quad (6.72)$$

where $B_j = \sqrt{\frac{\omega^2_{n0}}{v^2\gamma^2} + M^2_j}$.

For collisions at infinite γ the transition amplitude, given by (6.61), is valid for any impact parameter b. For collisions with light atoms, where $\frac{2Z_A}{c} \ll 1$, or at large impact parameters, where the condition $\frac{2Z_A}{c} \sum_j A_j K_0(M_j \mid \mathbf{r}_\perp - \mathbf{b}\mid) \ll 1$ holds for any atom, the transition amplitude (6.61) reduces to the first order amplitude (6.72). Note that in the limit $\gamma \to \infty$, in contrast to collisions with point-like charges, the transition amplitude (6.61) for a screened interaction does not result in infinite cross sections for dipole allowed transitions.

6.4.5 Collisions at High but Finite γ: Combination of the Light-Cone and First Order Approaches

For collisions with high but finite values of γ both transition amplitudes (6.61) and (6.72) are not exact. In such a case the expressions (6.61) and (6.72), in general, are better suited to describe the transition amplitude at small and large impact parameters, respectively. In a comparative analysis for these two amplitudes we first consider colliding systems where $b_0 = \frac{\gamma v}{\omega_{n0}} \gg a_0$. In such a case one has $B_j \simeq M_j$ since $M_j \gtrsim 1$. For large impact parameters $b \gg \frac{Z_A}{Z_{IC}}$, where the atomic field acting on the electron of the ion is weak compared to the interaction between the electron and the nucleus of the ion, the exponent in (6.61) can be expanded in series and one sees that the transition amplitude (6.61) is approximately equivalent to the first order transition amplitude for these impact parameters (if in the latter one neglects terms proportional to $\frac{1}{\gamma^2}$). For collisions with smaller impact parameters, where the atomic field can reach considerable magnitudes during the collisions, the first order transition amplitude (6.72) is inferior to the amplitude (6.61). Therefore, for colliding systems, which satisfy the condition $b_0 \gg a_0 \sim 1$, the light-cone amplitude (6.61) should be used for all impact parameters.

Let us now consider colliding systems where $b_0 \lesssim a_0$. One should note that in ultrarelativistic collisions such a condition can be fulfilled only for very heavy ions. If, in addition, $b_0 \gg a_I$, where $a_I \sim \frac{1}{Z_I}$ is the typical dimension of the ground state of the electron in the ion, then a simple method can be applied to calculate cross sections (see e.g. [138–141]). Namely, for collisions with small impact parameters $b \ll b_0$, where the atom-electron interaction can be strong, the transition probability is calculated according to the nonperturbative expression (6.61). For collisions with larger impact parameters $b \gg \frac{1}{Z_I} \gtrsim \frac{Z_A}{Z_I c}$, where the perturbation is already weak, the first order perturbation theory can be used to calculate the transition probability. This method of combining the light-cone and first-order treatments can be employed if there exists an overlap between the regions $b \ll b_0$ and $b \gg \frac{1}{Z_I}$, i.e. when $Z_I b_0 \gg 1$. Then, taking into account (6.61) and (6.72), the screening contribution to the cross section can be written as

$$\sigma_{0 \to n} = 2\pi \int_0^{b_1} db\, b \mid a_{0n}^{LC}(b) \mid^2 + 2\pi \int_{b_1}^{\infty} db\, b \mid a_{0n}^{1}(b) \mid^2, \qquad (6.73)$$

where a_{0n}^{LC} and a_{0n}^{1} are the light-cone and first order amplitudes, respectively, and the magnitude b_1 has to be in the range of the impact parameters where the transition probabilities calculated with these amplitudes are approximately equal. The existence of such an overlap region is very important. Indeed, only if this region exists does become the cross section (6.73) independent of a particular choice made for the value of b_1. This point is discussed in detail in Appendix B where it is shown that, for the electron loss in ultra-relativistic collisions with a point-like charged particle, one can always find a range of impact parameters where the light-cone and first order transition amplitudes are approximately equal and that in collisions with neutral atoms a similar range of impact parameters does exist for the loss from very heavy ions.

6.4.6 The Light-Cone Approximation for a Nonrelativistic Electron

The light-cone approximation assumes that the incident atomic particle moves with the speed of light but does not necessarily imply that the motion of the electron is relativistic. Indeed, as it was already remarked in Sect. 5.11.1, provided the electron is initially bound in a relatively light ion ($Z_I/c \ll 1$), its motion before, during and after the collision remains practically nonrelativistic.

One could start to analyze the motion of a nonrelativistic electron in collisions with a particle moving at the speed of light using the corresponding Schrödinger equation with the light-cone potentials taken, for instance, in the form (6.27). However, the exact solution (6.61) for the electron transition amplitude became possible because both the Dirac and Maxwell equations are covariant under the same space–time transformation (the Lorentz transformation) and thus possess the same symmetry. In particular, the derivation of the light-cone amplitude, which was discussed in Sect. 6.4.4, explicitly involved the fact that the time and space derivatives enter the Dirac equation symmetrically.

All this is of course not the case when one considers the nonrelativistic wave equation. Now the wave equation and the electromagnetic potentials are characterized by the different symmetries. In particular, if we were to calculate the transition amplitude by considering the Schrödinger equation, we would immediately encounter a difficulty with using the light-cone coordinates (6.40).

In order to derive the (analog of the) light-cone amplitude in the case of a nonrelativistic electron there is an alternative way. Instead starting with the Schrödinger equation we can simply begin with the relativistic light-cone amplitude (6.61) and perform a transition to the nonrelativistic description of the electron motion directly in (6.61).

Having in mind a possible extension of the light-cone amplitude to collisions with velocities $v < c$, in this subsection we shall denote the speed of light, which is related to the collision velocity, by a symbol v_c. In all other places the usual notation c will be kept. Note that it is the velocity v_c which enters the minimum momentum transfer $q_{min} = \omega_{n0}/c = \omega_{n0}/v_c$ and the factor G (given by (6.54)) in the exact amplitude (6.61). Besides, the term in (6.61), which is proportional to α_z, now gets multiplied by a factor v_c/c.

In the amplitude (6.61) we shall first present the initial and final electron states according to

$$\psi_{0,n} = \begin{pmatrix} \chi_{0,n} \\ \xi_{0,n} \end{pmatrix}, \tag{6.74}$$

where χ and ξ are the large and small two-component spinors, respectively, of the Dirac four-spinor ψ. According to the standard consideration of the transformation of the Dirac equation into the Schrödinger one (see, for instance, [10], [73]) these components are approximated by

$$\left(\frac{\hat{\mathbf{p}}^2}{2} - \frac{Z_I}{r} \right) \chi_m = \varepsilon_m \chi_m,$$

$$\xi_m = \frac{\boldsymbol{\sigma} \cdot \hat{\mathbf{p}}}{2mc} \chi_m, \tag{6.75}$$

where the first line is the Schrödinger equation for the electron moving in the (undistorted) ion with ε_m being its nonrelativistic energy, $\boldsymbol{\sigma} = (\sigma_x, \sigma_y, \sigma_z)$ are the Pauli matrices and $\hat{\mathbf{p}}$ is the electron momentum operator.

We now approximate the states ψ_0 and ψ_n in the exact amplitude (6.61) with the help of (6.74) and (6.75). Then, after performing some simple manipulations with the transition matrix elements, we arrive at the following approximate expression for the transition amplitude

$$a_{0 \to n}(\mathbf{b}) = \langle \chi_n \mid \exp(i\omega_{n0}z/v_c) \exp(-iG) \mid \chi_0 \rangle$$
$$- \frac{v_c}{2c^2} \langle \chi_n \mid \exp(i\omega_{n0}z/v_c) \exp(-iG) \hat{p}_z \mid \chi_0 \rangle$$
$$- \frac{v_c}{2c^2} \langle \chi_n \mid \hat{p}_z \exp(i\omega_{n0}z/v_c) \exp(-iG) \mid \chi_0 \rangle$$
$$+ i \frac{v_c}{2c^2} \langle \chi_n \mid \exp(i\omega_{n0}z/v_c) \exp(-iG) \, F \mid \chi_0 \rangle$$

$$= \left(1 - \frac{\omega_{n0}}{2c^2}\right) \langle \chi_n \mid \exp(i\omega_{n0}z/v_c) \exp(-iG) \mid \chi_0 \rangle$$

$$- \frac{v_c}{c^2} \langle \chi_n \mid \exp(i\omega_{n0}z/v_c) \exp(-iG)\hat{p}_z \mid \chi_0 \rangle$$

$$+ i \frac{v_c}{2c^2} \langle \chi_n \mid \exp(i\omega_{n0}z/v_c) \exp(-iG)\, F \mid \chi_0 \rangle. \tag{6.76}$$

In the above expression $G = G(\mathbf{r}_\perp - \mathbf{b})$ is defined in (6.54) (with $c \to v_c$) and the 2×2 matrix $F = F(\mathbf{r}_\perp - \mathbf{b})$ is given by

$$F = \sigma_x \frac{\partial G}{\partial y} - \sigma_y \frac{\partial G}{\partial x}. \tag{6.77}$$

Before we continue with further simplifications of the transition amplitude (6.76) it is of interest to attempt to trace the origin of the different terms in this amplitude by relating them to the interaction terms of the Schrödinger equation

$$i\frac{\partial \phi(t)}{\partial t} = \left(\frac{\hat{\mathbf{p}}^2}{2} - \frac{Z_I}{r} - \Phi + \frac{1}{2c}(\mathbf{A} \cdot \hat{\mathbf{p}} + \hat{\mathbf{p}} \cdot \mathbf{A}) + \frac{1}{2c^2}\mathbf{A}^2 - \boldsymbol{\mu} \cdot \mathbf{H}\right)\phi(t). \tag{6.78}$$

This equation describes a nonrelativistic electron which moves in the field of the ionic nucleus with a charge Z_I and is subjected in the collision to the field of the incident atom. The electromagnetic field of the atom is described by the scalar and vector potentials Φ and $\mathbf{A}$ given e.g. by Eqs. (6.27). The last term on the right hand side of (6.78) represents the interaction between the spin magnetic moment of the electron, $\boldsymbol{\mu} = -\boldsymbol{\sigma}/2c$, and the magnetic field

$$\mathbf{H} = (H_x, H_y, H_z) = \delta(v_c t - z)\left(\frac{\partial G}{\partial x}, \frac{\partial G}{\partial y}, 0\right), \tag{6.79}$$

which is generated by the atom moving with the speed of light in the rest frame of the ion.

Comparing the structure of (6.76) with that of (6.78) we see that the term in the fourth line of the amplitude (6.76) obviously describes transitions caused by the interaction between the spin of a nonrelativistic electron and the magnetic field of the moving atom. It is also plausible to assume that the two terms in the second and third lines of (6.76) are related to those interaction terms of the Schrödinger equation which are proportional to $\mathbf{A} \cdot \hat{\mathbf{p}}$ and $\hat{\mathbf{p}} \cdot \mathbf{A}$, respectively. Then, the remaining part of the amplitude (6.76) (the first line of (6.76)) should have its origin in the interaction terms $-\Phi$ and $A^2/2c^2$ of the Schrödinger equation.

Similarly to the amplitude (6.61), the light-cone amplitude for a nonrelativistic electron is expected to be a good approximation also for collisions with velocities $v < c$ provided the effective collision time $\frac{b}{v\gamma}$ is much less than the typical electron transition time $1/\omega_{n0}$. Then the amplitude (6.76) can be used if the replacement $v_c \to v$ is done in the expressions for the minimum momentum transfer and the function G.

Besides, similarly to the amplitude (6.61), the amplitude (6.76) is also expected to yield infinite cross sections for dipole allowed transitions occurring in collisions with atomic systems which are not electrically neutral.

It is known (see e.g. [142–144]) that a nonzero value of the minimum momentum transfer is responsible for the angular asymmetry in the calculated spectra of electrons emitted in ion–atom collisions. However, in high-energy collisions this asymmetry does not have a substantial impact on the total cross section for the ionization/loss of an electron which was initially not very tightly bound ($Z_I \ll v$). Therefore, when evaluating the total loss cross section in collisions with neutral atoms one can set $q_{min} = 0$ in (6.76).

Further, as a rough estimate one has $\frac{1}{c}\hat{p}_z \chi_0 \sim \frac{Z_I}{c}\chi_0$. Therefore, one can drop in the amplitude (6.76) the terms containing the electron momentum operator. Moreover, since the spin-flip transitions contribute very little to the loss process, the last term in (6.76) can also be omitted.

Finally, as a result of all these simplifications, we obtain

$$a_{0 \to n}(\mathbf{b}) = \langle \varphi_n \mid \exp(-iG(\mathbf{r}_\perp - \mathbf{b})) \mid \varphi_0 \rangle$$

$$= \left\langle \varphi_n \left| \exp\left(\frac{-2iZ_A}{v} \sum_j A_j K_0 \left(M_j \mid \mathbf{r}_\perp - \mathbf{b} \mid \right) \right) \right| \varphi_0 \right\rangle, \quad (6.80)$$

where φ_0 and φ_n are the space parts of the initial and final states of the electron. It is worth to point to the similarity between the above formula and the corresponding results for the transition amplitude which were obtained in Chap. 3 by using the (nonrelativistic) sudden approximation.

6.5 Collisions at Relatively Low Energies: Three-Body Distorted-Wave Models

In the previous section we considered the projectile-electron excitation and loss in collisions at extreme relativistic energies. Because of the effectively short-range force, which a neutral atom exerts on the electron of the projectile, the electron–atom interaction is restricted to relatively small impact parameters and this holds true at any impact energy. Therefore, provided the atom has a sufficiently high atomic number, the atomic field may be too strong leading to deviations from the first order predictions which can survive even in the limit $\gamma \to \infty$.

In general the deviations from the first order predictions increase when the impact energy decreases and now we shall consider the region of the collision parameters where the first order approximation, being applied to calculate cross sections for the projectile-electron excitation and loss, is expected to fail especially strongly. This region is represented by collisions of very highly charged ions with heavy atoms at relatively low impact energies

0.1-1 $\mathrm{GeV\,u^{-1}}$ [14]. The electron excitation of and loss from very heavy ions occurring in collisions with many-electron atoms at these impact energies are the topic of this section.

In order to treat these processes we shall use, following [145–147], a simplified picture of the collision which consists of the following main 'ingredients'. Firstly, the electron of the ion is regarded as the only particle having the dynamical degrees of freedom which are described by a wave equation. Secondly, the nuclei of the ion and the atom are treated as classical particles which move along given (straight-line) trajectories and are just the sources of the external electromagnetic field acting on the electron of the ion. Thirdly, as was already discussed in Sect. 5.11, in the range of the relatively low impact energies the influence of the electrons of the atom on the excitation and loss processes is of minor importance and, therefore, the presence of the atomic electrons can be ignored. Thus, within the simplified picture the process of the electron excitation/loss effectively reduces to a three-body problem of the motion of the electron in the electromagnetic fields generated by two point-like classical particles.

The basic assumption underlying our approach to this problem is that the motion of the electron in the collision is determined mainly by its interaction with the field of the ionic nucleus. This interaction, therefore, should be treated as accurate as possible whereas the interaction of the electron with the nucleus of the atom is supposed to be less important for the electron motion and thus can be taken into account in an approximate way. For calculations of the total loss cross section such an 'asymmetric' approach seems be certainly reasonable as long as the charge Z_I of the ionic nucleus noticeably exceeds the charge Z_A of the atomic nucleus and the collision velocity is not too low.

It is convenient to treat the electron loss process using a reference frame K_I in which the nucleus of the ion is at rest. We take the position of the nucleus as the origin and assume that in the frame K_I the nucleus of the atom moves along a straight-line classical trajectory $\mathbf{R}(t) = \mathbf{b} + \mathbf{v}t$, where $\mathbf{b} = (b_x, b_y, 0)$ is the impact parameter, $\mathbf{v} = (0, 0, v)$ is the collision velocity and t is the time.

The Dirac equation for the electron of the ion reads

$$i\frac{\partial \Psi}{\partial t} = \left(\hat{H}_0 + \hat{W}(t)\right)\Psi. \tag{6.81}$$

Here $\hat{H}_0$ denotes the electronic Hamiltonian for the undistorted ion and $\hat{W}(t)$ is the interaction between the electron of the ion and the nucleus of the atom given by (5.85).

The semi-classical transition amplitude can be written as

$$a_{\mathrm{fi}}(\mathbf{b}) = -\mathrm{i}\int_{-\infty}^{+\infty} dt \int d^3\mathbf{x} \left(\rho_{\mathrm{fi}}(\mathbf{x}, t)\Phi(\mathbf{x}, t) - \frac{1}{c}\mathbf{j}_{\mathrm{fi}}(\mathbf{x}, t) \cdot \mathbf{A}(\mathbf{x}, t)\right). \tag{6.82}$$

---

[14] One should note that the main bulk of the existing experimental data on the loss from highly charged ions is related exactly to this region.

Here $\rho_{\text{fi}}(\mathbf{x}, t)$ and $\mathbf{j}_{\text{fi}}(\mathbf{x}, t)$ are the transition charge and current densities, respectively, created by the electron of the ion at a space point $\mathbf{x}$ at a time t, and $\Phi(\mathbf{x}, t)$ and $\mathbf{A}(\mathbf{x}, t)$ are the scalar and vector potentials of the field of the nucleus of the atom. In what follows we shall again work with the transition amplitude in the momentum space, $S_{\text{fi}}(\mathbf{q}_\perp)$, which is related to the amplitude (6.82) by the two-dimensional Fourier transformation (3.13).

6.5.1 Symmetric Eikonal Approximation

In an attempt to improve the description of the interaction between the electron of the ion and the nucleus of the atom we first approximate the electron states within the symmetric eikonal approach (SEA). According to the SEA, the interaction $\hat{W}$ between the electron and the atomic nucleus is taken using the Lienard–Wiechert potentials (5.86) and the initial and final states of the electron are given by

$$\chi_i^{\text{eik}}(t) = \psi_i(\mathbf{r}) \exp(-i\varepsilon_i t) \times (vs + \mathbf{v} \cdot \mathbf{s})^{-i\nu_t}$$
$$\chi_f^{\text{eik}}(t) = \psi_f(\mathbf{r}) \exp(-i\varepsilon_f t) \times (vs - \mathbf{v} \cdot \mathbf{s})^{i\nu_t} \qquad (6.83)$$

with $\nu_t = Z_A/v$. The states χ_i^{eik} and χ_f^{eik} take fully into account the interaction between the electron and the nucleus of the ion whereas the effect of the interaction with the nucleus of the atom is treated in an approximate way by introducing the eikonal distortion factors. Correspondingly, the transition charge and current densities of the electron in the SEA read

$$\rho_{\text{fi}}^{\text{sea}}(\mathbf{x}, t) = (vs_\perp)^{-2i\nu_t} \psi_f^\dagger(\mathbf{x}) \psi_i(\mathbf{x}) \exp(i(\varepsilon_f - \varepsilon_i)t)$$
$$\mathbf{j}_{\text{fi}}^{\text{sea}}(\mathbf{x}, t) = (vs_\perp)^{-2i\nu_t} \psi_f^\dagger(\mathbf{x}) c\,\boldsymbol{\alpha}\psi_i(\mathbf{x}) \exp(i(\varepsilon_f - \varepsilon_i)t), \qquad (6.84)$$

where $s_\perp = \sqrt{s_x^2 + s_y^2}$.

According to the spirit of the SEA, one starts with the Dirac equation with the electromagnetic potentials Φ and $\mathbf{A}$ taken in the Lienard–Wiechert form (5.86) and chooses the distortion factors $(vs \pm \mathbf{v} \cdot \mathbf{s})^{\mp i\nu_t}$ in such a way as to eliminate from the Dirac equation the term with the scalar potential Φ (see e.g. [145]). This effectively corresponds to a gauge transformation from the field potentials in the Lienard–Wichert form to the field potentials given by (5.93). Therefore, when building the eikonal transition amplitude we shall couple the charge and current densities (6.84) with the potentials (5.93) that yields

$$a_{\text{fi}}^{\text{sea}}(\mathbf{b}) = \frac{icZ_A}{v} \int_{-\infty}^{+\infty} dt \exp(i(\varepsilon_f - \varepsilon_i)t)$$
$$\times \int d^3\mathbf{x} (vs_\perp)^{-2i\nu_t} \psi_f^\dagger(\mathbf{x}) \left(\frac{s_x \alpha_x + s_y \alpha_y}{s + s_z} + \frac{\alpha_z}{\gamma} \right) \psi_i(\mathbf{x}). \qquad (6.85)$$

The above expression and the relation (3.13) enable one, after some manipulations, to obtain the symmetric eikonal amplitude in the momentum space

$$
S_{\text{fi}}^{\text{sea}}(\mathbf{q}_\perp) = \frac{2iZ_A c}{v^2} \frac{1}{q'^2 q_z} \left(\frac{q'}{2}\right)^{2i\nu_t} \Gamma^2(1 - i\nu_t)
$$
$$
\times \Bigg((1 - i\nu_t)_2 F_1 \left(1 - i\nu_t, i\nu_t; 2; Q^2/q'^2\right)
$$
$$
\times \langle \psi_{\text{f}} \mid \exp(i\mathbf{q}\cdot\mathbf{r})(q_x \alpha_x + q_y \alpha_y) \mid \psi_{\text{i}} \rangle
$$
$$
+ {}_2F_1 \left(1 - i\nu_t, i\nu_t; 1; Q^2/q'^2\right)
$$
$$
\times \frac{1}{\gamma^2} \langle \psi_{\text{f}} \mid \exp(i\mathbf{q}\cdot\mathbf{r})q_z \alpha_z \mid \psi_{\text{i}} \rangle \Bigg), \qquad (6.86)
$$

where Γ and $_2F_1$ are the gamma and hypergeometric functions, respectively (see e.g. [108]).

Modified Eikonal Amplitude

The symmetric eikonal approximation is known to yield unphysical results for the electron capture occurring in relativistic collisions (see for discussions e.g. [3,4]). It has been pointed out [145] that the difficulties with the application of the SEA approximation to treat the capture are mainly caused by the problem of gauge dependence since in the case of capture the initial and final undistorted states of the electron in the symmetric eikonal approximation are described by wavefunctions belonging to different Hamiltonians.

Nevertheless, certain shortcomings remain in the SEA even if the initial and final undistorted states of the electron belong to the same Hamiltonian. Indeed, the recent application of the amplitude (6.86) to evaluate cross sections for the collision-induced excitation [145] has unveiled that the SEA may face difficulties in treating bound–bound electron transitions involving spin-flip even if exact eigenstates of the Hamiltonian $\hat{H}_0$ are employed in the calculation. After the analysis of these difficulties it was shown in [145] that the following 'modified' SEA amplitude,

$$
S_{\text{fi}}^{\text{msea}}(\mathbf{q}_\perp) = \frac{2iZ_A c}{v^2} \frac{1}{q'^2 q_z} \left(\frac{q'}{2}\right)^{2i\nu_t}
$$
$$
\times (1 - i\nu_t)\,\Gamma^2(1 - i\nu_t)_2 F_1 \left(1 - i\nu_t, i\nu_t; 2; Q^2/q'^2\right)
$$
$$
\times \left\langle \psi_{\text{f}} \mid \exp(i\mathbf{q}\cdot\mathbf{r}) \left(q_x \alpha_x + q_y \alpha_y + \frac{q_z}{\gamma^2}\alpha_z\right) \mid \psi_{\text{i}} \right\rangle, \quad (6.87)
$$

does not have problems with the description of spin-flip transitions and suggested that, on overall, the amplitude (6.87) may yield substantially better results for excitation cross sections. Below this amplitude will be applied to calculate the total electron loss cross sections.

6.5.2 Continuum-Distorted-Wave-Eikonal-Initial-State Approximation

In order to reflect the influence of the nucleus of the atom on the electron of the ion, within the symmetric eikonal approximation the initial and final electron states are distorted similarly. Besides, the final distortion factor is the same for all electron continuum states and does not depend on the electron velocity with respect to the nucleus of the atom. From the physical grounds this may be not very reasonable, especially, if one tries to calculate differential loss cross sections. Therefore, in an attempt to obtain a better description of the continuum, we now approximate the initial and final states in (6.82) in the spirit of the continuum-distorted-wave-eikonal-initial-state (CDW-EIS) approach by

$$\chi_i^{\text{eik}}(t) = \psi_i(\mathbf{r})\exp(-i\varepsilon_i t) \times (vs + \mathbf{v}\cdot\mathbf{s})^{-i\nu_t}$$

$$\chi_f^{\text{cdw}}(t) = \psi_f(\mathbf{r})\exp(-i\varepsilon_f t)$$
$$\times \Gamma(1+i\eta_t)\exp(\pi\eta_t/2)\,_1F_1\left(-i\eta_t, 1, -i(ps+\mathbf{p}\cdot\mathbf{s})\right). \quad (6.88)$$

Here, $\eta_t = Z_A/v_e'$ where v_e' is the electron velocity in the rest frame of the nucleus of the atom, $\mathbf{p}$ is the kinetic momentum of the electron in the rest frame of the atomic nucleus and $_1F_1$ is the confluent hypergeometric function (see e.g. [108]).

The electron transition charge and current densities entering the amplitude (6.82) are now constructed with the states (6.88) and are given by

$$\rho_{fi}^{\text{cdw-eis}}(\mathbf{x}, t) = G(\nu_t, \eta_t, \mathbf{v}, \mathbf{p}, \mathbf{s})\psi_f^\dagger(\mathbf{x})\,\psi_i(\mathbf{x})\,\exp(i(\varepsilon_f - \varepsilon_i)t)$$

$$\mathbf{j}_{fi}^{\text{cdw-eis}}(\mathbf{x}, t) = G(\nu_t, \eta_t, \mathbf{v}, \mathbf{p})\psi_f^\dagger(\mathbf{x})\,c\,\boldsymbol{\alpha}\psi_i(\mathbf{x})\,\exp(i(\varepsilon_f - \varepsilon_i)t), \quad (6.89)$$

where

$$G(\nu_t, \eta_t, \mathbf{v}, \mathbf{p}, \mathbf{s})$$
$$= \Gamma(1-i\eta_t)\exp(\pi\eta_t/2)\,_1F_1\left(i\eta_t, 1, i(ps+\mathbf{p}\cdot\mathbf{s})\right)(vs + \mathbf{v}\cdot\mathbf{s})^{-i\nu_t}. \quad (6.90)$$

Because of the same reason as in the case with the SEA transition amplitude, it is natural to combine in the CDW-EIS amplitude the charge and current densities (6.90) with the potentials (5.93) to obtain

$$a_{fi}^{cdw-eik}(\mathbf{b}) = \frac{icZ_A}{v}\int_{-\infty}^{+\infty} dt\,\exp(i(\varepsilon_f - \varepsilon_i)t)$$
$$\times \int d^3\mathbf{x}\, G(\nu_t, \eta_t, \mathbf{v}, \mathbf{p}, \mathbf{s})\psi_f^\dagger(\mathbf{x})\left(\frac{s_x\alpha_x + s_y\alpha_y}{s + s_z} + \frac{\alpha_z}{\gamma}\right)\psi_i(\mathbf{x}).$$
$$(6.91)$$

Taking into account (6.91) and (3.13) one can show that the CDW-EIS transition amplitude, written in the momentum space, is given by

$$S_{\mathrm{fi}}^{\mathrm{cdw\text{-}eis}}(\mathbf{q}_\perp) = \frac{2\mathrm{i}Z_A c}{\gamma v} \left(\frac{A}{C}\right)^{\mathrm{i}\nu_\mathrm{t}} \left(\frac{A+B}{A}\right)^{-\mathrm{i}\eta_\mathrm{t}} \langle\psi_\mathrm{f}|J_x\alpha_x + J_y\alpha_y + J_z\alpha_z|\psi_\mathrm{i}\rangle.$$

$$(6.92)$$

Here,

$$J_{x(y)} = \frac{\Gamma(-\mathrm{i}\nu_\mathrm{t})}{C}\left(\Omega_{x(y)}\ {}_2F_1\left(\mathrm{i}\nu_\mathrm{t},\mathrm{i}\eta_\mathrm{t},1,Z\right) + \Omega'_{x(y)}\ {}_2F_1\left(\mathrm{i}\nu_\mathrm{t}+1,\mathrm{i}\eta_\mathrm{t}+1,2,Z\right)\right),$$
$$J_z = \frac{\Gamma(1-\mathrm{i}\nu_\mathrm{t})}{A\gamma v}\ {}_2F_1\left(\mathrm{i}\nu_\mathrm{t},\mathrm{i}\eta_\mathrm{t},1,Z\right),$$

$$(6.93)$$

$$A = q'^2,\ \ B = -2\mathrm{i}\mathbf{q}'\cdot\mathbf{p},\ \ C = -2\mathrm{i}q'_z v,\ \ D = 2\mathrm{i}v(p_z - p),\ \ Z = \frac{BC - AD}{C(A+B)}$$

$$(6.94)$$

and

$$\Omega_{x(y)} = (\nu_\mathrm{t} + \eta_\mathrm{t})\frac{\partial\ln(A)}{\partial q_{x(y)}} - \eta_\mathrm{t}\frac{\partial\ln(A+B)}{\partial q_{x(y)}} + \eta_\mathrm{t}\frac{Z}{1-Z}\frac{\partial\ln(Z)}{\partial q_{x(y)}},$$
$$\Omega'_{x(y)} = \mathrm{i}\eta_\mathrm{t}Z\left(\Omega_{x(y)} + \frac{\nu_\mathrm{t}}{1-Z}\frac{\partial\ln(Z)}{\partial q_{x(y)}}\right).$$

$$(6.95)$$

6.5.3 The Relationship with the First Order Approximation and with Other Distorted-Wave Models

Using the known properties of the gamma and hypergeometric functions, it is not difficult to show that in the weak perturbation limit, when $\nu_\mathrm{t} \ll 1$ and $\eta_\mathrm{t} \ll 1$, all distorted-wave amplitudes (6.86), (6.87) and (6.92) reduce to the first order amplitude in the form given by (5.88). If, in addition, the initial and final states of the electron in the undistorted ion are described by the Coulomb–Dirac states, the amplitude (5.88) and other forms of the first order amplitude (for instance, (5.83), (5.89)–(5.90) become fully equivalent and, thus, in the limit $\nu_\mathrm{t} \ll 1$ and $\eta_\mathrm{t} \ll 1$ the distorted-wave amplitudes (6.86), (6.87) and (6.92) reduce also to all other forms of the first order amplitudes.

The amplitudes (6.86) and (6.92) can be derived also by using the 'standard' approach in which the distorted-wave transition amplitude is given (in the prior form) by

$$a_{\mathrm{fi}}(\mathbf{b}) = -\mathrm{i}\int_{-\infty}^{+\infty} dt\langle\chi_\mathrm{f}(t)\,|\,\left(\hat{H}_0 + \hat{W}(t) - \mathrm{i}\partial/\partial t\right)\chi_\mathrm{i}(t)\rangle,\qquad (6.96)$$

where $\chi_\mathrm{i}(t)$ and $\chi_\mathrm{f}(t)$ are the corresponding distorted states.

The CDW-EIS and SEA approximations were initially introduced to study the atomic ionization and excitation and the electron capture from atoms which occur in nonrelativistic collisions with point-like ions [33–36] (see also

[37] where a historic overview of the nonrelativistic distorted-wave models is presented). Attempts to extend and apply these approximations to the domain of relativistic collisions (see e.g. [148–153]) have been the subject of much controversy (see discussions in [3, 4, 160] and referenced therein).

To our knowledge, the SEA model employed in [147] had not been used previously to treat relativistic collisions in which electronic transitions involve continuum states. A very detailed discussion of the CDW-EIS approximation in the problem of the K-shell ionization of intermediately heavy atoms in relativistic collisions with highly charged ions can be found in [4]. There are two main differences between the CDW-EIS model for the ionization considered in [4] and the present CDW-EIS model for the electron loss which was employed in [147] and is used here.

First, in order to treat the initial and final undistorted electron states within the present CDW-EIS model one does not resort to the semi-relativistic Darwin and Sommerfeld–Maue–Furry approximations. Instead, these states are described by using the Coulomb–Dirac wave functions. The importance of such a fully relativistic description in the case of very heavy ions will be shown below.

Second, the present model does not contain the distortion terms containing the Dirac matrices α_x, α_y and α_z. Such terms would arise if the α-dependent parts of the Furry wave function and of its eikonal asymptotics would be kept in the distortion factors. In the present CDW-EIS model the distortions are also introduced with the help of the Furry wave function and its eikonal asymptotics, but their α_j-dependent parts were omitted. One should note, however, that the analysis of the distortion factors undertaken in [4] suggests that the inclusion of the matrix distortion terms does not have a noticeable impact on the calculated cross sections.

6.5.4 Comparison of Relativistic and Semi-Relativistic Electron Descriptions

In order to get an idea about the importance of the relativistic electron description, we have calculated the electron loss from U^{91+}(1s) by proton impact at the impact energies ranging between 0.1 and $1\,\mathrm{GeV\,u^{-1}}$. In these calculations the states ψ_i and ψ_f were described (1) by using the relativistic (Coulomb–Dirac) wave functions, (2) by employing the semi-relativistic approach in which the states ψ_i and ψ_f are approximated by the Darwin and Furry wave functions, respectively, and (3) by applying the purely nonrelativistic description of these states when the loss process is considered by describing the electron transitions using the Schrödinger–Pauli equation (see Fig. 6.7).

It is seen in Fig. 6.7 that the application of both nonrelativistic and semi-relativistic approximations to the uranium ion leads to a considerable overestimation of the loss cross section. The semi-relativistic results are about

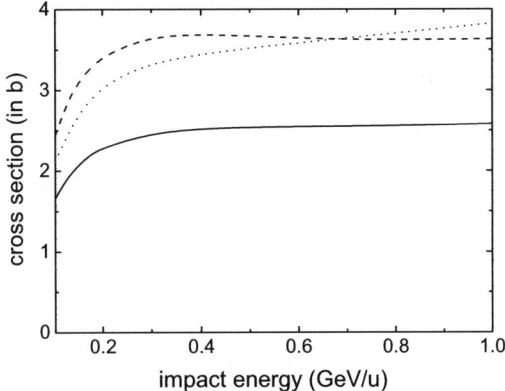

Fig. 6.7. The total cross sections for the electron loss from U $^{91+}(1s)$ ions by proton impact given as a function of the impact energy. *Solid, dash* and *dot curves* display results of the first order calculations which employ the relativistic, the semi-relativistic and the nonrelativistic wave functions, respectively, to describe the initial and final states of the undistorted ion. From [147].

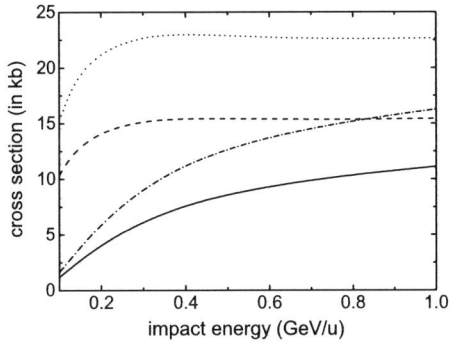

Fig. 6.8. The total cross sections for the electron loss from U $^{91+}(1s)$ ions in collisions with atoms of gold ($Z_A = 79$) given as a function of the impact energy. *Dash* (*solid*) and *dot* (*dash-dot*) *curves* display results of the first order (eikonal) calculations which employ the relativistic and semi-relativistic wave functions, respectively, to describe the initial and final states of the undistorted ion. From [146].

a factor 1.5 larger compared to those obtained by employing the Dirac states and this ratio remains basically a constant for the whole range of impact energies considered in the figure. Besides, for this range of impact energies the semi-relativistic description in fact does not represent an improvement over the purely nonrelativistic electron treatment. A similar relationship between results obtained with the relativistic and semi-relativistic descriptions remains also when the distorted-wave approaches are applied (see Fig. 6.8).

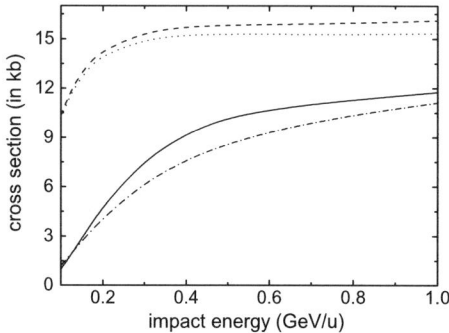

Fig. 6.9. Total cross section for the electron loss from $U^{91+}(1s)$ in collisions with atoms of gold given as a function of the collision energy. *Dash* and *dot curves* display results of the first order calculations without and with taking into account the screening effect of the atomic electrons, respectively. For more information, results of the calculation with the amplitudes (6.87) (*dash dot curve*) and (6.90) (*solid curve*) are also shown. From [147].

6.5.5 Higher Orders versus Screening

In collisions with heavy atoms the first order approximation for the electron loss becomes less inaccurate when the impact energy increases. However, as follows from Fig. 6.9, even at 1 GeV u^{-1} the first order approximation, compared to results of more elaborate treatments, is still likely to overestimate the electron loss cross section in collisions with atoms of gold by a factor of about 1.35–1.4. The magnitude of this factor as well as that of the difference between the loss cross sections calculated with the relativistic and semi-relativistic electron descriptions are to be compared with the screening effect of the atomic electrons which does not exceed 5%. Besides, because of a very large number of electrons in heavy atoms their antiscreening effect always remains weak. Therefore, one can expect that for a proper description of the electron loss from such very heavy ions, like hydrogen- and helium-like uranium, occurring in collisions with heavy atoms at impact energies below 1 GeV u^{-1} it is more important to treat with a necessary care the interaction of the electron of the ion with the bare atomic nucleus and to address the relativistic effects in the motion of this electron than to account for the presence of the atomic electrons.

6.6 The High-Energy Limit
   of the Distorted-Wave Models

In the previous section we have considered two distorted-wave models: the symmetric eikonal and the continuum-distorted-wave-eikonal-initial-state approximations. In particular, we have seen that at the relatively low impact energies these approximations result in cross section values which strongly differ from predictions of the first order calculations. In Chap. 8 we will also see that the application of these distorted-wave models yields a much better description of experimentally measured cross sections.

Here we shall briefly discuss the high-energy limit ($\gamma \gg 1$) of these distorted-wave approximations. This point is of great interest from the purely theoretical point of view. Besides, its discussion may be also quite useful having in mind possible applications of these models to study various electromagnetic processes occurring at extreme relativistic energies which are becoming experimentally accessible with the advent of Relativistic Super-Colliders.

At very high impact energies ($\gamma \gg 1$) the difference between the continuum-distorted-wave-eikonal-initial-state and symmetric eikonal models vanishes. Therefore, it is quite sufficient to consider the high-energy limit of the symmetric eikonal model.

At the collision energies of interest it is possible (and convenient) to start with the electromagnetic potentials in the form which they attain at $\gamma \to \infty$, i.e. with

$$\Phi(\mathbf{r}, t) = 2Z_A \delta(ct - z) G(|\mathbf{r}_\perp - \mathbf{b}|),$$
$$A_z(\mathbf{r}, t) = \Phi(\mathbf{r}, t),$$
$$A_x(\mathbf{r}, t) = A_y(\mathbf{r}, t) = 0 \tag{6.97}$$

(compare with (6.25)). Indeed, at extreme relativistic impact velocities the difference between the electromagnetic potentials and their light-cone approximation is proportional to $1/\gamma^2$ (see [154, 155]) and, thus, becomes quite small already at $\gamma \simeq 10$.

Within the symmetric eikonal approximation the standard representation of the transition amplitude is given by

$$a_{\mathrm{fi}}^{\mathrm{sea}}(\mathbf{b}) = -i \int_{-\infty}^{+\infty} dt \left\langle \Psi_{\mathrm{f}} \left| \left(\hat{H} - i\frac{\partial}{\partial t} \right) \right| \Psi_{\mathrm{i}} \right\rangle. \tag{6.98}$$

In (6.98) $\hat{H}$ is the total Hamiltonian of the electron, which moves in the field of the nucleus of the ion and is also affected by the field of the moving atom. Further, the initial, Ψ_{i}, and final, Ψ_{i}, states of the electron are approximated by

$$\Psi_{\mathrm{i}}(\mathbf{r}, t) = \psi_{\mathrm{i}}(\mathbf{r}) \exp(-i\varepsilon_{\mathrm{i}}t) \exp\left(i \int_{-\infty}^{t} dt' \Phi(\mathbf{r}, t') \right)$$

$$= \psi_{\mathrm{i}}(\mathbf{r}) \exp(-i\varepsilon_{\mathrm{i}}t) \exp\left(i\frac{2Z_A}{c} \theta(ct - z) G(|\mathbf{r}_\perp - \mathbf{b}|) \right)$$

$$\Psi_{\mathrm{f}}(\mathbf{r}, t) = \psi_{\mathrm{f}}(\mathbf{r}) \exp(-i\varepsilon_{\mathrm{f}}t) \exp\left(i \int_{+\infty}^{t} dt' \Phi(\mathbf{r}, t') \right),$$

$$= \psi_{\mathrm{f}}(\mathbf{r}) \exp(-i\varepsilon_{\mathrm{f}}t) \exp\left(i\frac{2Z_A}{c} (\theta(ct - z) - 1) G(|\mathbf{r}_\perp - \mathbf{b}|) \right)$$

$$\tag{6.99}$$

where ψ_i and ψ_f are the initial and final electronic states of the undistorted ion with the energies ε_i and ε_f, respectively.

Inserting (6.99) into (6.98) we obtain

$$
a_{\text{fi}}^{\text{sea}}(\mathbf{b}) = -2\mathrm{i}Z_A \int_{-\infty}^{+\infty} \mathrm{d}t \int \mathrm{d}^3\mathbf{r} \exp(\mathrm{i}\omega_{\text{fi}}t)\,\theta(ct - z)\,\psi_f^\dagger(\mathbf{r})\,\psi_i(\mathbf{r})
$$
$$
\times \left(\frac{\partial G(\mid \mathbf{r}_\perp - \mathbf{b} \mid)}{\partial x}\alpha_x + \frac{\partial G(\mid \mathbf{r}_\perp - \mathbf{b} \mid)}{\partial y}\alpha_y \right)
$$
$$
\times \exp\left(\mathrm{i}\frac{2Z_A}{c}G(\mid \mathbf{r}_\perp - \mathbf{b} \mid) \right), \tag{6.100}
$$

where $\omega_{\text{fi}} = \varepsilon_f - \varepsilon_i$. As before, we convert the symmetric eikonal amplitude into the momentum space which gives

$$
S_{\text{fi}}^{\text{sea}}(\mathbf{q}_\perp) = \frac{1}{2\pi} \int \mathrm{d}^2\mathbf{b}\, a_{\text{fi}}^{\text{sea}}(\mathbf{b}) \exp(\mathrm{i}\mathbf{q}_\perp \cdot \mathbf{b})
$$
$$
= \frac{1}{2\pi} \langle \psi_f \mid \exp(\mathrm{i}\mathbf{q} \cdot \mathbf{r})\,\boldsymbol{\alpha}_\perp \mid \psi_i \rangle \cdot \mathbf{J}_\perp(\mathbf{q}), \tag{6.101}
$$

where $\mathbf{q} = (\mathbf{q}_\perp, \omega_{\text{fi}}/c)$, $\boldsymbol{\alpha}_\perp = (\alpha_x, \alpha_y)$ and the two-dimensional vector $\mathbf{J}_\perp = (J_x, J_y)$ is given by

$$
\mathbf{J}_\perp(\mathbf{q}) = \int \mathrm{d}^3\mathbf{s} \exp(-\mathrm{i}\mathbf{q} \cdot \mathbf{s})\,\theta(-s_z)\,\nabla_{\mathbf{s}_\perp} \left[\exp\left(\mathrm{i}\frac{2Z_A}{c}G(\mathbf{s}_\perp) \right) - 1 \right]. \tag{6.102}
$$

In (6.102) the integration runs over the three-dimensional space ($\mathbf{s} = (\mathbf{s}_\perp, s_z)$ and $\mathbf{s}_\perp \cdot \mathbf{v} = 0$) and $\nabla_{\mathbf{s}_\perp}$ denotes the two-dimensional gradient operator.

The integration over s_z in (6.102) is performed to yield the factor $\mathrm{i}c/\omega_{\text{fi}}$.[15] Further, assuming that $\lim_{\xi \to \infty} G(\xi) = 0$ the integral over $\mathbf{s}_\perp$, after the integration by parts, can be transformed according to

$$
\boldsymbol{\alpha}_\perp \cdot \int \mathrm{d}^2\mathbf{s}_\perp \exp(-\mathrm{i}\mathbf{q}_\perp \cdot \mathbf{s}_\perp)\,\nabla_{\mathbf{s}_\perp} \left[\exp\left(\mathrm{i}\frac{2Z_A}{c}G(\mathbf{s}_\perp) \right) - 1 \right]
$$
$$
= \mathrm{i}\boldsymbol{\alpha}_\perp \cdot \mathbf{q}_\perp \int \mathrm{d}^2\mathbf{s}_\perp \exp(-\mathrm{i}\mathbf{q}_\perp \cdot \mathbf{s}_\perp) \exp\left(\mathrm{i}\frac{2Z_A}{c}G(\mathbf{s}_\perp) \right)
$$
$$
\equiv \mathrm{i}\boldsymbol{\alpha}_\perp \cdot \mathbf{q}_\perp \int \mathrm{d}^2\mathbf{s}_\perp \exp(-\mathrm{i}\mathbf{q}_\perp \cdot \mathbf{s}_\perp) \left[\exp\left(\mathrm{i}\frac{2Z_A}{c}G(\mathbf{s}_\perp) \right) - 1 \right]. \tag{6.103}
$$

Taking all this into account we obtain that

$$
S_{\text{fi}}^{\text{sea}}(\mathbf{q}_\perp) = \frac{1}{2\pi} \int \mathrm{d}^2\mathbf{s}_\perp \exp(-\mathrm{i}\mathbf{q}_\perp \cdot \mathbf{s}_\perp) \left[\exp\left(\mathrm{i}\frac{2Z_A}{c}G(\mathbf{s}_\perp) \right) - 1 \right]
$$
$$
\times \frac{c}{\omega_{\text{fi}}} \langle \psi_f \mid \exp(\mathrm{i}\mathbf{q} \cdot \mathbf{r})\,\boldsymbol{\alpha}_\perp \mathbf{q}_\perp \mid \psi_i \rangle. \tag{6.104}
$$

---

[15] We suppose that $\omega_{\text{fi}} \neq 0$.

According to the continuity equation (see (5.56)) the last line in (6.104) is equal to $\langle \psi_f | \exp(i\mathbf{q} \cdot \mathbf{r})(1 - \alpha_z) | \psi_i \rangle$ and we finally arrive at the following symmetric eikonal amplitude:

$$S_{fi}^{sea}(\mathbf{q}_\perp) = \frac{1}{2\pi} \int d^2 \mathbf{s}_\perp \, \exp(-i\mathbf{q}_\perp \cdot \mathbf{s}_\perp) \left[\exp\left(i\frac{2Z_A}{c} G(\mathbf{s}_\perp)\right) - 1 \right]$$
$$\times \langle \psi_f | \exp(i\mathbf{q} \cdot \mathbf{r})(1 - \alpha_z) | \psi_i \rangle. \tag{6.105}$$

By comparing this expression with (6.62)–(6.63) we see that the amplitude (6.105) coincides with the light-cone amplitude. Thus, in the limit of very high impact energies the symmetric eikonal model yields the same results as the 'exact' light-cone transition amplitude. The same can also be said about the continuum-distorted-wave-eikonal-initial-state model because at $\gamma \gg 1$ the latter becomes essentially identical to the symmetric eikonal model.

6.7 Nonperturbative Approaches

6.7.1 Classical Description

As was already mentioned in Sect. 3.6, in the case of nonrelativistic ion–atom collisions the classical-trajectory Monte Carlo (CTMC) approach can, under certain conditions, yield results which are in reasonable agreement with experimental.

The CTMC approach can also be applied to relativistic collisions in which, for instance, a heavy hydrogen-like ion collides with a high-energy nucleus [156]. In such collisions the nuclei move along straight-line trajectories and their fields can be regarded as external. Therefore, a relativistic version of the CTMC approach appears after the implementation of the necessary changes which account for the relativistic electron dynamics and the relativistic character of the external fields acting on the electron.

If both colliding particles carry initially active electrons, the electromagnetic field transmitting the interactions in general can no longer be considered as an external field. In such a case, in addition to the equations for the electrons and nuclei, a classical treatment of relativistic collisions should also include the description of the degrees of freedom of the electromagnetic field.

We are not aware about any classical calculation performed for the projectile-electron transitions occurring in relativistic collisions with atoms.

6.7.2 Collisions at Relatively Low Energies: Nonperturbative Quantum Descriptions

When the projectile-electron excitation and loss may be reduced to a three-body problem of the motion of a single electron in the combined fields of two nuclei, these processes can be described by considering the Dirac equation

$$i\frac{\partial \Psi(\mathbf{r},t)}{\partial t} = \left(\hat{H}_0 + W(t)\right)\Psi(\mathbf{r},t). \tag{6.106}$$

In this equation

$$\hat{H}_0 = c\boldsymbol{\alpha}\cdot\hat{\mathbf{p}} - \frac{Z_A}{r} + \beta c^2 \tag{6.107}$$

is the electronic Hamiltonian of the undistorted initial hydrogen-like ion which has a nucleus with a charge Z_A, $\boldsymbol{\alpha} = (\alpha_x, \alpha_y, \alpha_z)$ and β are the Dirac matrices, and $\mathbf{r}$ are the coordinates of the electron with respect to the nucleus Z_A, which is taken as the origin. Further,

$$W(t) = -\gamma\frac{Z_B}{s}\left(1 - \frac{v}{c}\alpha_z\right) \tag{6.108}$$

is the interaction between the electron and the incident nucleus. This nucleus has a charge Z_B and moves in the rest frame of the nucleus Z_A along a classical straight-line trajectory $\mathbf{R}(t) = \mathbf{b} + \mathbf{v}t$, where $\mathbf{b}$ is the impact parameter and $\mathbf{v}$ the collision velocity. In (6.108) $\mathbf{s}$ represents the electron coordinates with respect to the nucleus Z_B (given in the rest frame of this nucleus) and is defined similarly as in (5.87). In (6.108) the scalar and vector potentials of the field generated by the nucleus Z_B are chosen to be in the Lienard–Wiechert form (see (5.86)).

The (6.106) can be solved by using two basically nonperturbative quantum approaches, (1) coupled channel methods and (2) numerical integrations of this equation on a lattice.

Coupled Channel Methods

Coupled channel methods have been used in a number of articles devoted to the study of the ionization, excitation, charge exchange and pair production occurring in relativistic ion–ion collisions (see e.g. [157]– [162]).

There are two main types of coupled channel expansions which have been used in the literature to treat relativistic collisions between a hydrogen-like ion and a bare nucleus. They are considered in some detail in [3] and [5] and here we restrict ourselves just to a very brief discussion.

One type of the coupled channel calculations involves the expansion of the time-dependent wave function $\Psi(\mathbf{r},t)$ using eigenstates of both $(Z_A + e^-)$ and $(Z_B + e^-)$ hydrogen-like ions. In addition to the description of the excitation and loss/ionization, such an expansion enables one also to address the problem of the charge exchange, in which the electron is finally captured by the nucleus Z_B forming a bound (or low-lying continuum) state of the ion $(Z_B + e^-)$.

Within the other type, the expansion is restricted to using only eigenstates $\varphi_j(\mathbf{r},t)$ of the initial ion $(Z_A + e^-)$:

$$\Psi(\mathbf{r},t) = \sum_j a_j(t)\varphi_j(\mathbf{r},t). \tag{6.109}$$

Concerning the choice between these two types in our case, one has to keep in mind that we have arrived at the three-body problem by simplifying the consideration of the projectile-electron transitions occurring in collisions with neutral atoms. The capture of an electron by a neutral atom is unlikely.[16] Therefore, in our case the expansion of the type of (6.109) is relevant.

Inserting the expansion (6.109) into the Dirac equation (6.106) and assuming that $i\partial_t \varphi_j(\mathbf{r}, t) = \hat{H}_0 \varphi_j(\mathbf{r}, t)$ we obtain

$$i\frac{da_j}{dt} = \sum_k a_k \langle \varphi_j | W | \varphi_k \rangle,$$

$$a_j(t \to -\infty) = \delta_{j0}. \tag{6.110}$$

Provided all the terms are kept in the expansion (6.109), (6.110) are equivalent to the Dirac equation. In practice, however, the expansion (6.109) of course must be truncated which limits the accuracy of coupled channel calculations. Besides, in such calculations there also exists the problem of how to incorporate continuum states into the consideration. These points are discussed in some detail in [3,5] where the interested reader can also find further references.

In the case of applications of coupled channel approaches it is in general not very easy to estimate the accuracy of calculated results. The same can also be said about results which are obtained when the distorted wave approaches are used. The main ideas lying behind the distorted wave and coupled channel approaches are quite different and these approaches also face quite different difficulties. Therefore, it is of interest to compare results of coupled channel calculations with those obtained when the distorted wave approaches are employed.

To this end we show in Table 6.1 results of various calculations for the excitation of $U^{91+}(1s_{1/2}(+1/2))$ by $1\,\text{GeV}\,\text{u}^{-1}$ U^{92+} nuclei. In the first column the excited states are displayed. The second column contains results of the first Born calculation performed with the amplitude (5.88). The next two column present results obtained with the eikonal amplitudes (6.86) and (6.87). The last two columns show results of the coupled channel calculations of [161] (see also Table 8.3 in [3]).

The latter calculations were performed using the expansions into the unperturbed or boundary corrected states and in the table the corresponding results are referred to as CC-1 and CC-2 results, respectively.[17] The

---

[16] A direct electron capture leading to the formation of a negative ion is either impossible or has very small cross section. The latter can also be said about the more complicated process, in which the removal of an electron from the atom is accompanied by the consequent capture of the electron from the ion.

[17] The authors of [161] actually used a two-center expansion. However, at a collision energy of $1\,\text{GeV}\,\text{u}^{-1}$ the capture is already much weaker that the excitation and, therefore, the expansion states centered on the incident nucleus are not expected to substantially influence the excitation cross sections.

Table 6.1. Theoretical cross sections (in **b**) for the excitation of $U^{91+}(1s_{1/2}(+1/2))$ by $1\,\mathrm{GeV\,u^{-1}}\,U^{92+}$ ions. The *first column*: final states of U^{91+}. The *second column*: first Born results. The *third* and *fourth columns*: results of the SEA and the modified SEA approximations, respectively. The *fifth* and *sixth columns*: results of the coupled-channel calculations performed in [161] with the undistorted and boundary corrected basis sets, respectively. The last line contains the sum of the corresponding excitation cross sections

Final state	1B	SEA	SEA-mod	CC1	CC2
$2s_{1/2}(+1/2)$	2,133	1,646	1,682	4,950	1,660
$2s_{1/2}(-1/2)$	64.3	105	58.2	394	114
$2p_{1/2}(+1/2)$	427	583	216	3,120	648
$2p_{1/2}(-1/2)$	5,682	3,827	3,748	5,830	4,710
$2p_{3/2}(+3/2)$	6,152	4,002	3,979	5,830	4600
$2p_{3/2}(+1/2)$	951	569	437	7540	1,480
$2p_{3/2}(-1/2)$	1,363	864	883	1,160	1,010
$2p_{3/2}(-3/2)$	117	143	94.4	106	99.4
$2s+2p_{1/2}+2p_{3/2}$	16,889	11,739	11,097	28,610	13,821

undistorted and boundary corrected expansion states differ by a phase-factor $\exp(-i\nu_B \ln(R' - vt'))$, $\varphi_j^{BC} = \exp(-i\nu_B \ln(R' - vt'))\varphi_j$, where $R' = \sqrt{b^2 + v^2 t'^2}$ is the asymptotic distance (at $t \to -\infty$) between the electron and the nucleus Z_B as it is viewed in the rest frame of this nucleus and $t' = \gamma(t - vz/c^2)$ is the electron time measured in the rest frame of the nucleus Z_B (see [3]).

The introduction of the distortion factor has an important advantage enabling one to replace the Lienard–Wiechert potentials of the field of the incident ion in the Dirac equation by the scalar and vector potentials which fall off more rapidly with the distance s. It is obvious that such an introduction can be also regarded as a gauge transformation.

The table shows that there exists a large discrepancy between the results of the coupled channel calculations. This suggests that the calculations with the restricted basis sets, performed in [161], are strongly gauge dependent. Comparing the cross sections yielded by the coupled channel and the first Born approximations we see that, with respect to the first Born results, the CC-1 (CC-2) on overall gives substantially larger (lower) values. Taking into account that, compared to the first Born predictions, the inclusion of the higher order effects normally leads to a decrease in the values of calculated cross sections, the CC-1 results seem to be very questionable and we may safely rule out these results from the further comparison.

The difference between the two sets of the eikonal results becomes substantial only for electronic transitions which are characterized by relatively small cross sections. On overall, the difference turns out to be just about several per cent. On average, the eikonal results also differ by about 20% from the CC-2 results. Compared to the first Born results, the eikonal cross sections are on overall by about 45% lower.

Numerical Integration on a Lattice

Processes occurring with an electron in relativistic collisions between a bare nucleus and an hydrogen-like ion have been also considered by integrating numerically the Dirac equation on a lattice. There are various grid methods and such integrations have been performed for the Dirac equation written in the configuration space as well as in the momentum space (see, for instance, [163–167]).

Numerical Integration on a Lattice in Configuration Space

In a numerical integration of the Dirac equation on a lattice in the configuration space one starts with replacing the whole space by a finite volume, to which the processes of interest are supposed to be mainly restricted, and introducing in this volume an equidistant grid.

Within the *finite-element* method one defines basis functions with respect to each coordinate on a grid. For instance, in the simplest case of one-dimensional Cartesian space the interval of interest $[x_{min} \leq x \leq x_{max}]$ is divided into subintervals (x_j, x_{j+1}) of length a $(j = 1, \ldots, N_x$ with $N_x = (x_{max} - x_{min})/a)$. The basis functions on this interval may be chosen according to

$$f_j(x) = \frac{x - x_{j-1}}{a}, \quad \text{if } x_{j-1} \leq x \leq x_j,$$

$$f_j(x) = \frac{x_{j+1} - x}{a}, \quad \text{if } x_j \leq x \leq x_{j+1},$$

$$f_j(x) = 0, \quad \text{for all other } x. \tag{6.111}$$

This definition, in particular, implies that the functions $f_j(x)$ satisfy the condition

$$f_{j'}(x_j) = \delta_{jj'}. \tag{6.112}$$

Note that (6.111) represents the simplest choice which is not necessarily the best one and other definitions for the basic functions, fulfilling the condition (6.112), are possible (see e.g. [167]).

In the case of three-dimensional Cartesian space (x, y, z) the corresponding basis functions are written as products $f_j(x)f_k(y)f_l(z)$ $(j \leq N_x, k \leq N_y, l \leq N_z)$. The time dependent components $\Psi^\alpha(t)$ $(\alpha = 1, 2, 3, 4)$ of the four-spinor $\Psi(t)$, which satisfies the Dirac equation (6.106), are expanded in terms of the basis functions

$$\Psi^\alpha(t) = \sum_{j,k,l} B^\alpha_{jkl}(t) \, f_j(x) f_k(y) f_l(z), \tag{6.113}$$

where $B^\alpha_{jkl}(t)$ are unknown time-dependent coefficients to be determined. Inserting this expansion into (6.106), and projecting onto the basis functions one obtains for these coefficients the following equation: [167]

$$\mathbf{iM}\frac{d\mathbf{B}}{dt} = \mathbf{H}(t)\mathbf{B}(t). \tag{6.114}$$

Here,

$$\mathbf{B} = \begin{pmatrix} B_{111}^1 \\ B_{111}^2 \\ B_{111}^3 \\ B_{111}^2 \\ \cdot \\ \cdot \\ \cdot \\ B_{N_x,N_y,N_z}^4 \end{pmatrix} \tag{6.115}$$

and

$$\mathbf{M} = \mathbf{M}^x \otimes \mathbf{M}^y \otimes \mathbf{M}^z,$$
$$\mathbf{H} = c\,P^x \otimes \mathbf{M}^y \otimes \mathbf{M}^z \otimes \alpha_x + c\,\mathbf{M}^x \otimes P^y \otimes \mathbf{M}^z \otimes \alpha_y$$
$$+ c\,\mathbf{M}^x \otimes \mathbf{M}^y \otimes P^z \otimes \alpha_z + mc^2\,\mathbf{M}^x \otimes \mathbf{M}^y \otimes \mathbf{M}^z \otimes \beta$$
$$+ (V_t + V_p) \otimes I_4 - \frac{v}{c}V_p\alpha_z, \tag{6.116}$$

where the sign $\otimes$ denotes matrix multiplication. The matrices $\mathbf{M}^n$ and P^n ($n = x, y, z$) are the overlap and momentum operator matrices, respectively, and V_t and V_p are the matrices containing the transition matrix elements of the scalar potentials of the target and projectile nuclei, respectively. For instance,

$$\mathbf{M}_{jj'}^x = \int dx\, f_j(x) f_{j'}(x),$$
$$V_{jkl,\,j'k'l'} = \int dx \int dy \int dz\, f_j(x) f_l(x) f_k(x) V f_{j'}(x) f_{k'}(y) f_{l'}(z), \tag{6.117}$$

where $V = V_t$ or $V = V_p$.[18]

If one knows $\mathbf{B}(t)$ for a given value of t, then the formal solution of (6.114) for $t + \Delta t$, where Δt is sufficiently small such that $\mathbf{H}$ can be regarded as a constant during the interval Δt, reads:

$$\mathbf{B}(t + \Delta t) = \exp\left(-i\mathbf{M}^{-1}\mathbf{H}\Delta t\right)\mathbf{B}(t). \tag{6.118}$$

Equation (6.118) can be solved by using appropriate numerical techniques [167].

---

[18] It worth noting that the method briefly outlined above has a certain similarity to a coupled-channel approach in which the initial Dirac equation is also replaced by the first order differential equation (6.110) for unknown time-dependent coefficients of the corresponding coupled channel expansion.

One of the main problems of a numerical integrations on a grid in the configuration space is the fact that the wave function in this space is poorly localized and may spread over the whole integration volume. If during the propagation time a part of the wave function reaches the boundaries, it will be reflected back leading to unphysical solutions. Therefore, once a noticeable part of the wave function reaches the boundaries, the propagation in time has to be stopped (or some absorbing potential should be introduced on the boundaries).

Such a problem does not arise if the Dirac equation is considered on a lattice in the momentum space [166]. This advantage appears because in the momentum space the wave function is very well localized around the origin. Being transformed into the momentum space, the Dirac equation reads

$$i\frac{\partial\psi(\mathbf{p},t)}{\partial t} = \left(c\boldsymbol{\alpha}\cdot\mathbf{p} + \beta c^2\right)\psi(\mathbf{p},t)$$
$$-\frac{1}{2\pi^2}\int d^3\mathbf{p}' \left(\frac{Z_A}{q^2} + \left(1 - \frac{v}{c}\alpha_z\right)\frac{Z_B\exp(-i\mathbf{q}\cdot\mathbf{R}(t))}{q_x^2 + q_y^2 + q_z^2/\gamma^2}\right)\psi(\mathbf{p}',t),$$

$$(6.119)$$

where $\mathbf{q} = \mathbf{p} - \mathbf{p}'$, $\mathbf{R}$ is the trajectory of the nucleus Z_B and

$$\psi(\mathbf{p},t) = \frac{1}{(2\pi)^{3/2}}\int d^3r\, \Psi(\mathbf{r},t)\exp(-i\mathbf{p}\cdot\mathbf{r}) \qquad (6.120)$$

is the Dirac spinor in the momentum space. While in the momentum space the wave function $\psi(\mathbf{p},t)$ does not tend to escape from the integration volume, the calculation of the momentum integrals is more demanding to the quality of a coordinate mesh and takes more computing time.

Like in the case with coupled channel calculations it is of interest to compare cross sections computed by means of numerical integrations of the Dirac equation on a lattice with results of the distorted-wave models discussed in Sect. 6.5.

In [166] collisions of 0.93 GeV u^{-1} Au^{79+} with uranium ions U^{91+}(1s) were studied by using a numerical integration on a lattice in the momentum space. In particular, the authors of [166] calculated cross section values for the excitations into the $2s_{1/2}$ and $3s_{1/2}$ states of U^{91+} as well as for the electron loss. It was reported in [166] that their nonperturbative excitation cross sections are 1.37 kb (for $1s_{1/2} \to 2s_{1/2}$) and 0.25 kb (for $1s_{1/2} \to 3s_{1/2}$). The nonperturbative loss cross section, obtained in [166], is 14.4 kb.

Note that the first order predictions for the bound–bound transitions amount to 1.65 and 0.3 kb, respectively, and the first order result for the electron loss is 15.3 kb.

Calculations with the distorted-wave amplitudes (6.86) and (6.87) for the same bound–bound transitions yield 1.36 and 0.245 and 1.35 and 0.243, respectively. A calculation with the CDW-EIS amplitude (6.92) results in the loss cross section of 12.2 kb.

We thus see that in the case of excitations the results of the distorted-wave models and of the integration on a grid are mutually supportive yielding quite close cross section values. However, in the case of the electron transitions to the continuum the agreement between the results of these approaches is not that good with the CDW-EIS ionization cross section being noticeably lower than the nonperturbative ionization cross section reported in [166].

Note that a similar correspondence between results of the distorted-wave approaches and the numerical integrations on a grid holds also for the collision system $0.466\,\mathrm{GeV\,u^{-1}}$ $U^{92+} + U^{91+}(1s)$. The nonperturbative results for excitation and ionization cross section for this system were reported in [167] where (6.118) was solved numerically. Here again the distorted-wave cross sections for the excitation turned out to be rather close to those obtained with the nonperturbative calculation whereas in the case of the electron loss the distorted-wave models, compared to the grid calculation, predict a noticeably lower cross section.

At present it is not clear whether the discrepancy observed for the loss cross section should be interpreted as a deficiency of the distorted-wave models for transitions which involve continuum states,[19] or it simply reflects the lack of a necessary accuracy in the nonperturbative calculations for the ionization cross sections performed in [166, 167]. This discrepancy, of course, may also be a signature that both these points are simultaneously present.

---

[19] In particular, we remind the reader that in the derivation of the CDW-EIS amplitude (6.92) the influence of the nuclei on the electron in the continuum was not treated symmetrically arguing that we consider the case when Z_A is noticeably larger than Z_B. The latter, however, is certainly not true for U^{92+} on $U^{91+}(1s)$ collisions.

7

Impact Parameter Dependence of Projectile-Electron Excitation and Loss in Relativistic Collisions

7.1 Preliminary Remarks

For a collision in which the electron of the projectile ion makes a transition $0 \to n$ and those of the target atom make a transition $0 \to m$, the semi-classical transition probability[1] reads

$$P_{0\to n}^{0\to m}(\mathbf{b}) = \mid a_{0\to n}^{0\to m}(\mathbf{b}) \mid^2, \tag{7.1}$$

where the transition amplitude $a_{0\to n}^{0\to m}(\mathbf{b})$ is given by (5.46).

In experiments on the projectile-electron excitation and loss in relativistic collisions performed until now final internal states of the atom were not observed. In order to describe an experiment in such a case one has to sum over all possible states of the atom. The total probability for the ion to make a transition $0 \to n$ in the collision then reads

$$P_{0\to n}(\mathbf{b}) = \sum_m \mid a_{0\to n}^{0\to m}(\mathbf{b}) \mid^2 . \tag{7.2}$$

This transition probability can be split into the sum of the contributions given by the elastic,

$$P_{0\to n}^{\mathrm{s}}(\mathbf{b}) = \mid a_{0\to n}^{0\to 0}(\mathbf{b}) \mid^2, \tag{7.3}$$

and inelastic,

$$P_{0\to n}^{\mathrm{a}}(\mathbf{b}) = \sum_{m\neq 0} \mid a_{0\to n}^{0\to m}(\mathbf{b}) \mid^2, \tag{7.4}$$

target modes.

---

[1] If the final state of the projectile or of the target is a continuum state, then (7.1) represents the probability density.

7.2 Transition Amplitudes

Except Sect. 7.4, throughout this chapter we shall use only the first order description of the ion–atom collisions. In what follows the probability $P_{0\to n}(\mathbf{b})$ will be evaluated within the nonrelativistic atom approximation. Within this approximation the semi-classical first order transition amplitude (5.46) is substantially simplified and reduces to

$$a_{0\to n}^{0\to m}(\mathbf{b}) = -\frac{i}{\pi v}\int d^2\mathbf{q}_\perp \exp(-i\mathbf{q}_\perp \mathbf{b}) \left\langle u_m \left| Z_A - \sum_{j=1}^{N_A}\exp(i\mathbf{q}_A \cdot \boldsymbol{\xi}_j) \right| u_0 \right\rangle$$

$$\times \frac{\left\langle \psi_n \left|\left(1 - \frac{v}{c}\alpha_z\right)\exp(i\mathbf{q}_I \cdot \mathbf{r})\right| \psi_0 \right\rangle}{q_\perp^2 + \frac{(\varepsilon_n-\varepsilon_0+\epsilon_m-\epsilon_0)^2}{v^2\gamma^2} + 2(\gamma-1)\frac{(\varepsilon_n-\varepsilon_0)(\epsilon_m-\epsilon_0)}{v^2\gamma^2}} . \tag{7.5}$$

We remind the reader that $\mathbf{q}_I = (\mathbf{q}_\perp, q_{\min}^I)$ and $\mathbf{q}_A = (-\mathbf{q}_\perp, -q_{\min}^A)$ are the momenta transferred to the ion (in the ion frame) and to the atom (in the atom frame), respectively, with $q_{\min}^I$ and $q_{\min}^A$ given by (5.31) and (5.32). Besides, $\boldsymbol{\xi}_j$ are the coordinates of the atomic electrons with respect to the atomic nucleus (given in the atomic frame).

7.2.1 Elastic Target Mode

In this case $m = 0$ and the corresponding transition amplitude is given by

$$a_{0\to n}^{0\to 0}(\mathbf{b}) = -\frac{i}{\pi v}\int d^2\mathbf{q}_\perp \exp(-i\mathbf{q}_\perp \mathbf{b})Z_{A,\text{eff}}(\mathbf{q}_A)$$

$$\times \frac{< \psi_n \mid \left(1 - \frac{v}{c}\alpha_z\right)\exp(i\mathbf{q}_I \cdot \mathbf{r}) \mid \psi_0 >}{q_\perp^2 + \frac{(\varepsilon_n-\varepsilon_0)^2}{v^2\gamma^2}}, \tag{7.6}$$

where the effective charge $Z_{A,\text{eff}}$ of the atom in the ground state is defined by (5.70).

With the effective charge given by (5.70) it is easy to perform the integration over the transverse part $\mathbf{q}_\perp$ of the momentum transfer in expression (7.6) and obtain that the elastic transition amplitude is given by

$$a_{0\to n}^{0\to 0}(\mathbf{b}) = \frac{2iZ_A}{v}\sum_{j=1}^3 A_j$$

$$\times \left\langle \psi_n \left|\exp\left(i\frac{\varepsilon_n-\varepsilon_0}{v}z\right)\left(1-\frac{v}{c}\alpha_z\right)K_0\left(B_{0,n}^j \mid \mathbf{b} - \mathbf{r}_\perp \mid\right)\right| \psi_0 \right\rangle, \tag{7.7}$$

where K_0 is the modified Bessel function, $\mathbf{r} = (\mathbf{r}_\perp, z)$ with $\mathbf{r}_\perp \cdot \mathbf{v} = 0$ are the coordinates of the electron of the ion,

$$B_{0,n}^j = \sqrt{\frac{(\varepsilon_n - \varepsilon_0)^2}{v^2\gamma^2} + \kappa_j^2}$$

and A_j and κ_j are the screening parameters. Note that if we neglect in (7.7) the screening effect of the atomic electrons by setting all $\kappa_j = 0$ then, taking into account that $\sum_j A_j = 1$, we obtain the amplitude for a transition $0 \to n$ of the electron of the ion in collisions with a bare nucleus with a charge Z_A

$$a_{0 \to n}^{0 \to 0}(\mathbf{b}) = \frac{2\mathrm{i}Z_A}{v}$$
$$\times \left\langle \psi_n \left| \exp\left(\mathrm{i}\frac{\varepsilon_n - \varepsilon_0}{v}z\right) \left(1 - \frac{v}{c}\alpha_z\right) K_0\left(B_{0,n}^C \mid \mathbf{b} - \mathbf{r}_\perp \mid\right) \right| \psi_0 \right\rangle,$$

$$\tag{7.8}$$

where

$$B_{0,n}^C = \frac{(\varepsilon_n - \varepsilon_0)}{\gamma v}.$$

7.2.2 Inelastic Target Mode

In this case $m \neq 0$. Taking into account that

$$\int \mathrm{d}^2\mathbf{q}_\perp \frac{\exp(-\mathrm{i}\mathbf{q}_\perp(\mathbf{b} - \mathbf{r}_\perp + \boldsymbol{\xi}_{\perp,j}))}{\mathbf{q}_\perp^2 + B_{m,n}^2}$$
$$= 2\pi K_0(\mid \mathbf{b} - \mathbf{r}_\perp + \boldsymbol{\xi}_{\perp,j} \mid B_{mn}), \quad B_{m,n} > 0, \tag{7.9}$$

where $\boldsymbol{\xi}_j = (\boldsymbol{\xi}_{\perp,j}, \xi_{z,j})$ with $\boldsymbol{\xi}_{\perp,j} \cdot \mathbf{v} = 0$, the transition amplitude can be written as

$$a_{0 \to n}^{0 \to m}(\mathbf{b}) = \left\langle u_m \left| \sum_{j=1}^{N_A} \mathrm{e}^{-\mathrm{i}\left(\frac{\epsilon_m - \epsilon_0}{v} + \frac{\varepsilon_n - \varepsilon_0}{v\gamma}\right)\xi_{z,j}} \alpha_{n0}^j(\mathbf{b} + \boldsymbol{\xi}_{\perp,j}, m0) \right| u_0 \right\rangle. \tag{7.10}$$

In the above formula

$$\alpha_{n0}^j(\mathbf{b} + \boldsymbol{\xi}_{\perp,j}, m0) = \frac{2\mathrm{i}}{v} \left\langle \psi_n \left| \mathrm{e}^{-\mathrm{i}\left(\frac{\varepsilon_n - \varepsilon_0}{v} + \frac{\epsilon_m - \epsilon_0}{v\gamma}\right)z} \left(1 - \frac{v}{c}\alpha_z\right) \right. \right.$$
$$\left. \left. \times K_0\left(\mid \mathbf{b} - \mathbf{r}_\perp + \boldsymbol{\xi}_{\perp,j} \mid B_{m,n}\right) \right| \psi_0 \right\rangle, \tag{7.11}$$

where

$$B_{m,n} = \sqrt{\frac{(\varepsilon_n - \varepsilon_0 + \epsilon_m - \epsilon_0)^2}{v^2\gamma^2} + 2(\gamma - 1)\frac{(\varepsilon_n - \varepsilon_0)(\epsilon_m - \epsilon_0)}{v^2\gamma^2}}.$$

It worth noting that the form of the expression (7.11) resembles the semi-classical transition amplitude for the electron of the ion in collisions with a point-like particle with a charge -1 which moves along a classical straight-line trajectory with the velocity v and the impact parameter $\mathbf{b} + \boldsymbol{\xi}_{\perp,j}$.

In order to find the total probability of the electron transition in the ion from all collisions, where the atom can finally be in any of its excited states including the atomic continuum, one has to perform the summation in (7.4). This can be done by using the closure method (see [22] and references therein). In the simplest form of this method the same averaged energy $\Delta\epsilon$ is assumed for all possible transitions of the atomic electrons. In nonrelativistic collisions this approximation yields good results for the electron loss at collision velocities well above the energy threshold for the projectile ionization by a beam of free electrons. Therefore, one can expect this approximation to give reasonable results for relativistic collisions when the kinetic energy T of an equivelocity electron is much larger than the transition energy of the electron in the ion: $T = m_e c^2 (\gamma - 1) \gg \varepsilon_n - \varepsilon_0$. Starting with $\gamma \sim 2 - 3$ the latter condition is fulfilled even for the heaviest single-electron ions. Since we already have the condition $\gamma \geq 3 - 4$ imposed by the application of the nonrelativistic atom approximation for the antiscreening mode, no additional restrictions on the collision energies are introduced here.

Within the closure approximation the closure relation for the electron states of the atom

$$\sum_m | u_m > < u_m | = \mathbf{I} \tag{7.12}$$

is applied in order to perform the summation over the final states of the atom. In addition, if the antisymmetrization in the ground state of the atom is ignored and the wavefunction of the ground state is expressed as

$$u_0 = \prod_\lambda \phi_\lambda(\boldsymbol{\xi}_\lambda), \tag{7.13}$$

where $\phi_\lambda(\boldsymbol{\xi})$ are the single electron orbitals, the antiscreening probability takes the much simpler form [92]

$$P^a_{0 \to n}(\mathbf{b}) = \sum_{m \neq 0} | a^{0 \to m}_{0 \to n}(\mathbf{b}) |^2$$

$$= \sum_\lambda \left\langle \phi_\lambda \left| | e^{-i(\frac{\Delta\epsilon}{v} + \frac{\varepsilon_n - \varepsilon_0}{v\gamma})\xi_z} \alpha_{n0}(\mathbf{b} + \boldsymbol{\xi}_\perp) |^2 \right| \phi_\lambda \right\rangle$$

$$- \sum_\lambda \left| \left\langle \phi_\lambda \left| e^{-i(\frac{\Delta\epsilon}{v} + \frac{\varepsilon_n - \varepsilon_0}{v\gamma})\xi_z} \alpha_{n0}(\mathbf{b} + \boldsymbol{\xi}_\perp) \right| \phi_\lambda \right\rangle \right|^2. \tag{7.14}$$

Here $\alpha_{n0}(\mathbf{b} + \boldsymbol{\xi}_\perp)$ is defined by (7.11) with the replacements $\epsilon_m - \epsilon_0 \to \Delta\epsilon$ and

$$B_{mn} \to \sqrt{\frac{(\varepsilon_n - \varepsilon_0 + \Delta\epsilon)^2}{v^2\gamma^2} + 2(\gamma - 1)\frac{(\varepsilon_n - \varepsilon_0)\Delta\epsilon}{v^2\gamma^2}}.$$

The expression (7.14) still contains the sixfold integration over the electronic coordinates and an additional threefold integration needs to be performed if one considers the electron loss.

7.3 Excitation of Bi^{82+}(1s) in Collisions with Cu and He

In this section we shall discuss probabilities for the electron excitation in Bi^{82+} in relativistic collisions with two neutral atoms, Cu and He. For a comparison the excitation of Bi^{82+} in collisions with the corresponding bare nuclei, Cu^{29+} and He^{2+}, will also be considered. In addition, results will be presented for the antiscreening probability of the projectile-electron excitation in collisions with helium. Helium as a target was chosen because of three main reasons. First, helium is a few-electron system where the contribution from the antiscreening mode is expected to be comparable in magnitude with that of the screening mode. Second, helium target is widely used in experiments on atomic collision physics. Third, in the helium case orbitals $\phi_\lambda(\boldsymbol{\xi})$ are 1s-orbitals and the sixfold integral in (7.14) can be reduced analytically to a fourfold integration [25]. The latter, however, has to be done numerically.

In contrast to helium, a copper atom has many electrons. Therefore, in collisions of Bi^{82+}, which has a very tightly bound electron, with copper the antiscreening mode is of minor importance and will not be considered here.

Throughout this section relativistic units $\hbar = m_e = c = 1$ are used except in Fig. 7.5 where the impact parameter is given in fermi (1 rel. unit. $\simeq 386\,\mathrm{fm}$).

7.3.1 Screening in Ultrarelativistic Collisions with Moderately Heavy Atoms

In Fig. 7.1 weighted probabilities $bP(b)$ are shown for the excitation of a $1s_{1/2}(m_j = -1/2)$ electron of a Bi^{82+} projectile incident on Cu at a collision energy corresponding to $\gamma = 10$. The different full curves show results for excitation to different states of Bi^{82+} where the screening effect has been included. The dashed curves show the excitation without any screening, i.e. in collisions with a bare nucleus Cu^{29+}. It can be seen from this figure that the main effect of the screening is to reduce the transition probabilities at larger impact parameters. For transitions to the $2s_{1/2}(m_j = -1/2)$-state the screening effect plays almost no role. This suggests that these transitions occur effectively at very small impact parameters where the electrons of the neutral copper atom cannot screen their nucleus.

In Fig. 7.2 results of similar calculations are displayed for the same projectile–target system but at a collision energy corresponding to $\gamma = 100$. Because of the retardation effect, in collisions with Cu^{29+} the probabilities of transitions to p-states have considerably longer tails at large b compared to the previous case. However, in collisions with a neutral atom the screening of the nucleus of Cu by its electrons reduces the transition probabilities in collisions

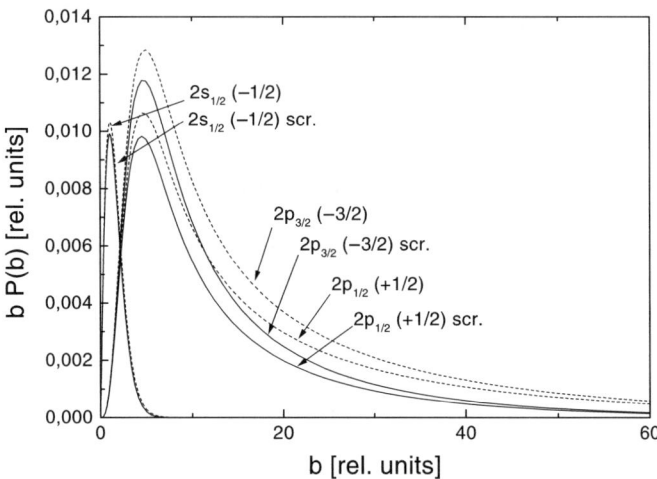

Fig. 7.1. Weighted probabilities for projectile excitation in collisions of a Bi^{82+} projectile with Cu at a collision energy corresponding to $\gamma = 10$. *Solid lines*: the screening mode, *dashed lines*: collisions with a bare atomic nucleus Cu^{29+}. The numbers in *brackets* denote the magnetic quantum numbers of the final electron states in the Bi^{82+} ion. From [92].

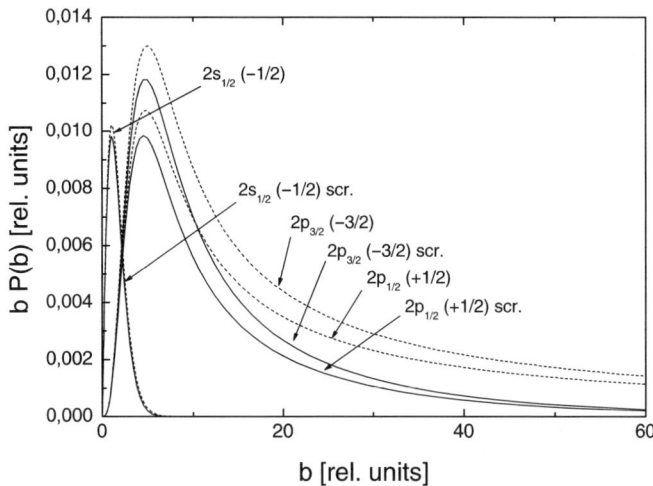

Fig. 7.2. As in Fig. 7.1 but at a collision energy corresponding to $\gamma = 100$. From [92].

with larger impact parameters. Thus, one obtains almost the same results for the screened probabilities at $\gamma = 10$ and $\gamma = 100$. For a many-electron atom like Cu the antiscreening mode is not expected to play a noticeable role. Therefore, one may conclude that at $\gamma \gtrsim 10$ the corresponding cross sections

for the excitation, considered as a function of collision energy, are very close
to or already have entered the 'saturation' region where the cross sections
become γ-independent (for more discussion of the cross section saturation
see 8.9).

7.3.2 Screening and Antiscreening in Ultrarelativistic Collisions with Very Light Atoms. 'Separation' of the Screening and Antiscreening Modes in the Impact Parameter Space

In Fig. 7.3 the weighted probabilities are depicted for the excitation of a
$1s_{1/2}(m_j = -1/2)$ electron of a Bi^{82+} projectile incident on He at a collision
energy corresponding to $\gamma = 10$. Similarly to the case of Bi^{82+}–Cu collisions,
the full and dashed lines represent results of calculations with and without the
screening, respectively. The helium atom is much lighter than copper and the
orbits of helium electrons are much larger than the orbits of inner electrons in
copper. Therefore, in contrast to collisions with Cu, in collisions with He the
screening effect plays a very modest role at $\gamma = 10$ for all transitions shown
in the figure.

The situation changes drastically for Bi^{82+}–He collisions at $\gamma = 100$ (see
Fig. 7.4). In collisions with He^{2+} at $\gamma = 100$, larger impact parameters (compared to the case with $\gamma = 10$) considerably contribute to transitions to
the p-states in Bi^{82+}. These impact parameters are already comparable in
magnitude with the dimension of the electron orbits in the ground state of
neutral He. Therefore, in collisions at such impact parameters, electrons of
He are able to effectively screen their nucleus and considerably reduce the
transition probabilities.

In Fig. 7.4 we compare the screening and antiscreening effects in the probabilities of the electron transitions in Bi^{82+} in collisions with He at a collision

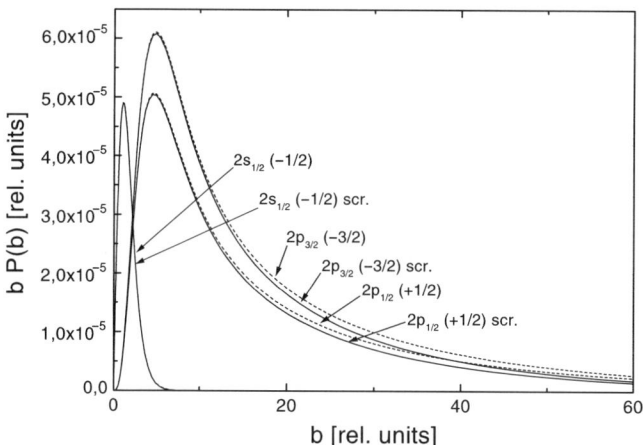

Fig. 7.3. As in Fig. 7.1 but for collisions with He. From [92].

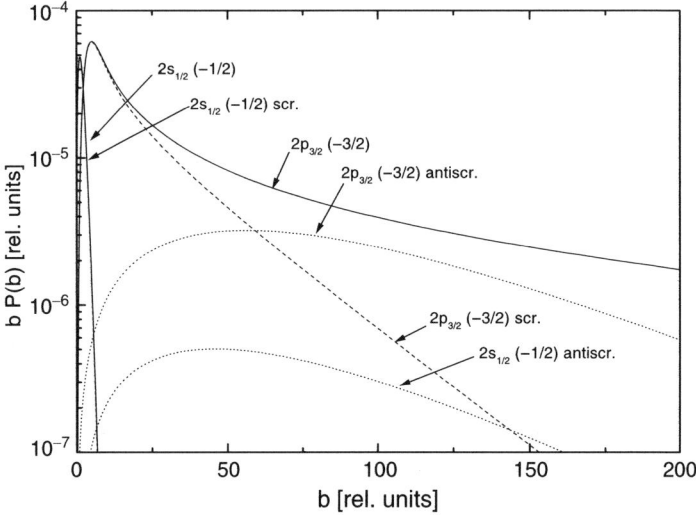

Fig. 7.4. Weighted probabilities for projectile excitation in collisions of a Bi^{82+} projectile with He. *Dashed lines*: the screening mode, *dotted lines*: the antiscreening mode, *solid lines*: collisions with He^{2+}. The numbers in brackets denote the magnetic quantum numbers of the final states of the Bi^{82+} ion. From [92].

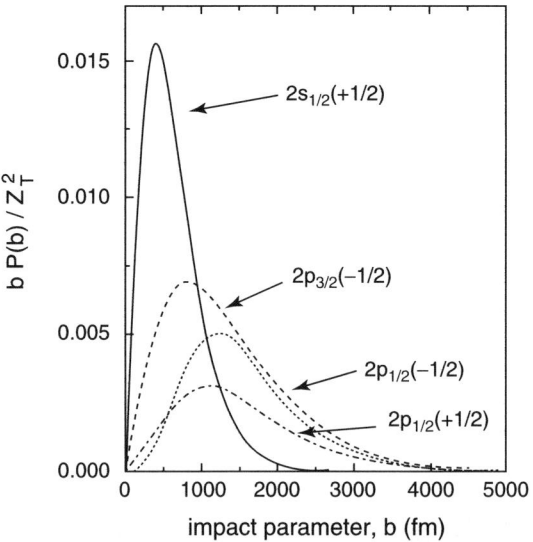

Fig. 7.5. Weighted probabilities for excitation of a $119\,MeV\,u^{-1}$ Bi^{82+} divided by Z_T where Z_T is the target atomic number. The numbers in *brackets* denote the magnetic quantum numbers of the final states of the Bi^{82+} ion. From [96].

energy corresponding to $\gamma = 100$. There are some interesting features in the antiscreening probabilities, which should be mentioned. First, at small impact parameters these probabilities are much lower than the screening probabilities. Second, the antiscreening probabilities spread to much larger impact parameters. At $b \sim 100$ the antiscreening probabilities for the $1s$–$2p$ transitions are comparable in magnitude to the probabilities in collisions with the unscreened helium nucleus. And only at $b \gtrsim 300$ (which is not shown in the figure) the antiscreening probabilities become much smaller than the probabilities in collisions with the unscreened helium nucleus. Thus, it turns out that the screening and antiscreening contributions to the excitation are to large extent separated in the b-space.

This relationship between the screening and antiscreening contributions can be understood by noting that, whereas the atomic nucleus is point-like (on a typical atomic scale), the atomic electrons spread over a large volume. Because of this reason in collisions with small impact parameters the action of the atomic electrons on the electron, which is bound in a highly charged ion, cannot effectively compete with that of the atomic nucleus. However, due to the same reason, the atomic electrons become more effective, compared to the atomic nucleus, at larger impact parameters.

Estimates show that, because of the long tails at large impact parameters, the antiscreening mode contributes considerably (about 25–30%) to the total cross sections for the (electric) dipole allowed electron transitions in Bi^{82+} in collisions with He at $\gamma = 100$. The relative contribution of the antiscreening mode to the total cross section for the $1s$–$2s$ transition is about 15%, i.e. it is considerably smaller. The latter point can be understood by noting that the contribution of large impact parameters, where the antiscreening mode could become more important, to the $1s_{1/2}(-1/2) \to 2s_{1/2}(-1/2)$ transition is strongly suppressed compared to the case of the excitation of the dipole allowed transitions.

7.3.3 Comparison between Excitation of Heavy Ions in Collisions with Neutral Atoms at Low and High γ

Excitation of hydrogen-like Bi ions in collisions with copper at a collision energy of $119\,\mathrm{MeV\,u^{-1}}$ corresponding to $\gamma = 1.13$ was studied in [95, 96]. Since the excitation energies of Bi^{82+} are very big and the value of γ is quite low, only collisions with momentum transfers which are large on the atomic scale of copper can effectively excite the ion. Therefore, under these conditions the screening effect is expected to be very weak and, as our calculations show, can be neglected. In addition, the collision velocity corresponding to the energy $119\,\mathrm{MeV\,u^{-1}}$ is below the threshold for the ionization of Bi^{82+} by a free electron having the same velocity in the ion frame as the atom. Therefore, under the experimental conditions of [95, 96] the antiscreening effect is very weak as well and the main contribution to the excitation is given by the interaction with the unscreened target nucleus. Thus, the physics of the excitation of $119\,\text{MeV}\,\text{u}^{-1}$ Bi^{82+} in collisions with a neutral copper atom is basically reduced to that in collisions with a bare copper nucleus.

Probabilities for the excitation of $119\,\text{MeV}\,\text{u}^{-1}$ Bi^{82+} in collisions with a point-like copper nucleus Cu^{29+} were calculated in [95, 96] within the first order of the perturbation theory (see Fig. 7.5). As was just discussed above, at this collision energy these results can be directly applied for collisions with neutral atoms of copper. It is of interest to state briefly the main differences between the excitation of very heavy hydrogen-like ions in collisions at low γ and in ultrarelativistic collisions. First, in contrast to collisions at low γ, at high values of γ the screening effect of the atomic electrons becomes important even for very light atomic targets. Second, the antiscreening mode is always of considerable importance in ultrarelativistic collisions with few-electron targets. Third, in collisions at low γ the $1s$–$2s$ transition in Bi^{82+} were shown to dominate over the transitions to $2p$-states [96]. In ultrarelativistic collisions this situation is changed. Now the $1s$–$2p$ transitions contribute most to the excitation with the transition $1s_{1/2}(m_j = -1/2) \rightarrow 2p_{3/2}(m_j = -3/2)$ being the most probable one. Compared to collisions with $\gamma = 1.13$, in collisions with $\gamma = 10$ the maximum of the distribution $bP(b)$ for transition to $2s$ state is reduced by a factor of about 4 and the position of the maximum and the width of this distribution are practically unchanged. In contrast, the distributions $bP(b)$ for the main transitions to $2p$ states are not reduced in height but are shifted towards larger b and acquire larger widths. This behaviour of the probabilities for transitions to $2s$ and $2p$ states is connected with the increase of the collision velocity (energy). At $\gamma = 1.13$ the collision velocity is considerably less than the speed of light ($v/c = 0.46$) and the increase of this velocity leads to a decrease of all the transition probabilities at small impact parameters. However, when the collision velocity approaches the light velocity and cannot be noticeably increased further, the retardation effect is the only important effect and it increases the transition probabilities for the dipole-allowed transitions. For large enough values of γ the retardation effect, which would lead to longer and longer tails for the probabilities of the dipole-allowed transitions, is neutralized by the screening effects discussed earlier in this section.

Since a $2s$-state can be effectively excited in collisions with very small impact parameters only, the influence of the retardation effect on the excitation to a $2s$-state is quite weak.

7.4 Higher-Order Effects in the Loss Probability in Collisions at Asymptotically Large γ

The asymptotic limits of a theory are a matter of general interest. In the case of ion–atom collisions one of the natural asymptotic limits is represented by the domain of (infinitely) high impact energies.

The nonrelativistic theories of ion–atom collisions assume that $c \to \infty$. According to such theories the high-energy limit of ion–atom collisions ($E_{col} \to \infty$), which is this case coincides with the high-velocity collision limit $v \to \infty$, is well described by the first order approximation in the ion–atom interaction. This, in particular, means that with increasing the impact energy the differential and total loss probabilities for any ion–atom collision pair will ultimately be very well described by applying merely a first order theory.

In the relativistic description the impact energy can in principle take on any value but the collision velocity has the upper limit given by the speed of light in vacuum. Therefore, the high-energy limit of relativistic theories is given by $E_{col} \to \infty$ ($\gamma \to \infty$) which corresponds to $v \to c$. Below we shall see that the existence of the upper limit for the collision velocity in general does not enable one to get a satisfactory description of the ion–atom collisions within the first order of the perturbation theory, no matter how high the impact energy is.

The electromagnetic field acting on the electron of the ion is stronger in collisions with heavier atoms. Therefore, in order to see whether the higher-order effects in the ion–atom interaction can 'survive' in the limit $\gamma \to \infty$, we shall consider only collisions with very heavy atoms in which these effects are strongest. At the same time, in the case of very heavy atoms the inelastic atomic mode of the collision is of minor importance and shall be simply neglected.

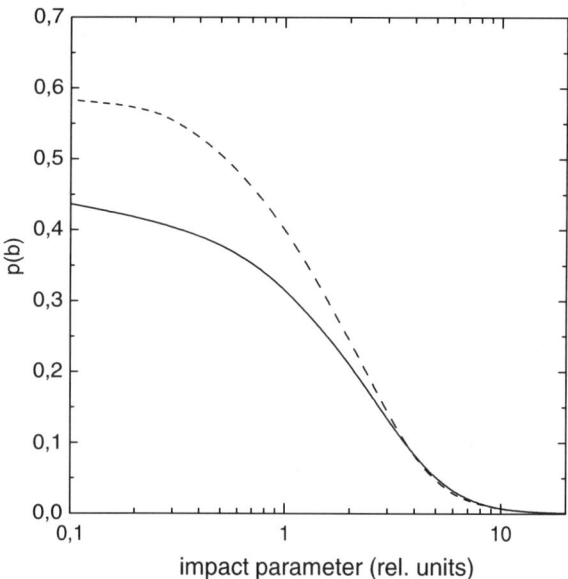

Fig. 7.6. Collisions at $\gamma \to \infty$. The probability for the total electron loss from $Pb^{81+}(1s)$ projectiles colliding with gold atoms ($Z_A = 79$). *Dashed line*: first order result, *solid line*: the light-cone result.

Figure 7.6 shows the probability for the electron loss from Pb^{81+} in collisions with Au at $\gamma = \infty$. Two results for the loss probability are displayed in this figure. One was obtained using the first order theory and considering only the elastic contribution. The second was calculated within the light-cone approximation, discussed in Sect. 6.4. The latter represents an 'exact' solution for $\gamma \to \infty$ and can serve as a reference for the first order calculation.

The deviation from the first order prediction is clearly seen at the impact parameters of the order or smaller than the typical dimension of the electron orbit in the ground state of Pb^{81+}. Yet, these impact parameters are so small that they actually contribute very little to the total loss cross section whose 'exact' and first values are thus very close.

This observation, however, by no means imply that at the asymptotically high energies ($\gamma \to \infty$) the first order and exact loss cross sections always tend to coincide. In the case considered in Fig. 7.6 the closeness in the values of the loss probabilities, observed for almost all impact parameters, is simply the direct consequence of the fact that at $\gamma \to \infty$ even such a heavy target, like Au, does not represent a sufficiently strong perturbation for the electron which is initially very tightly bound by the lead nucleus. Therefore, for a fixed atomic number of the projectile-ion, it is merely a question of how far one has to (or one can) increase the target atomic number in order to see that a noticeable deviation from the first order prediction is present at any impact energy.

The latter point is illustrated in Fig. 7.7. Now the difference between the first order and light-cone loss probabilities has 'survived' also at larger impact parameters such that the corresponding cross sections differ roughly by a factor of 1.16. It is also worth noting that, as is seen in this figure, even at the infinite impact energy the first order approximation still violates the unitarity in collisions with very small impact parameters.

Another possibility to observe a noticeable deviation from the first order predictions is not to increase the atomic number of the target but to decrease the atomic number of the projectile. This is because in a lighter ion the electron is weaker bound which makes the electron behavior more 'vulnerable' to the higher order effects in its interaction with the atomic field.[2]

In Fig. 7.8 the probability for the electron loss from a much lighter ion, $Kr^{35+}(1s)$, in collisions with gold atoms at $\gamma \to \infty$ is presented. Compared to the electron in $Pb^{81+}(1s)$ the electron in $Kr^{35+}(1s)$ is much weaker bound. As a result, now the deviation from the first order prediction for the loss probability in collisions with atoms of gold becomes more substantial which

---

[2] Important to note that, compared to a tighter bound electron, a weaker bound electron is relatively more exposed to the higher order effects only provided the condition of the suddenness of the collision is well fulfilled for both electrons. For instance, in the case of relatively slow collisions considered in Sect. 6.5 the situation is just opposite: the first order approximation is less suitable for the description of a tighter bound electron.

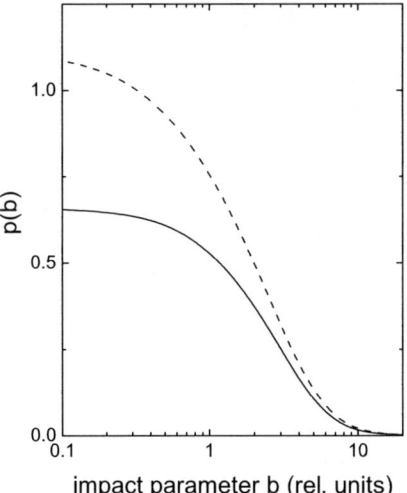

Fig. 7.7. Collisions at $\gamma \to \infty$. The probability for the total electron loss from $\mathrm{Au}^{78+}(1s)$ projectiles colliding with neutral atomic targets having $Z_A = 112$. *Dashed line*: the first order result, *solid line*: the light-cone result.

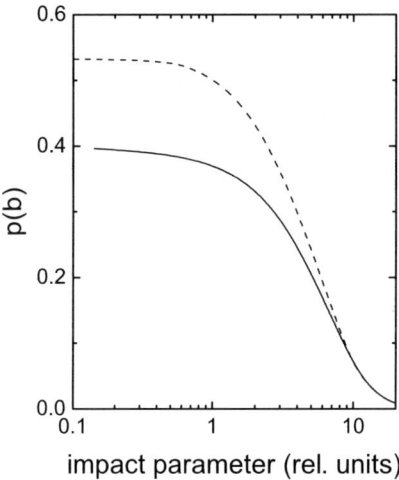

Fig. 7.8. Collisions at $\gamma \to \infty$. The probability for the total electron loss from $\mathrm{Kr}^{35+}(1s)$ projectiles colliding with gold atoms. *Dashed line*: the first order result, *solid line*: the light-cone result.

implies a noticeable difference between the first order and the light-cone cross sections.

By comparing Figs. 7.6 and 7.8 one can also infer that the loss probability in collisions with very small impact parameter is larger for the more tightly bound electron. The latter is suggested by both the first order and the light-cone calculations. It is plausible to assume that the lower values of the loss probability, observed in Fig. 7.8 for the lighter ion,[3] are caused by the larger screening effect of the atomic electrons. Indeed, since the dimension of the electron orbit increases with decrease of the ion charge, the screening of the atomic nucleus by the atomic electrons in collisions with small impact parameters should become more effective in the case of a lighter ion.

One should note that, although we considered the case when $\gamma \to \infty$, the above discussion is not just of academic interest. This is because for the projectile-electron excitation and loss occurring in collisions with neutral atoms the high-energy limit is actually reached at finite (and sometimes relatively modest) values of the impact energy.

In the case of the elastic target mode the latter can be clearly seen in the form of the transition amplitude (7.7). Indeed, at sufficiently high (but finite) values of γ the factor $B_{0,n}^j$ reduces to the screening parameter κ_j. Besides, starting with $\gamma \sim 5$–10 the collision velocity becomes practically equal to the speed of light and does not change when the impact energy increases further. As a result, provided the conditions $\gamma \gg (\varepsilon_n - \varepsilon_0)/(c\kappa_j)$ and $\gamma > 5$–10 are simultaneously fulfilled, the elastic target mode already 'functions' in the asymptotic regime.

Compared to the elastic amplitude (7.7), the form of the inelastic amplitude (7.10) is much more cumbersome and does not enable one to get the idea about the asymptotic limit so easily. However, in the case of the inelastic target mode one can apply the knowledge which we have gained when the two-center dielectronic interaction was considered (see Sect. 5.14.1). In particular, it was shown there that at sufficiently high (and finite) impact energies the cross sections 'saturate' and do no change any more with a further increase in the collision energy. The saturation in the cross sections, of course, also implies that the corresponding probabilities become energy-independent. Thus, both for the elastic and inelastic target contributions to the probabilities of the projectile-electron excitation and loss the asymptotic high-energy limit $\gamma \to \infty$ is in fact reached at finite impact energies.

---

[3] It is of interest to compare the loss probabilities at $b \to 0$ in the case of collisions with neutral atoms and bare atomic nuclei. It was predicted in [103, 158] that for 'ionization' of a hydrogen-like ion with a charge $Z_I > 12$–15 by a nucleus with a charge Z_A at collision energies corresponding to $\gamma > 5$ the ionization(loss) probability at $b = 0$ is very weakly dependent on γ and Z_I.

8

Cross Sections and Comparison with Experiment

8.1 Electron Loss in Collisions at Low γ

As was discussed in Chap. 5 (see Sect. 5.11), under certain conditions the projectile excitation and loss in collisions with neutral atoms can be well approximated as occurring solely due to the interaction with the (unscreened) atomic nucleus since the atomic electrons play just a minor role. This, for instance, is the case for the electron loss from $105\,\text{MeV}\,\text{u}^{-1}$ $\text{U}^{90+}(1s^2)$ projectiles for which the total loss cross sections were measured in [168–171].[1]

Indeed, according to Fig. 6.9, at this impact energy the screening effect of the atomic electrons has a negligible impact on the electron loss process. Besides, since the effective energy threshold for the inelastic target mode in the electron loss from $\text{U}^{90+}(1s^2)$ is about $240\,\text{MeV}\,\text{u}^{-1}$, the antiscreening effect of the atomic electrons in these collisions is also very weak and can safely be neglected not only for many-electron but also for few-electron atomic targets. Thus, the process of the electron loss can really be regarded as an effectively three-body problem and, therefore, one may apply the three-body approximations discussed in Sects. 5.11 and 6.5.

Results of calculations for the total cross section for the single electron loss from $105\,\text{MeV}\,\text{u}^{-1}$ $\text{U}^{90+}(1s^2)$ ions in collisions with different atomic targets ranging between beryllium and gold are presented in Fig. 8.1. The calculations were performed using the first order transition amplitude (5.83) and the distorted-wave amplitudes (6.86), (6.87) and (6.90). In this figure the theoretical results are also compared with experimental data for the loss cross section reported in [168] for collisions of $105\,\text{MeV}\,\text{u}^{-1}$ $\text{U}^{90+}(1s^2)$ with solid state targets of beryllium, carbon, aluminum, copper, silver and gold. In the calculation the electron loss was considered as occurring from the ground state of a hydrogen-like ion whose effective nuclear charge was determined from the binding energy of the electrons in $\text{U}^{90+}(1s^2)$.

---

[1] For a discussion of the electron loss cross sections measured in collisions with relatively light atoms see [172, 173].

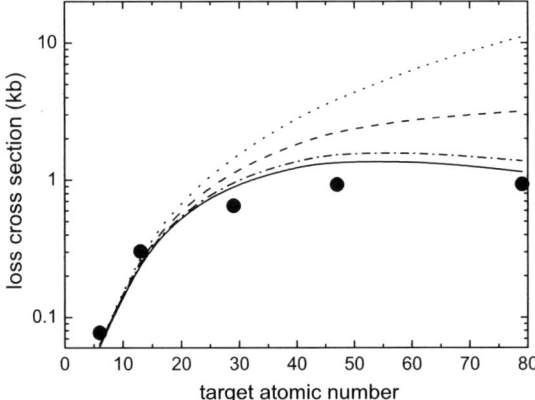

Fig. 8.1. The total cross section (per electron) for the single electron loss from $105\,\mathrm{MeV\,u}^{-1}$ $U^{90+}(1s^2)$ ions colliding with different targets. Circles show experimental results for the loss in collisions with solid state targets of carbon, aluminum, copper, silver and gold which were measured in [168]. *Dot, dash, dash–dot* and *solid curves* display results of calculations using the amplitudes (5.83), (6.86), (6.87) and (6.90), respectively.

It is seen in Fig. 8.1 that in collisions with atomic targets having not very large atomic numbers, for which one has $Z_A/v \ll 1$, all the theoretical models yield very similar loss cross section values. When the ratio Z_A/v increases, the difference between the results of the first order approximation and the distorted-wave models rapidly grows. Compared to the first order result, for collisions with atoms of gold the SEA and CDW-EIS models yield the loss cross sections which are smaller by about a factor of 3.7 and 10, respectively.

A comparison with the experimental data clearly shows a complete failure of the first order approximation. This approximation, predicting the dependence $\sim Z_A^2$ for the loss cross section, does not reproduce the 'saturation' of this cross section, which is clearly visible in the experimental data at $Z_A \gtrsim 40$, and overestimates these data by more than the order of magnitude in collisions with the gold target.

The results obtained with the symmetric eikonal amplitude (6.86) are also not in agreement with the experimental data. Although compared to the first order approximation this amplitude predicts a much weaker growth of the loss cross section with increasing Z_A, at very large Z_A this cross section does substantially overestimate the experimental results.

In contrast, the application of the modified eikonal amplitude (6.87) leads to the results which are much closer to the experimental data. The application of the CDW-EIA model in this case is also quite successful yielding cross section values which are closer to the experimental data than all the other theoretical results.

Taking into account that the experimental data shown in Fig. 8.1 possess a possible systematic error of up to 20% (see [3]), one can say that the results

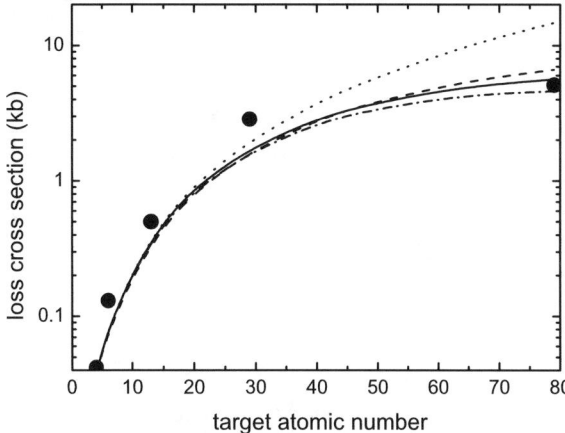

Fig. 8.2. The total cross section for the electron loss from $220\,\mathrm{MeV\,u^{-1}}$ $\mathrm{U^{91+}}(1s)$ ions colliding with different targets. Circles show experimental results measured in [168] for the loss in collisions with solid state targets of beryllium, carbon, aluminum, copper and gold. *Dot, dash, dash–dot* and *solid curves* display results of calculations with the amplitudes (5.83), (6.86), (6.87) and (6.90), respectively.

for the loss obtained with the amplitudes (6.87) and (6.90) are in reasonable agreement with the experiment.

Figure 8.2 displays results for the total cross section for the electron loss from $220\,\mathrm{MeV\,u^{-1}}$ $\mathrm{U^{91+}}(1s)$ ions. The impact energy of $220\,\mathrm{MeV\,u^{-1}}$ is not yet sufficiently large to make the screening effect of the atomic electrons important for the electron loss process. Besides, this impact energy is still below the effective threshold of $240\,\mathrm{MeV\,u^{-1}}$ for the antiscreening collision mode. Therefore, like in the case with $105\,\mathrm{MeV\,u^{-1}}$ $\mathrm{U^{90+}}(1s^2)$ ions, the presence of the atomic electrons can be neglected and again the projectile-electron loss process can be considered using the three-body models.

Similarly to the electron loss from $105\,\mathrm{MeV\,u^{-1}}$ $\mathrm{U^{90+}}(1s^2)$ ions, all the theoretical models again yield very close results for collisions with targets having low atomic numbers $(Z_A/v \ll 1)$. When the atomic number of the target increases, the difference between the results of the first order approximation, on one hand, and the distorted-wave amplitudes, on the other, rapidly increases and reaches about a factor of 2.5–3 for collisions with the gold target. Compared to the case at $105\,\mathrm{MeV\,u^{-1}}$, the smaller differences between the first order and the other results are probably related to the fact that at an impact energy of $220\,\mathrm{MeV\,u^{-1}}$ the effective strength of the interaction between the electron of the ion and the nucleus of the atom is weaker. This point also seems to be responsible for the smaller difference between the predictions of the SEA and CDW-EIS models.

Taking into account the possible systematic error of up to 20% in the experimental data one may conclude the following. First, the first order

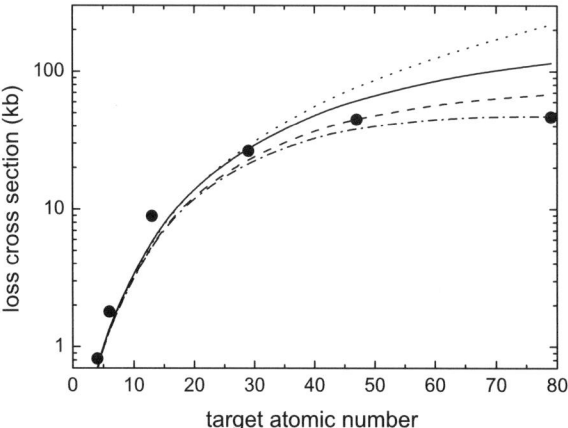

Fig. 8.3. The total cross section for the electron loss from $105\,\mathrm{MeV\,u^{-1}}\,U^{89+}(1s^2 2s)$ ions colliding with different targets. *Circles* show experimental results measured in [168] for the loss in collisions with solid state targets of beryllium, carbon, aluminum, copper, silver and gold. *Dot, dash, dash–dot* and *solid curves* display results of calculations with the amplitudes (5.83), (6.86), (6.87) and (6.90), respectively.

approximation strongly fails also in the case of the loss from $220\,\mathrm{MeV\,u^{-1}}$ $U^{91+}(1s)$ ions. Second, the results obtained with all distorted-wave amplitudes (6.86), (6.87) and (6.90) are in reasonable agreement with the experiment.

The total cross sections for the single electron loss from $105\,\mathrm{MeV\,u^{-1}}$ $U^{89+}(1s^2 2s)$ are shown in Fig. 8.3. Note that at this relatively low impact energy there is a very large difference between the cross sections for the electron loss from the K and L shells. Therefore, following [146], it was assumed in the calculations, results of which are presented in Fig. 8.3, that the loss occurs only from the L shell. The $2s$-electron in the initial and final states of the undistorted ion was described by considering this electron as moving in the Coulomb field of the ionic core (the nucleus plus the two K-shell electrons) whose effective charge was determined from the binding energy of the $2s$-electron in $U^{89+}(1s^2 2s)$.

Compared to the K-shell electrons, the $2s$-electron of the uranium ion is substantially less tightly bound. However, due to the relatively low value of the impact energy, the typical minimum momentum transfer to this electron, which is necessary to remove it from the ion, is still very large on the atomic scale. Therefore, although the screening effect is now substantially larger than in the case of the electron loss from the K-shell, it nevertheless still remains quite modest and can be neglected.[2]

---

[2] According to the first order calculations, the reduction of the loss cross section caused by the screening effect reaches about 11%. This effect is to be compared with the difference of a factor of 2–4 between the first order result and the

The important difference between the previous cases and the electron loss from 105 $U^{89+}(1s^22s)$ ions is that the effective threshold for the antiscreening mode is about 60 MeV u^{-1} and, thus, this mode is now open.

In theoretical results shown in Fig. 8.3 the contribution of the antiscreening mode was estimated by treating this mode within the first order perturbation theory. Note that, since the relative contribution of this mode to the loss cross section scales approximately as $1/Z_A$, the antiscreening effect has to be taken into account in collisions with light targets, like beryllium and carbon, but may be simply neglected in collisions with atoms of silver and gold.

It is seen in Fig. 8.3 that, similarly to the case with 105 MeV u^{-1} $U^{90+}(1s^2)$ ions, both the first order and distorted-wave calculations yield very close cross section values for collisions with atomic targets having low atomic numbers. When the ratio Z_A/v increases the difference between the results of these models rapidly starts to grow. Yet, compared to the case of the electron loss from 105 MeV u^{-1} $U^{90+}(1s^2)$, this difference remains substantially smaller reaching 'merely' a factor of 2–4 for collisions with atoms of gold. The smaller difference could be attributed to the fact that the loss from the L-shell occurs in collisions with smaller momentum transfers corresponding to larger impact parameters, where the interaction between the electron of the ion and the nucleus of the atom is weaker.

An interesting peculiarity in the theoretical data shown in Fig. 8.3 is that now the CDW-EIS model overestimates the loss cross section by up to a factor of 2. Compared to the other distorted-wave calculations, this model now yields the worst agreement with the experiment. This is rather surprising, especially taking into account the good results obtained with the CDW-EIS amplitude for the loss from 105 MeV u^{-1} $U^{90}(1s^2)$ ions. The reasons for this failure are not clear.

In contrast, the application of the amplitude (6.87) yields good agreement with the experiment also in the case of the electron loss from the lithium-like uranium ion.

Compared to the simple first order approach, the more elaborated distorted-wave models are supposed to improve the treatment of the interaction between the electron and the nucleus of the atom. As a result, these models are expected to yield better descriptions for the loss process. As we have just seen, this is indeed the case.

What, however, is important to keep in mind is that, similarly to the first order theory, the distorted-wave models represent in essence high-energy (or high-velocity) approximations which in general have solid grounds only provided the impact energy is 'sufficiently high'.

Indeed, such models, like the CDW-EIS and SEA, were first proposed (and turned out to be very successful) to treat ionization and excitation of very light

---

predictions of the distorted-wave models which do not take the screening effect into account but instead attempt to describe the interaction between the electron of the ion and the nucleus of the atom in a better way.

atomic systems (hydrogen and helium) occurring in collisions with fast bare nuclei whose impact velocities are much larger than the typical velocities of the 'active' atomic electron. In such collisions the range of the 'sufficiently high' impact energies is of course reached.

When we consider the excitation or loss of an electron, which is initially bound by a very strong field (like e.g. in a few-electron uranium ions), even collision energies about $\sim100\,\mathrm{MeV\,u}^{-1}$ can hardly be considered as 'sufficiently high' since the typical collision time is not yet (much) smaller than the typical electron transition time. In such a case results of the distorted-wave models should be taken with due caution.

8.2 Excitation and Simultaneous Excitation-Loss in Collisions at Low γ

8.2.1 Excitation

Compared to the electron loss, the process of the projectile-electron excitation involves smaller changes in the electron energy. This means that the excitation is characterized by smaller values of the momentum transfer or, viewed from a different perspective, occurs in collisions with effectively larger impact parameters. As a result, the excitation process is in general more sensitive to the presence of the atomic electrons.

However, if the projectile-ion has a very high charge and the collision energy is relatively low, the screening effect of the atomic electrons is weak and can be neglected (see for illustration Fig. 8.4). At the same time at such energies the higher-order effects in the projectile–target interaction may become quite important leading to the failure of the first order approximation.

Experimental data for the excitation of relativistic heavy hydrogen-like ions were reported in [95,96] (see also [112]) for collisions of 82 and 119 MeV u^{-1} Bi^{82+}(1s) with solid state targets of carbon, aluminum and nickel. As was mentioned in [96], the accuracy of the experimental data for the 82 MeV u^{-1} projectiles was substantially affected by the electron capture process which, because of the very high projectile charge, is still very strong at this impact energy. Therefore, in what follows we restrict a comparison of the theoretical results only to the excitation of 119 MeV u^{-1} Bi^{82+}(1s).

At this energy the screening effect of the atomic electrons is quite weak. The effective energy threshold for the antiscreening mode in collisions with Bi^{82+}(1s) is about 140 MeV u^{-1}. Therefore, the antiscreening effect is very weak as well and the influence of the atomic electrons on the excitation process can be safely neglected. All this enables one to calculate cross sections by treating the excitation process as a three-body problem.

In Fig. 8.5 results are shown for the excitation of 119 MeV u^{-1} Bi^{82+}(1s) and are given as a function of the atomic number of the target. Figure 8.5 consists of two parts. In part (a) dash and dot curves show results obtained

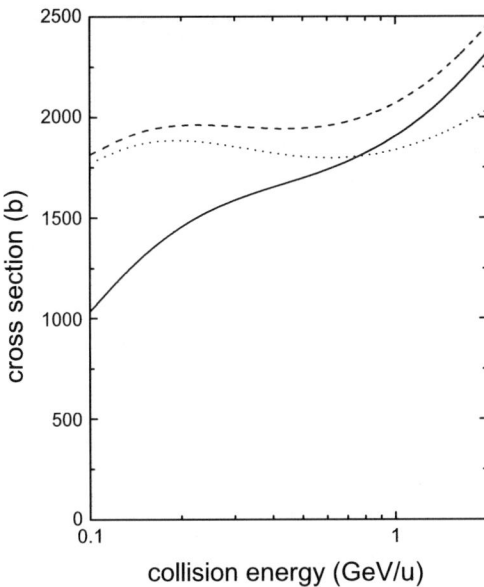

Fig. 8.4. Cross section for the total excitation into the $2p_{3/2}$-states of $Bi^{82+}(1s_{1/2})$ in collisions with krypton ($Z_A = 36$) given as a function of the collision energy. *Dash and dot curves* display results of the first order calculations without and with taking into account the screening effect of the atomic electrons, respectively. Results of the calculation with the amplitude (6.87) are shown by the *solid curve*. From [145].

with the first order amplitude for the excitation of the Bi ion into the states with $n = 2, j = 1/2$ and $n = 2, j = 3/2$, respectively, where n is the principal quantum number and j the total angular momentum of the electron. Dash and dot curves in Fig. 8.5b display results for the same transitions but calculated with the eikonal amplitude (6.87). In both parts of Fig. 8.5 solid curves show the total cross section for the transitions to all the states with $n = 2$ obtained with the corresponding amplitudes. Both these parts also display the (same set of) experimental data from [95, 96].

Similarly to the case with the electron loss, it is seen in the figure that the first order and eikonal transition amplitudes yield very close results for the excitation cross sections in collisions with very light atoms ($Z_A \lesssim 10$) where the interaction of the electron with the nucleus of the atom is weak. In collisions with atoms having larger atomic numbers the difference between the predictions of the first order and eikonal approaches starts to appear. When the atomic number of the target increases further, this difference rapidly increases and reaches almost a factor of 2 at $Z_A \simeq 50$.

Amongst targets, for which experimental data are available, nickel ($Z_A = 28$) has the highest atomic number. For the excitation of $119\,\mathrm{MeV\,u^{-1}}$ $Bi^{82+}(1s)$ in collisions with this target the eikonal calculation predicts the reduction of the excitation cross section approximately by 30% compared to the

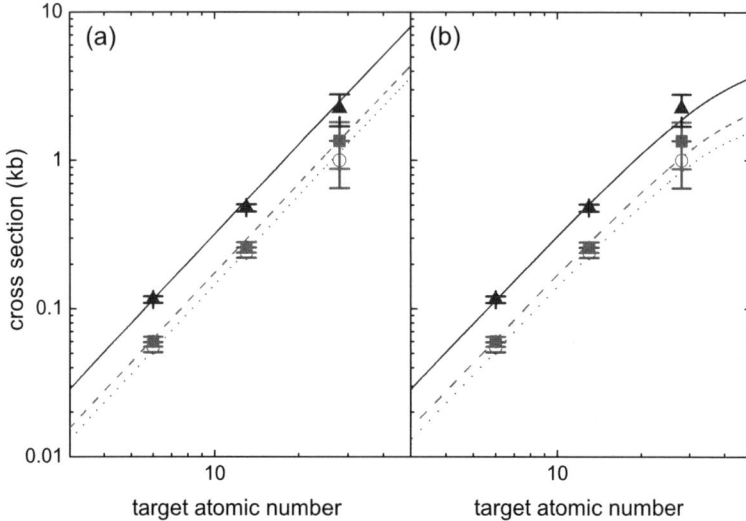

Fig. 8.5. Cross sections for the excitation into states with $n = 2, j = 1/2$, with $n = 2, j = 3/2$ and for the total excitation into $n = 2$-states of $119 \, \mathrm{MeV \, u^{-1}} \, \mathrm{Bi^{82+}}(1s_{1/2})$ in collisions with atomic targets whose atomic numbers run between 1 and 54. **(a)** Results of the first order calculation. *Dashed curve*: $n = 2, j = 1/2$; *dot curve*: $n = 2, j = 3/2$ and *solid curve*: $n = 2$. **(b)** Results obtained with the amplitude (6.87). *Dashed curve*: $n = 2, j = 1/2$; *dot curve*: $n = 2, j = 3/2$ and *solid curve*: $n = 2$. *Circles*, *squares* and *triangles* (with the corresponding *error bars*) in **(a)** and **(b)** display experimental results from [95, 96] for the excitation into the states with $n = 2$ and $j = 1/2$, $n = 2$ and $j = 3/2$, and $n = 2$, respectively, which were measured in collisions with solid state targets of carbon ($Z_A = 6$), aluminum ($Z_A = 13$) and nickel ($Z_A = 28$). From [145].

results of the first order model. However, the uncertainty in the experimental data is also about 30% and both the eikonal and first order calculations are in good overall agreement with the experiment. Thus, the accuracy of the experimental data does not enable one to conclude which calculation describes better the experiment.

In Fig. 8.6 we compare results for the ratio $\sigma(n = 2, j = 3/2)/\sigma(n = 2, j = 1/2)$ where $\sigma(n = 2, j = 1/2)$ and $\sigma(n = 2, j = 3/2)$ are the cross sections for the excitation to the states with $n = 2, j = 1/2$ and $n = 2, j = 3/2$, respectively. The figure shows that, for the excitation to different final states, the deviations between the results, obtained with the first order and eikonal transition amplitudes, are accumulating at a different pace. In particular, this deviation is somewhat stronger for the excitation into the states with $n = 2, j = 3/2$. Note also that, in contrast to the case of the absolute cross sections, for which the available experimental data do not allow us to prefer one of the two calculations, the experimental data for the cross section ratio seem to be more in favor of the results obtained using the amplitude (6.87).

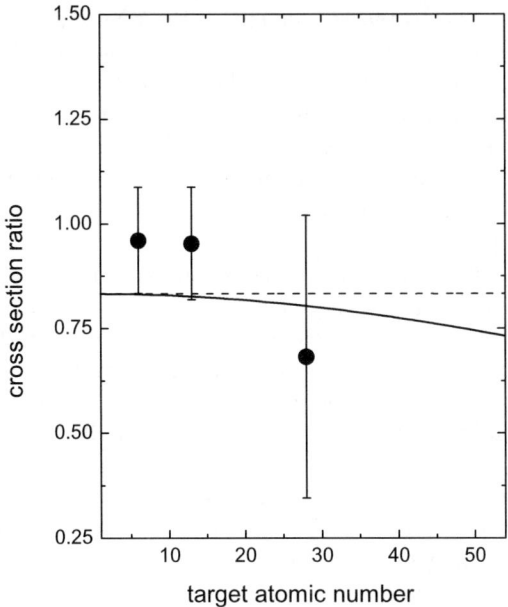

target atomic number

Fig. 8.6. The ratio $\sigma(n = 2, j = 3/2)/\sigma(n = 2, j = 1/2)$ between the cross sections for the excitation of $119\,\mathrm{MeV\,u^{-1}}$ $\mathrm{Bi^{82+}}(1s_{1/2})$ into the states with $n = 2, j = 1/2$ and $n = 2, j = 3/2$. The ratio is given as a function of the atomic number of the target. *Dashed line*: results of the first order calculation. *Solid curve*: results obtained with the amplitude (6.87). *Circles with error bars*: experimental data from [95] and [96]. From [145].

8.2.2 Simultaneous Excitation-Loss

If a heavy ion initially carries several electrons, then more than one electron of the ion can be simultaneously excited and/or lost in a collision with a neutral atom. In [97] simultaneous excitation and loss of the projectile electrons was investigated experimentally for $223.2\,\mathrm{MeV\,u^{-1}}$ $\mathrm{U^{90+}}$ ions impinging on atomic targets of Ar, Kr and Xe. In the collision between the projectile-ion and the target-atom one of the two electrons of $\mathrm{U^{90+}}$ was ejected and the other was simultaneously excited into the L-subshell states of $\mathrm{U^{91+}}$. Note that such a process represents one of the simplest and basic processes which can occur with projectiles having initially more than one electron.

There are essentially two qualitatively different possibilities to get the simultaneous two-electron transitions. The first is that the atom in the collision effectively interacts with only one electron inducing its transition while the second electron undergoes a transition due to the electron–electron-correlations in the ion and/or due to rearrangement in the final state of the ion.[3] These processes are often called the two-step-1 and shake-off, respectively (see e.g. [2]).

The other possibility is that the field of the atom has enough power to interact simultaneously with each of the two electrons and to become the main driving force for both electrons to undergo transitions. In case when the interaction with the atomic nucleus involves the exchange of only two virtual photons (one per electron) such a process is often referred to as the two-step-2 process (see e.g. [2]).

It was shown in [174] that, provided the condition $\frac{Z_I Z_A}{v} > 0.4$ is fulfilled (where Z_I and Z_A are the charges of the ionic and atomic nuclei, respectively, and v the collision velocity), two-electron transitions in a heavy helium-like ion occurring in collisions with an atom are governed mainly by the 'independent' interactions between the atom and each of the electrons of the ion. This condition, in particular, was very well fulfilled in the experiment [97].

When the electron transitions are governed by the independent interactions the application of the independent electron model often yields reasonable results. According to this model the cross section for the simultaneous loss-excitation is evaluated as

$$\sigma = 2\pi \int_0^\infty \mathrm{d}b b P(b), \tag{8.1}$$

where the probability $P(b)$ for the two-electron process is given by

$$P(b) = 2\, P_{\mathrm{exc}}(b)\, P_{\mathrm{loss}}(b). \tag{8.2}$$

Here, $P_{\mathrm{exc}}(b)$ and $P_{\mathrm{loss}}(b)$ are the single-electron excitation and loss probabilities, respectively, in a collision with a given value of the impact parameter b.

Considering transitions of a single electron, which was initially very tightly bound by the field of a highly charged nucleus, we have already seen that in collisions with many-electron atoms, provided the collision energy is not too large, these transitions are caused mainly by the interaction between the electron of the ion and the nucleus of the atom. Compared to single-electron transitions the two-electron processes are characterized by larger momentum and energy transfers and, thus, by even smaller impact parameters. Therefore, when considering the simultaneous loss-excitation one can also neglect the presence of the atomic electrons.

Thus, in the simplified picture of the ion–atom collision, sketched in the above paragraphs, the theoretical treatment of the process of the simultaneous loss and excitation is reduced to the finding of the single-electron transition

---

[3] The second electron 'tries' to adjust its wave function to the Hamiltonian which was 'suddenly' changed because of the rapid removal of the first electron which leads to the population of excited states of the new Hamiltonian.

probabilities within the three-body problem in which a relativistic electron is moving in the external electromagnetic fields generated by the nuclei of the colliding ion and atom.

In addition to the experimental data the authors of [97] also reported results of their calculations for the simultaneous excitation-loss cross sections. These calculations were performed by using (8.1) and (8.2) and evaluating the single-electron transition probabilities $P_{exc}(b)$ and $P_{loss}(b)$ in the first order interaction between the electron of the ion and the nucleus of the atom.

However, this two-electron process occurs at effectively very small impact parameters, where the field of the atomic nucleus acting on the electrons of the ion may be quite strong and where the first order results for the transition probabilities may be not very reliable. Indeed, the calculations and the experimental data of [97] were in agreement only for collisions with Ar atoms while for collisions with Kr and, especially, Xe target the theoretical results of [97] very substantially overestimated the experimental data.

Recently, the problem of the simultaneous excitation and loss of the projectile electrons in collisions of $223.2 \, \text{MeV} \, \text{u}^{-1}$ U^{90+} ions with atoms was considered theoretically in [175]. The consideration of [175] was also based on the independent electron model. However, in addition to using the first order theory, the transition probabilities were also evaluated in [175] by using the distorted-wave models. The results of [175] are shown in Fig. 8.7 where they are compared with the experimental data from [97]. It is clearly seen in the figure that, compared to the first order calculations, the distorted-wave models are capable of a much better description of the experimental data.

8.3 Electron Loss in Collisions at Moderately High γ

Unless the range of asymptotically high energies is reached, where the higher order and screening effects 'saturate' (see Sects. 8.9 and 8.11), the increase in the impact energy diminishes the influence of the higher-order effects in the interaction between the projectile-electron and a neutral atomic target on the electron transitions but increases the role of the atomic electrons (the screening effect).

In experiment of Claytor et al. [176] the loss cross section was measured for incident $10.8 \, \text{GeV} \, \text{u}^{-1}$ ($\gamma = 12.6$) $Au^{78+}(1s)$ ions penetrating different solid state targets. At this collision energy the typical impact parameters $b \lesssim b_{max} = \gamma v / \omega_{eff}$ (v is the collision velocity and $\omega_{eff} \sim Z_I^2$ is the averaged transition frequency), which would give the main contribution to the cross sections for the total electron loss in collisions with an unscreened atomic nucleus, are already much larger than the size of the inner shells in very heavy elements.[4] This means that in collisions with very heavy targets (like

---

[4] Although still substantially smaller than 1 a.u.

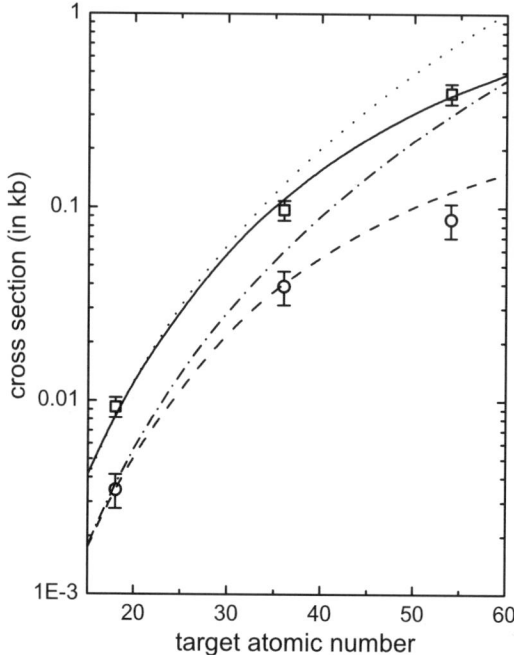

Fig. 8.7. Total cross sections for the reactions $223.2\ \mathrm{U}^{90+}(1s^2) + Z_t \rightarrow \mathrm{U}^{91+}(n =$
$2, j) + e^- + \cdots$, where $j = 1/2$ and $j = 3/2$ are the angular momentum of the
L-shell states of the hydrogen-like uranium ion. The cross sections are given as a
function of the atomic number Z_t of the target atom. *Circles* and *squares* with
the corresponding error bars are experimental data for the $j = 3/2$ and $j = 1/2$
cases, respectively, reported in [97] for collisions with argon, krypton and xenon
gas targets. *Dot* $(j = 1/2)$ and *dash-dot* $(j = 3/2)$ *curves* show the cross sections
calculated with the single-electron transition probabilities obtained in the first order
of perturbation theory in the interaction between the electron and the nucleus of the
atom. *Solid* $(j = 1/2)$ and *dash* $(j = 3/2)$ curves display theoretical results obtained
using the symmetric eikonal approximation to estimate the excitation probability
and the CDW-EIS approximation to calculate the loss probability.

e.g. gold) the influence of the screening effect of the atomic electrons on the
electron loss process has to be taken into account and the three-body models,
used in the previous sections, should no longer be applied.

The targets used in the experiment [176] were ranging from carbon to
gold. Taking into account that the influence of the higher order effects on
the total electron loss in collisions with these targets is obviously strongest
for the gold target, the upper boundary for this influence can be estimated
by considering $10.8\,\mathrm{GeV\,u^{-1}}\ \mathrm{Au}^{78+}(1s) + \mathrm{Au}$ collisions and by comparing
the first order result with that obtained by using the method described in
Sect. 6.4.5. The 'exact' cross section obtained in such a way turned out to be
just several percent less than the first order result.

Simple estimates show that an impact energy of $\simeq 10\,\mathrm{GeV\,u^{-1}}$ is not yet sufficiently high to expect the method of Sect. 6.4.5 to be a very good approximation for considering the loss from very heavy ions like $\mathrm{Au^{78+}}(1s)$. As a result, such a comparison cannot yield a very precise description of the change in the first order loss cross section caused by the higher order effects.

According to the first order calculation, the screening effect reduces the loss cross section by $\simeq 30\%$ compared to the result obtained for the unscreened atomic nucleus. In collisions with heavy atoms at this impact energy the screening effect of the atomic electrons is obviously more important than the higher order effects in the interaction between the electron of the projectile and the nucleus of the atom. Therefore, in what follows in this section, we shall simply restrict our attention just to the first order results.

The first order consideration is based on (5.72). In order to compare theoretical results with the experiment, the cross section (5.72) has to be summed over all possible states of the target. The resulting loss cross section can be split into the elastic (screening, $m = 0$) and inelastic (antiscreening, all $m \neq 0$) target contributions. In the gauge, where the scalar potential of the atomic field in the rest frame of the ion is chosen to be zero, the screening part reads

$$\sigma^{\mathrm{s}} = \frac{4\,c^2}{v^2 \omega_{k0}^2} \int d^2 q_\perp \frac{\left|Z_{\mathrm{A,eff}}(q_0^{\mathrm{A}})\right|^2}{\left(\left(q_0^{\mathrm{I}}\right)^2 - \frac{\omega_{n0}^2}{c^2}\right)^2}$$

$$\times \left|\left\langle \psi_k(r)\left|\exp\left(iq_0^{\mathrm{I}} \cdot r\right)\left(q_x \alpha_x + q_y \alpha_y + \frac{\omega_{k0}}{v\gamma^2}\alpha_z\right)\right|\psi_0(r)\right\rangle\right|^2, \quad (8.3)$$

where $\omega_{k0} = \varepsilon_k - \varepsilon_0$, $q_0^{\mathrm{I}} = (q_\perp, \omega_{k0}/v) = (q_x, q_y, \omega_{k0}/v)$, $q_0^{\mathrm{A}} = (-q_\perp, -\omega_{k0}/v\gamma)$. The final continuum states are normalized according to $\langle \psi_k | \psi_{k'} \rangle = \delta^3(k - k')$. It is implied in (8.3) and throughout the section that an averaging over the initial and sum over the final spin states of the electron of the ion is performed. The effective charge $Z_{\mathrm{A,eff}}$ of the atom in the ground state is defined by (5.70).

Using the closure approximation in order to sum over all excited final states of the atom (including the atomic continuum), the antiscreening contribution to the loss cross section can be written as

$$\sigma^{\mathrm{a}} = \frac{4\,c^2}{v^2 \omega_{k0}^2} \int d^3 k \int d^2 q_\perp \frac{S(q_1^{\mathrm{A}})}{\left(q_\perp^2 + \frac{(\omega_{k0}+\Delta\epsilon)^2}{v^2\gamma^2} + 2(\gamma - 1)\frac{\omega_{k0}+\Delta\epsilon}{v^2\gamma^2}\right)^2}$$

$$\times \left|\left\langle \psi_k(r)\left|\exp(iq_1^{\mathrm{I}} \cdot r)\left(q_x \alpha_x + q_y \alpha_y + \frac{1}{\gamma}q_{\mathrm{min},1}^{\mathrm{A}}\alpha_z\right)\right|\psi_0(r)\right\rangle\right|^2. \quad (8.4)$$

In (8.4)

$$q_1^{\mathrm{I}} = \left(q_\perp, \frac{\omega_{k0}}{v} + \frac{\Delta\epsilon}{v\gamma}\right) = \left(q_x, q_y, \frac{\omega_{k0}}{v} + \frac{\Delta\epsilon}{v\gamma}\right),$$

$$q_1^{\mathrm{A}} = \left(-q_\perp, -q_{\mathrm{min},1}^{\mathrm{A}}\right)$$

and

$$q^{A}_{min,1} = \frac{\Delta\epsilon}{v} + \frac{\omega_{k0}}{v\gamma},$$

where $\Delta\epsilon$ is the mean excitation energy for transitions of atomic electrons, and

$$S(q) = \sum_{m\neq 0} \left| \langle u_m(\tau) \left| \sum_{j=1}^{Z_A} \exp(i q \cdot \xi_j) \right| u_0(\tau) \rangle \right|^2 \tag{8.5}$$

is the so called incoherent scattering function. These functions are tabulated in [177, 178] for all atomic elements. The mean energy, which is used in calculations of the stopping power and is tabulated for a variety of atoms (see e.g. [179]), has been taken as the mean excitation energy, $\Delta\epsilon$.

Figure 8.8 shows a comparison between the experimental data of [176] and different theoretical results. The theoretical results include those of [84, 86] as well as results obtained by using (8.3) and (8.4).

The results of [84] are based on the theory for the projectile-electron loss proposed in [80, 81, 84] and the semi-relativistic description of the initial and

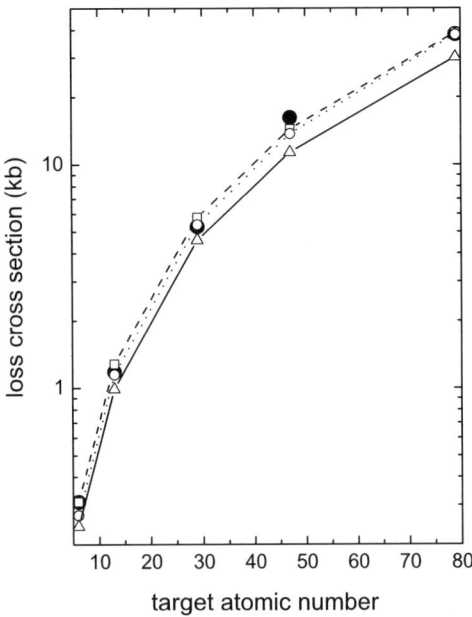

Fig. 8.8. Cross section for the electron loss from $10.8\,\mathrm{GeV\,u^{-1}}$ $Au^{78+}(1s)$ in collisions with neutral atoms of C, Al, Cu, Ag and Au. *Solid circles*: experimental data from [176]. *Open squares* connected by *dash lines*: results of [84]. *Open circles* connected by *dot lines* are results of [86]. *Open triangles* connected by *solid lines*: results of the calculation with (8.3) and (8.4). Lines connecting the theory points are intended just to guide the eye. For more explanations see the text.

final electronic states in the gold ion. The results of [86] are cross section estimates using the simple method which was briefly outlined in Sect. 5.1. In the calculations with (8.3) and (8.4) the ground and continuum states of the electron in Au^{78+} were described by the relativistic (Coulomb–Dirac) wave functions, the atomic screening coefficients for the effective charge $Z_{A,\text{eff}}$ were taken from [102].

It is seen in the figure that the results of [84, 86] are rather close to each other and agree well with the experimental data. At the same time the results obtained using (8.3) and (8.4) are noticeably lower than the other theoretical predictions and the experiment.

Concerning the comparison of the different calculations, it is quite expected that the results obtained with (8.3) and (8.4) turned out to be noticeably lower than those of [84].[5] Indeed, compared to the semi-relativistic description of the electron, its fully relativistic treatment yields smaller loss cross sections. Besides, the theory of [84] predicting the dependence $\sigma_{\text{loss}} \sim A + B \ln \gamma$ largely ignores the relativistic peculiarities (discussed in Sect. 5.9.1) in the screening effect which also leads to an overestimation of the cross section.

What is more surprising is that the most rigorous set of theoretical data has the largest deviations from the experimental results. The most probable reason for this is that the calculations have been done for the projectiles colliding with atoms whereas in the experiment the projectiles were stripped in solids. When penetrating solids the projectile may suffer multiple collisions. As a result, the electron transitions to the continuum states can proceed not only directly from the ground state but also via intermediate excited bound states. Since cross sections for the electron loss from excited states are larger than that for the loss from the ground state, the population of the intermediate states effectively enhances the loss process.

We shall postpone the discussion of the influence of the multiple collisions with atoms inside solids on the charge state of the projectiles to Sect. 8.6, where much higher impact energies are considered for which this influence is already rather strong.

8.4 Collisions at High γ: Electron Loss and Capture Cross Sections

In this and the next sections we consider in some detail the cases studied experimentally in [85, 180], where a much higher collision energy was considered. In these experiments the loss and capture cross sections were measured for Pb^{81+} and Pb^{82+} ions incident on solid and gas targets at an impact energy of 33 TeV, where the corresponding collisional Lorentz factor is very high ($\gamma = 168$).

---

[5] As was discussed in Sect. 5.1, the method of [86] involves a number of approximations whose accuracy is not very clear. Therefore, it is not easy to find out why results of this method are (or are not) in agreement with other calculations.

Explorations of ion–atom collisions at such very high impact energies are of special interest because during the interaction between a projectile-ion and a target-atom both these particles can be exposed to extremely intense and extraordinarily short pulses of the electromagnetic field.

For instance, in collisions of 33 TeV $Pb^{81+}(1s)$ ions Au atoms the typical duration of the electromagnetic pulses, generated by the atoms in the rest frame of the ions, are $\lesssim 10^{-21}$ s. The peak intensities of these pulses reach $\sim 10^{31}$ W cm^{-2}.[6] These intensities are so large that such pulses, despite the very short interaction time, not only enable one to induce transitions between electron states with positive energies leading to the excitation of and the electron loss from very tightly bound $Pb^{81+}(1s)$ ions but also cause transitions between negative and positive states resulting in the electron–positron pair production with quite noticeable cross sections.

8.4.1 Electron Loss Cross Sections

In Fig. 8.9a shown are experimental data from [180] on the electron loss cross sections measured for 33 TeV Pb projectiles ($\gamma = 168$) which were penetrating three gas targets (Ar, Kr and Xe). The loss cross sections were measured for the 'ionization' scenario, in which the incident projectiles were $Pb^{81+}(1s)$ ions. These cross sections were also obtained in the 'capture' scenario, in which a beam of initially bare Pb^{82+} nuclei was traversing the same targets: in the latter case the electron had to be captured into a bound state of the projectile before it could be lost in a consequent collision. The experimental loss data extracted in both scenarios are rather close to each other.

Figure 8.9a also shows results of calculations for the loss from 33 TeV $Pb^{81+}(1s)$ projectiles for a more extended set of atomic targets (Be, C, Al, Ar, Cu, Kr, Ag, Sn, Xe and Au). These calculations were performed by using (8.3) and (8.4) in which the initial and final electron states were described by the Coulomb–Dirac wave functions.

In addition to the results, already shown in Fig. 8.9a, 8.9b contains also experimental data from [85] as well as theoretical results from [84] for the electron loss from the ground state of the hydrogen-like lead ions. In the experiment of [85] beams of incident 33 hydrogen-like ('ionization' scenario) and bare ('capture' scenario) lead ions were penetrating solid state targets. The figure presents results for the loss cross sections extracted from the measurements carried out in both scenarios.

Two main conclusions can be drawn from the Fig. 8.9. First, according to Fig. 8.9a, there is a good agreement between the experimental data obtained for the gaseous targets and the results of the calculation using (8.3) and (8.4). Second, the experimental loss cross sections measured in collisions with solid

---

[6] These field parameters may be compared with the parameters of state of the art laser systems whose shortest pulse lengths are about $\sim 10^{-16}$–10^{-15} s and whose peak intensities do not exceed $\sim 10^{21}$–10^{22} W cm^{-2}.

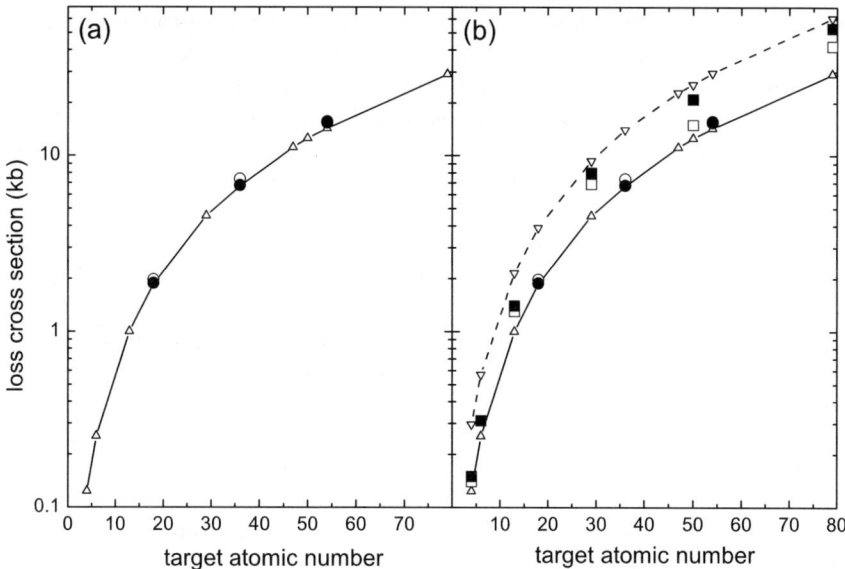

Fig. 8.9. Cross sections for the electron loss from 33 TeV $Pb^{81+}(1s)$ projectiles. (**a**) *Circles* represent experimental data from [180] on the electron loss in gas targets (Ar ($Z_A = 18$), Kr ($Z_A = 36$) and Xe ($Z_A = 54$)) where the *open* and *solid* symbols refer to the 'ionization' and 'capture' experimental scenarios, respectively. *Up triangles* connected by guiding *solid line* are results of the calculation with (8.3) and (8.4) for $Z_A = 4$, 6, 13, 18, 29, 36, 47, 50, 54 and 79. (**b**) *Circles* and *up triangles*: same as in the part (**a**) of the figure. *Squares* show the experimental data from [85] on the electron loss in solid state targets (Be, C, Al, Cu, Sn and Au). *Down triangles* connected by guiding *dash line* display theoretical results of [84]. Note that both in (**a**) and (**b**) the *open* and *solid* symbols denoting the experimental data refer to the 'ionization' and 'capture' experimental scenarios, respectively. See the text for more explanations.

targets are substantially larger that those obtained for gas targets. Besides, concerning the two theoretical calculations presented in the figure, one has to note that at this very high impact energy the results of [84] are already about a factor of 2 larger than those of the more rigorous treatment (and, thus, the difference is even much more pronounced than for the loss from $10.8\,\mathrm{GeV\,u^{-1}}$ $Au^{78+}(1s)$).

8.4.2 Electron Capture from Pair Production

In the Dirac sea picture the electron–positron pair production is considered as a transition of a negative-energy electron into a state with a positive energy. If this final state is in the continuum, the process is termed as free pair production. The final state can also be a bound state of the ion which means that the electron is created directly in the bound state changing the net charge of the ion. Such a process is called bound-free pair production.

In contrast to the free pair production, not much attention has been paid to this process until the eighties when it had began to attract great interest. The latter was caused, in particular, because of the importance of this process for the design and operation of relativistic heavy-ion colliders.[7]

The pair production processes have been considered in very many papers. For detailed discussions of the different aspects of these processes occurring in collisions between bare nuclei one can refer, for instance, to [3,5,75,182–184] where also many references to original papers can be found. In particular, starting with the first studies [185,186] the bound-free pair production in high-energy nuclear–nuclear collisions has been a subject of extensive theoretical research.

There have also been several experimental studies on the bound-free pair production, in which the total cross sections for this process were reported [85,180,187–190]. In these experiments bare ions (La^{57+}, Au^{79+}, Pb^{82+}, U^{92+}) were incident on solid foils of different chemical elements ranging from beryllium to gold. In these experiments also a very broad interval of impact energies was considered starting with relatively low energy collisions ($\sim 0.5\,\mathrm{GeV\,u^{-1}}$) up to extreme relativistic collisions ($\sim 160\,\mathrm{GeV\,u^{-1}}$ Pb^{82+}).

Ions penetrating foils 'see' atomic nuclei surrounded by electrons and the process of the pair production may be better regarded as occurring in collisions with neutral atoms rather than with the bare atomic nuclei. In ion–atom collisions the bound-free pair production may be influenced by the presence of atomic electrons. Since this influence increases when the impact energy increases, below we shall restrict our attention to the discussion of the bound-free pair production by incident 33 TeV Pb^{82+} ions.

Taking into account the close analogy between the projectile-electron excitation/loss and the pair production, which was briefly discussed in Sect. 5.13, the cross section for the pair production occurring in ion–atom collisions can be calculated using the straightforward modifications of (8.3) and (8.4). For instance, within the elastic atomic mode the cross section for the pair production in the rest frame of the ion is given by

$$\sigma^s = \frac{4\,c^2}{v^2 \omega_{k0}^2} \int \mathrm{d}^2 q_\perp \frac{\left| Z_{\mathrm{A,eff}}\left(q_0^{\mathrm{A}} \right) \right|^2}{\left(\left(q_0^{\mathrm{I}} \right)^2 - \frac{\omega_{k0}^2}{c^2} \right)^2}$$

$$\times \left| \left\langle \psi_0 \left| \exp\left(i q_0^{\mathrm{I}} \cdot r \right) \left(q_x \alpha_x + q_y \alpha_y + \frac{1}{\gamma^2} q_{\mathrm{min},0}^{\mathrm{I}} \alpha_z \right) \right| \psi_k^{(p)} \right\rangle \right|^2 , \quad (8.6)$$

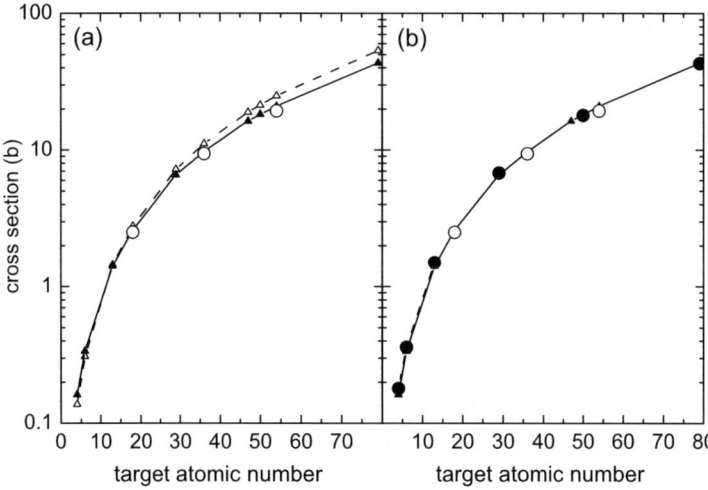

Fig. 8.10. Cross sections for the electron capture from the pair production in collisions of 33 TeV Pb^{82+} projectiles with gas and solid state targets given as a function of the target atomic number. **(a)** *Open circles* are experimental data from [180] for collisions with Ar, Kr and Xe gas targets. *Solid triangles* connected by *solid curve* are results of the calculations for collisions with atoms having atomic numbers $Z_A = 4$, 6, 13, 18, 29, 36, 47, 50, 54 and 79. *Open triangles* connected by *dash curve* are results for the pair production in collisions with the bare atomic nuclei. The curves are just to guide the eye. **(b)** *Open circles* and *solid triangles* connected by *solid curve* represent the same results as in **(a)**. *Solid circles* are data from [85] obtained for collisions with solid state targets (Be, C, Al, Cu, Sn and Au). From [191].

where ψ_0, as before, refers to a bound state of the electron having a total energy ε_0 but now $\psi_{\boldsymbol{k}}^{(p)}$ is an outgoing continuum state of the positron with a momentum $\boldsymbol{k}$ and total energy $\varepsilon_k > 0$, the transition frequency is given by $\omega_{k0} = \varepsilon_k + \varepsilon_0$ and the minimum momentum transfer in the rest frame of the ion reads $q_{\min,0}^{I} = \omega_{k0}/v$.

In Fig. 8.10 we show results for the total cross section for the electron capture via the pair production by 33 TeV Pb^{82+} projectiles incident on different targets. At this very high impact energy the pair production represents the main capture mechanism: its contribution to the capture amounts from $\simeq 60\%$ in collisions with Be atoms up to $\simeq 96\%$ in collisions with Au atoms. Experimental data for these cross sections were reported in [85, 180] for collisions with solid state and gas targets, respectively.

The theoretical cross sections shown in the figure were obtained using (8.6) and also the analog of (8.4) for the pair production. These cross sections were calculated in [191] using the Coulomb–Dirac wave functions for the states of the electron and positron. In the rest frame of the ion the energy spectrum of the positrons extends to tens of mc^2. Such positrons carry in general a big amount of angular momentum which makes it necessary to take into account continuum states with large values of the angular quantum number κ.[8] Theoretical results shown in the figure were obtained by directly integrating over the interval of the total positron energies $[mc^2; 30\,mc^2]$ and summing over the positron angular momenta corresponding to $\kappa \leq \kappa_{\max} = 30$. The contributions from the positron states with higher energy and/or larger κ were evaluated by an extrapolation. The capture cross sections were calculated to all bound states with the principal quantum number $n \leq 6$. According to our estimates, the capture into the states with larger n is negligible and can safely be neglected. In the figure only the total capture cross section is displayed.

Calculations in the light-cone approximation show that the pair production in collisions between 33 TeV Pb^{82+} and atoms ranging between Be and Au can be well described within the first order theory in the interaction between the lepton transition current and the field of the atom. The reasons for this are the very high values of the impact energy and of the projectile charge so that even such a heavy atom like gold still represents an effectively weak perturbation.

The calculations for the capture cross section were done assuming the single-collision condition. Therefore, they should be first of all compared with the experimental results obtained for collisions with gas targets. Such a comparison is shown in Fig. 8.10a where a good agreement is seen between the experiment and theory. Moreover, in contrast to the experimental loss cross sections, the experimental capture cross sections in collisions with both solids and gases fall on the same curve. Therefore, in contrast to the calculated loss cross sections, the calculated cross sections are in agreement also with the experimental data for solid targets.

Comparing calculated results for the pair production with neutral atoms and the corresponding bare atomic nuclei we see that the atoms are more effective at smaller values of Z_A. This reflects the relative importance of the contribution from the antiscreening mode. In the case of collisions with atoms having small Z_A the antiscreening overcompensates a small decrease in the cross section caused by the screening of the atomic nucleus by the atomic electrons inherent to the elastic atomic mode. For heavy targets the screening effect becomes stronger while the antiscreening contribution decreases (in relative terms) which leads to the reduction of the capture cross section compared to the case with the bare nuclei.

Compared to the electron loss, the pair production with capture involves much larger momentum transfers. Therefore, the screening effect of the atomic electrons is much weaker. For instance, for the pair production in collisions of 33 TeV Pb with Be and Au atoms this effect reduces the cross section for the capture to the ground state by about 5% and 24%, respectively.

Note also that the magnitude of the screening effect obtained in the calculations, which employ the exact Coulomb–Dirac wave functions, turns out

---

[8] The quantum number κ is an eigenvalue of the operator defined in (4.54), see Sect. 4.3.

to be somewhat larger compared to that reported in the calculation of [192], where the Darwin and Furry wave functions were used. This increase can be attributed to the fact that, compared to the semi-relativistic, the fully relativistic treatment of the pair production predicts an enhancement in the emission of positrons with the intermediate energies ($\sim$1–20 mc^2 in the rest frame of the ion), whereas according to the semi-relativistic description there should be more positrons with higher energies. Since the creation of more energetic positrons implies larger values of the momentum transfers in the collision, the screening effect tends to be weaker in the semi-relativistic consideration.

The magnitude of the screening effect for the pair production may be compared to the corresponding reduction by a factor of about 1.4 and 2 due to the screening effect of the atomic electrons in the elastic target mode in the case of the electron loss from the ground state of Pb^{81+} in collisions with the same atoms. It is obvious that such large differences in the magnitude of this effect are caused by the fact that, compared to the electron loss process, the pair production involves much larger momentum transfers (or, in other words, proceeds at much smaller impact parameters).

8.5 Screening Effects in Free–Free Pair Production

Compared to the bound-free pair production, the free (or free–free) electron–positron pair production, in which both the leptons are created in the continuum states, involve larger momentum transfers. Therefore, if this process occurs in collisions between a nucleus and a neutral atom, one may expect that the screening effect of the atomic electrons should be smaller.

The value of this effect was estimated in [75] by using the Weizsäcker–Williams approximation of equivalent photons [88, 89]. The main conclusions of [75] were: (a) the screening effect is important at all energies (where the Weizsäcker–Williams approximation is valid) reducing the free pair production by at least a factor of 1.5–2 and (b) the screening effect decreases when the collision energy increases.

Both the size and the energy dependence of the screening effect predicted in [75] were rather unexpected. They were not confirmed by a later theoretical study performed in [193]. In the latter paper the free pair production in nucleus-atom collisions was also calculated in the lowest order of perturbation theory (but without using the Weizsäcker–Williams approximation). The authors of [193] found that the screening effect increases with increasing the collision energy and, unless one considers extreme high impact energies, is rather modest.

In particular, the reduction of cross sections by the screening effect found in [193] is substantially lower compared to what we discussed for the bound-free pair production. According to [193] in Au^{79+} on Au^0 collisions the screening effect reduces the cross section for the free pair production by 4.5% at a collision energy of $E = 200\,GeV\,u^{-1}$, by 13.5% at $E = 2\,TeV\,u^{-1}$ and by 31.4% at $E = 200\,TeV\,u^{-1}$.

8.6 Charge States of 33 TeV Pb Projectiles Penetrating Solid Targets: Multiple Collision Effects

As we have seen in Sect. 8.4, the theoretical results for the projectile-electron loss cross sections agree quite well with the experimental data obtained for collisions with gas targets but substantially underestimate the experimental data reported for collisions with solid state targets.

A possible reason for this might be excitations suffered by the projectiles when they penetrate solids. Compared to the cross section for the loss from the ground state, the cross sections for the loss from excited states are larger. Therefore, if the beam of the hydrogen-like projectiles can attain, due to collisions with atoms of the foil, a noticeable fraction of excited-state ions, the cross section for the electron loss from such a beam may effectively become larger.

Yet, for quite a long time such a possibility was not considered seriously. Indeed, in the high-energy experiments [85, 180] quite noticeable differences between the loss cross sections measured in gases and solids were found already for collisions with light elements (see Fig. 8.9). However, according to the experience (see e.g. [198]) which was gained in the investigations of relativistic collisions at comparatively low energies, excitations of very heavy hydrogen-like ions inside thin foils of relatively light elements were expected to have no substantial impact on the electron loss process.

The reasons for this expectation were: (i) in such collisions the excitation cross sections are quite small and (ii) the lifetimes of the excited states of such ions with respect to the radiative decay are very short. Both (i) and (ii) certainly play not in a favor of the role of the excitations. In particular, these two points strongly diminish the possible role of the excitations of the heavy ions compared to the case when much lighter projectile-ions would be involved. Moreover, even in the latter case the effect caused by the excitations was quite noticeable but not dramatic.[9]

The different aspects of the penetration of fast ions through solids have been studied for several decades [195, 196]. In particular, there has been quite an extensive research, both experimental and theoretical, aimed at exploring the influence of solid state effects on the charge states of and the electron emission from energetic ions penetrating solid targets and the differences between the processes of the penetration of solid and gas targets (for recent references see [194, 197–203]). One has to note, however, that all these studies dealt with the domain of relatively low impact energies where the corresponding collisional Lorentz factor was still rather close to 1 ($\lesssim 1.5$).[10]

---

[9] For instance, in the experimental–theoretical study [194] on $200\,\mathrm{MeV\,u^{-1}}$ $Ni^{27+}(1s)$ ions incident on gaseous and solid targets it was found that the fraction of the ions excited inside the solids is about 5–6%.

[10] We let apart numerous papers which explore various aspects of the penetration of high-energy ions through a crystal where a periodic structure of the crystal has

In order to clarify the reasons for the substantial differences between the calculated and experimental cross sections observed in Fig. 8.9b a detailed theoretical analysis was performed in [208] for the electron loss from 33 TeV $Pb^{81+}(1s)$ and Pb^{82+} ions incident on Al and Au foils, respectively. In this section we shall follow this analysis.

The consideration of the change in the charge states of the projectiles, performed in [208], assumes that the foil materials are amorphous (not crystals) and consists of two main steps.

First, the basis of the consideration is represented by calculations of cross sections for the projectile-electron excitation (de-excitation) and loss occurring in the ion–atom collisions. Cross sections for the bound-free pair production are also calculated. In all these calculations the Dirac–Coulomb wave functions are employed to describe bound and continuum states of the electron (and the positron) in the field of the bare lead nucleus. The ion–atom interaction is described in the first order approximation with taking into account the shielding of the atomic nucleus by the atomic electrons. The antiscreening contributions of the atomic electrons to these processes is included as well.

In addition to the electron capture via the pair production, the radiative and kinematic capture channels are also considered [208]. At this very high impact energy the radiative capture, whose cross section scales as $1/\gamma$ at $\gamma \gg 1$, is much weaker than the bound-free pair production: in collisions with aluminum and gold atoms the radiative capture cross section, compared to the capture cross section via the pair production, is smaller roughly by a factor of 5 and 30, respectively. The kinematic capture is even weaker than the radiative one and can simply be neglected.

Besides, within the basic atomic physics analysis, rates for the spontaneous radiative decay of excited hydrogen-like lead ions to all possible internal states with lower energies are also calculated [208].

The second step of the consideration consists of solving kinetic equations. These equations describe the population of the internal states of the ion inside the foil given as a function of time t or of the ionic coordinate $z = vt$ inside the foil (z and t are measured in the laboratory frame, $\boldsymbol{v} = (0,0,v)$ is the projectile velocity). These equations read

$$\frac{dP_0}{dt} = -\frac{P_0}{\tau^{capt}} + \sum_{j=1}^{N_{max}} \frac{P_j}{\tau_j^{loss}},$$

$$\frac{dP_j}{dt} = \frac{P_0}{\tau_j^{capt}} - \frac{P_j}{\tau_j^{loss}} - P_j \sum_{i=1}^{i<j} \frac{1}{\tau_{j\to i}^{sp}}$$

$$- P_j \sum_{i=1(i\neq j)}^{N_{max}} \frac{1}{\tau_{j\to i}} + \sum_{i=1(i\neq j)}^{N_{max}} \frac{P_i}{\tau_{i\to j}}. \tag{8.7}$$

a very profound effect on the projectile charge states and the electron emission (for recent references see [204–207]).

Here, P_0 is the number of the bare ions, P_j is the number of the lead ions with one electron in the jth internal state $(j = 1, 2, .., N_{\text{max}})$ and N_{max} is the total number of the involved bound states. Further, τ_j^{capt} is the mean time for the electron capture into the jth state, τ^{capt} is the mean time for the electron capture into any state $\left(1/\tau^{\text{capt}} = 1/\tau_1^{\text{capt}} + 1/\tau_2^{\text{capt}} + ...\right)$, τ_j^{loss} is the mean time for the electron loss from the state j into the continuum, $\tau_{j \to i}$ is the mean time for the collision induced transition from the internal state i to the internal state j and $\tau_{j \to i}^{\text{sp}}$ is the lifetime of the state j with respect to the spontaneous radiative transition to any possible state i.

Note that the beam losses due to the nuclear reactions are not included in (8.7). Although cross sections for such losses are *per se* even somewhat larger (see [85, 180, 213]) than the corresponding capture cross sections, these losses, as calculations show, start to affect the populations P_0 and P_j at much larger foil thicknesses than those which are considered below.

The elementary cross sections and spontaneous decay rates obtained during the first step of the consideration enable one to get the above mean excitation/de-excitation loss and capture times in the usual way. For instance, $\tau_{j \to i}^{\text{sp}} = \gamma/\Gamma_{j \to i}^{\text{sp}}$ where $\Gamma_{j \to i}^{\text{sp}}$ is the spontaneous decay rate for the transition $j \to i$ calculated in the rest frame of the ion and $\gamma = 1/\sqrt{1 - v^2/c^2}$ is the collision Lorentz factor, and $\tau_j^{\text{loss}} = 1/\left(n_a \sigma_j^{\text{loss}} v\right)$, where σ_j^{loss} is the cross section for the electron loss from the jth internal state of the ion and n_a is the atomic density of the target.

8.6.1 Fraction of Hydrogen-Like Ions

In the case of 33 TeV $\text{Pb}^{81+}(1s)$ ions incident on a gold foil calculated results for the fraction of the hydrogen-like lead ions, $P_h = \sum_{j=1}^{N_{\text{max}}} P_j$, which were obtained in [208], are shown in Fig. 8.11. The dash curve corresponds to the situation when all bound states with the principal quantum numbers n larger than 1 were ignored in the analysis, the dot curve displays the results obtained when only the bound states with $n = 1$ and $n = 2$ were taken into account, and so on. It is seen in the figure that the inclusion of the excited states into the analysis reduces the calculated values for the fraction of the hydrogen-like lead ions. This reduction is caused by the fact that, compared to the ground state, the excited states are characterized by larger loss cross sections.

Normally excited states in such very heavy ions like Pb^{81+} do not play a noticeable role in the loss process, in particular, because of the spontaneous radiative decay of these states, which is very strong. In the case under consideration, however, the situation drastically changes because of the relativistic time dilatation. The latter effectively increases by a factor of ≈ 170 the lifetime of the excited states with respect to the radiative decay. As a result, the population of the excited states very substantially increases having a much stronger impact on the charge states of the projectiles.

Note also that for 33 TeV Pb projectiles the capture cross sections are about three orders of magnitude smaller than the loss and excitation cross

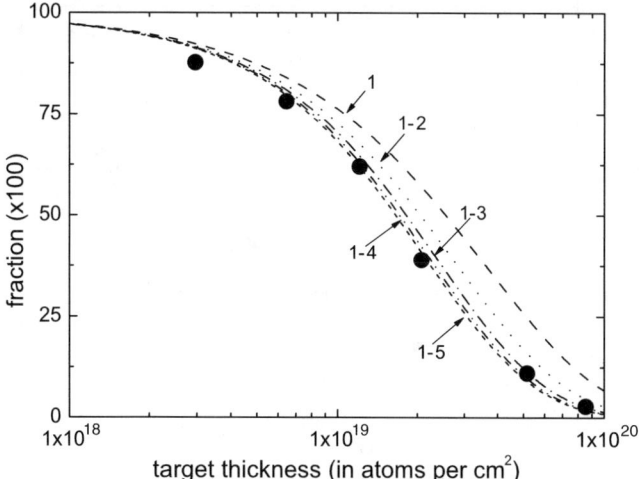

Fig. 8.11. The fraction of the hydrogen-like ions given as a function of the target thickness for 33 TeV $Pb^{81+}(1s)$ projectiles incident on a gold foil. The different curves correspond to taking into account different numbers of bound states in the theoretical analysis. *Dash curve*: only states with the principal quantum number $n = 1$ are included. *Dot curve*: the states with $n = 1$ and $n = 2$ are included. *Dash–dot curve*: states with $n = 1$–3 are included. *Dash–dot–dot curve*: states with $n = 1$–4 are included. *Short-dash curve*: states with $n = 1$–5 are included. *Circles*: experimental data from [85]. For more explanations see the text. From [208].

sections. Therefore, under the conditions of the 'ionization' experiment $(P_{1s}(t = 0) = 1, P_{j \neq 1s}(t = 0) = 0)$ the capture may become important only if rather thick foils are used in which the dynamical equilibrium is effectively reached for the transitions between the bound and continuum states. This is not the case under the conditions of Fig. 8.11 where, according to the theoretical analysis of [208], the electron capture channels play just a very minor role.

In Fig. 8.11 the theoretical results are also compared with the experimental data measured in [85]. It is seen that the inclusion of the excited states in the theoretical analysis results in a noticeably better agreement with the experiment.

Data, similar to those displayed in Fig. 8.11 but obtained for the case of 33 TeV Pb^{82+} projectiles incident on a gold foil, are shown in Fig. 8.12. We again observe that the inclusion of the excited states into the theoretical analysis leads to the reduction of the calculated results for the fraction of the hydrogen-like ions and yields a better agreement with the experimental data.

8.6.2 Effective Loss Cross Section

In addition to the data on the fractions of the hydrogen-like ions, the authors of [85] presented also values for the loss and capture cross sections. It is

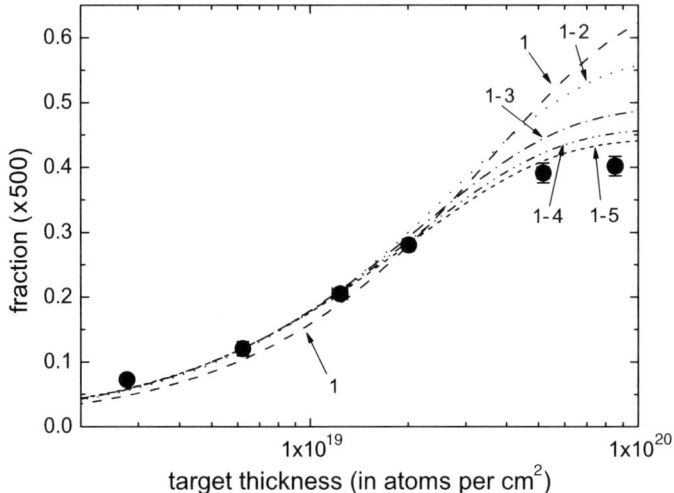

fraction (x500)

target thickness (in atoms per cm^2)

Fig. 8.12. Same as in Fig. 8.11 but for the case of 33 Pb^{82+} bare nuclei incident on a gold foil. *Circles*: experimental data from [85]. *Curves*: results of [208]. From [208].

important to keep in mind, however, that the ion fractions, given as a function of the foil thickness, were the only data which had been directly measured in the experiment [85].

Values for the loss cross sections, which were reported in [85] and which are shown in Fig. 8.9, were obtained in [85] by fitting the measured data for the fraction of the hydrogen-like ions using solutions of the two-charge-state model. This very simplified model considers just two populations, the population of the bare ions and that of the hydrogen-like ions without separating the latter ones over their internal states. The capture and (effective) loss cross sections were considered in this model as independent of the foil thickness.

Indeed, by replacing in (8.7) the coefficients P_j by their sum the system of equations (8.7) can formally be reduced to merely two equations

$$\frac{\mathrm{d}P_0}{\mathrm{d}t} = -\frac{P_0}{\tau^{\mathrm{capt}}} + \frac{P_{\mathrm{h}}}{\tau_{\mathrm{eff}}^{\mathrm{loss}}},$$
$$\frac{\mathrm{d}P_{\mathrm{h}}}{\mathrm{d}t} = \frac{P_0}{\tau^{\mathrm{capt}}} - \frac{P_{\mathrm{h}}}{\tau_{\mathrm{eff}}^{\mathrm{loss}}}. \tag{8.8}$$

Here

$$P_{\mathrm{h}} = \sum_{j=1}^{N_{\mathrm{max}}} P_j \tag{8.9}$$

is the total population of the bound states,

$$\tau_{\mathrm{eff}}^{\mathrm{loss}} = \frac{1}{n_{\mathrm{a}} \sigma_{\mathrm{eff}}^{\mathrm{loss}} v} \tag{8.10}$$

and

$$\sigma_{\text{eff}}^{\text{loss}} = \frac{\sum_{j=1}^{N_{\text{max}}} P_j \sigma_j^{\text{loss}}}{\sum_{j=1}^{N_{\text{max}}} P_j} \qquad (8.11)$$

is the effective loss cross section.

In [85] the system of two equations (similar to (8.8)) was considered assuming that the corresponding cross sections are independent of the thickness of the target foil. This assumption enabled the authors of [85] to get analytical solutions for P_0 and P_{h}. For instance, in the 'ionization' scenario such solutions read

$$P_0(t) = \frac{\sigma_{\text{eff}}^{\text{loss}}}{\sigma^{\text{c}} + \sigma_{\text{eff}}^{\text{loss}}} \left(1 - \exp(-t/T)\right),$$

$$P_{\text{h}}(t) = \frac{\sigma^{\text{c}}}{\sigma^{\text{c}} + \sigma_{\text{eff}}^{\text{loss}}} + \frac{\sigma_{\text{eff}}^{\text{loss}}}{\sigma^{\text{c}} + \sigma_{\text{eff}}^{\text{loss}}} \exp(-t/T),$$

where σ^{c} is the total capture cross section (i.e. the sum of the capture cross sections to all bound states) and $T = \tau^{\text{capt}} \tau_{\text{eff}}^{\text{loss}} / \left(\tau^{\text{capt}} + \tau_{\text{eff}}^{\text{loss}}\right)$. However, the above solutions, of course, are sensible only if the effective loss cross section, given by (8.11), can indeed be well approximated as independent of time and thus of the target thickness.

In Fig. 8.13 results are shown for the effective loss cross section calculated in [208] for 33 TeV u^{-1} Pb^{81+}(1s) (Fig. 8.13a) and Pb^{82+} (Fig. 8.13b) ions incident on an aluminum foil. It follows from the figure that the calculated value of the effective loss cross section in general depends on (a) the number of bound states involved in the analysis, (b) the target thickness and (c) it may be quite sensitive to the scenario ('ionization' or 'capture') used in an experiment.

In Fig. 8.14 results are shown for the effective loss cross section calculated for 33 TeV Pb^{81+}(1s) and Pb^{82+} projectiles incident on a gold foil. The above conclusions drawn from Fig. 8.13 for the electron loss under the penetration of the aluminum foil are also valid in the case of the gold foil. At the same time, there are also important differences between these two cases. First, the changes in the effective loss cross section, which occur when one allows for more bound states in the theoretical analysis, are accumulating at a different pace. The largest change in the effective cross section now comes from the inclusion of the states with $n = 2$ and with further addition of bound states the effective cross section converges faster. Second, the total influence of the excitations on the effective loss cross section in collisions with the gold foil turns out to be even larger. All this is caused by the fact that in collisions with atoms of gold, which are much heavier and therefore are a source of the much stronger field acting on the ion, all the elementary collision cross sections are much larger whereas the radiative decay rates remain exactly the same.

Summarizing the above consideration one can conclude that the use of the simplified two-state model in [85] had, in fact, introduced quite a substantial

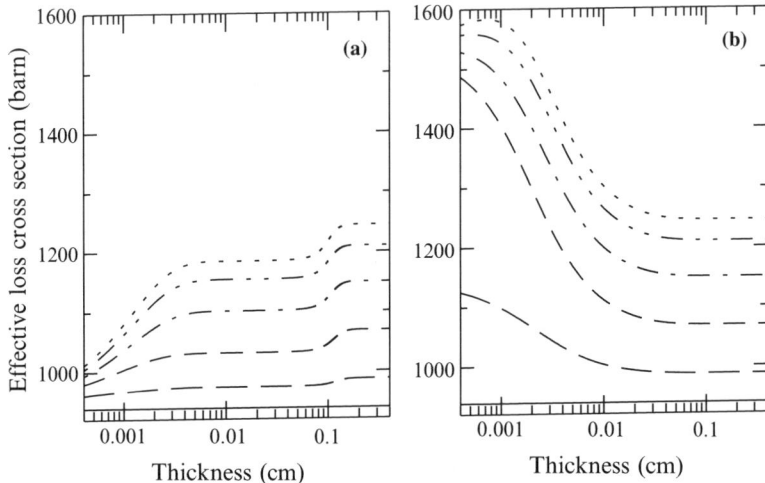

Fig. 8.13. The effective cross section (8.11) for the electron loss from 33 TeV lead projectiles penetrating an aluminum foil: (**a**) incident $Pb^{81+}(1s)$ ions; (**b**) incident Pb^{82+} ions. The cross section is given as a function of the foil thickness. The different curves correspond to taking into account different numbers of bound states in the analysis. *Solid curve*: bound states with $n = 1$. *Dash curve*: $n = 1$ and $n = 2$. *Short dash curve*: $n = 1$–3. *Dash–dot curve*: $n = 1$–4. *Dash–dot–dot curve*: $n = 1$–5. *Dot curve*: $n = 1$–6. From [208].

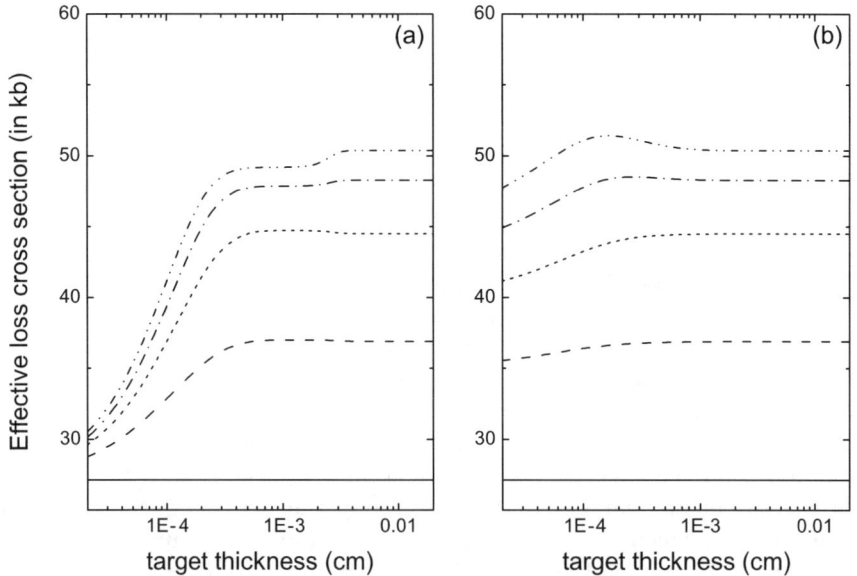

Fig. 8.14. Same as in Fig. 8.13 but for 33 TeV lead projectiles penetrating a gold foil. From [208].

error into the values of the loss cross sections reported there. This also explains the very substantial deviations between the theoretical and experimental results for the loss cross sections observed in Fig. 8.9b.

8.7 Differential Loss Cross Sections in Collisions at High γ

In general, much more information about the projectile-electron loss process can be obtained by considering differential loss cross sections. Energy spectra of electrons, emitted when extreme relativistic highly charged projectiles penetrate matter, possess important information about the collision physics and represent the very topic of this section.

Results of the first (and, up to now, the only) measurement of the spectrum of electrons emitted by 33 TeV ($\gamma = 168$) lead projectiles penetrating thin foils were reported in [209] (see also [181]). Having in mind these experimental results, in this section we restrict our attention to projectiles with $\gamma = 168$. Our consideration will be based on the first order theory in the projectile–target interaction. Electrons emitted in such collisions from the target atoms will not be taken into account since they do not really contribute to the range of electron energies considered below.

8.7.1 Energy Spectra of Electrons Emitted by Projectiles Under the Single-Collision Conditions

In this subsection we shall discuss spectra of electrons emitted from relatively light and very heavy hydrogen-like projectiles, which move with velocities corresponding to $\gamma = 168$ and collide with neutral atoms of aluminum. We shall assume that the single-collision conditions are fulfilled which corresponds to the projectiles penetrating gas targets or extremely thin solid foils. The first order perturbation theory for treating the loss process in such collisions is justified and will be used below.

In order to obtain the differential cross section in the laboratory frame K_A it is convenient to calculate first the differential cross section in the rest frame K_I of the projectile ion. Then, by using the relation $\frac{d^2\sigma'}{d\varepsilon' d\Omega'} = \frac{k'}{k}\frac{d^2\sigma}{d\varepsilon d\Omega}$ (compare with formula (4.20)) one can transform the results into the laboratory frame. In these expression ε, k, $d\Omega$ and $\frac{d^2\sigma}{d\varepsilon d\Omega}$ are the total energy of the electron, the absolute value of the electron momentum, the solid electron emission angle and the cross section, respectively. The primed and unprimed quantities refer to the laboratory and projectile frames, respectively.

Figure 8.15a shows the cross section differential in energy in the laboratory frame for the electron loss from the Pb^{81+} projectiles colliding with Al atoms. The following main features of the calculated spectrum can be noted. First, the electron energy distribution has a maximum at an electron

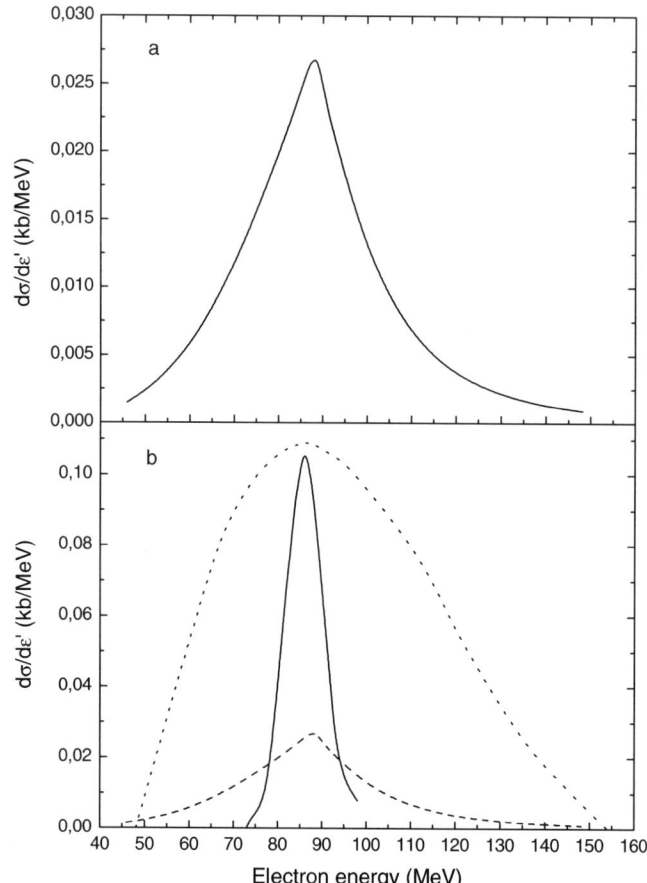

Fig. 8.15. Cross section differential in energy for the electron loss from $160\,\mathrm{GeV\,u^{-1}}$ $\mathrm{Pb}^{81+}(1s)$ colliding with Al atoms. The cross section is given in the laboratory frame, where the atoms are at rest. **(a)** Calculations of [210]. **(b)** *Full curve*: experimental results of [209] which we normalised according to the total cross section for the electron loss from $33\,\mathrm{TeV\,Pb}^{81+}$ colliding with Al solid target reported in [85]; *dashed curve*: our calculation; *dotted curve*: the Compton profile of $\mathrm{Pb}^{81+}(1s)$ mapped into the laboratory frame [209]. From [210].

energy $\varepsilon'_{max} = mc^2\gamma$ which corresponds to the emitted electron moving with a velocity equal to the velocity of the projectile. Second, this distribution is asymmetric with the majority of the lost electrons having energies lower than ε'_{max}. Third, the width of the distribution is much larger (about a factor 2.5–3) than it was measured experimentally in [209] for $33\,\mathrm{TeV\,Pb}^{81+}$ projectiles penetrating an aluminum foil (Fig. 8.15b). Fourth, this distribution also differs rather strongly from that calculated in [209] (see Fig. 1b) where, as the authors of [209] state, a $\mathrm{Pb}^{81+}(1s)$ Compton profile was mapped into

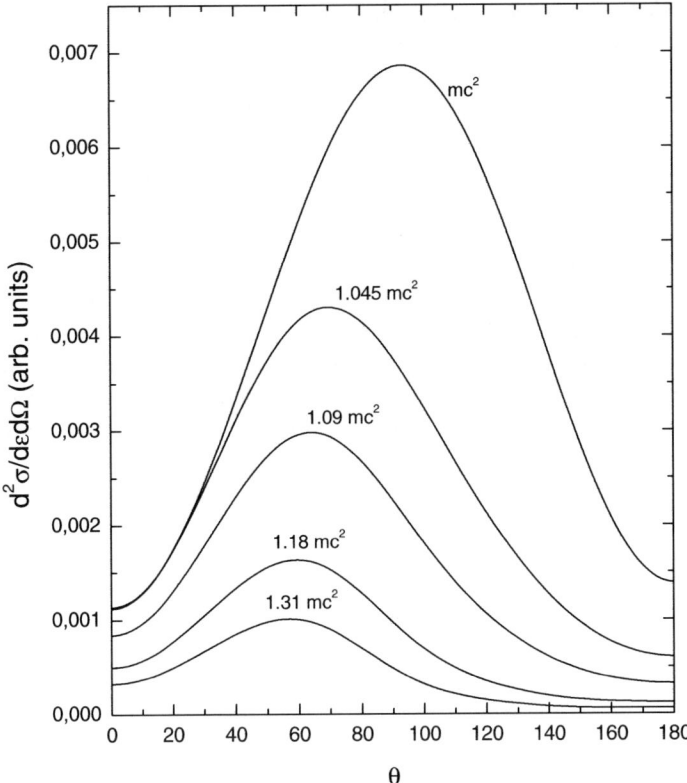

Fig. 8.16. Doubly differential cross section for the electron emission from $160\,\mathrm{GeV\,u^{-1}}$ $\mathrm{Pb^{81+}}$ colliding with Al atoms. The cross section is given in the projectile frame as a function of the electron emission angle for several electron energies. The zero angle corresponds to the direction of the velocity of the incident atom. From [210].

the laboratory frame assuming that the angular emission distribution in the projectile frame is of a dipole form.

In order to obtain some insight into the origin of the shape of the calculated loss peak in the laboratory frame, results for the double differential loss cross sections for $160\,\mathrm{GeV\,u^{-1}}$ $\mathrm{Pb^{81+}}$ in the projectile rest frame are shown in Fig. 8.16. The following points are worth to mention. First, the angular distribution of the emitted electrons in the projectile frame is rather asymmetric: the main part of the electrons in this frame is emitted into the forward semi-sphere (in the direction of the motion of the incident neutral atom). The angular asymmetry in the emission increases with increasing the electron kinetic energy. Second, the number of the emitted electrons rapidly decreases with increasing this energy and the main part of the emitted electrons has kinetic energies not substantially higher than the electron binding energy ($\approx 0.2mc^2$ in $\mathrm{Pb^{81+}}$).

The first point allows one to understand the asymmetry in the electron energy spectra in the laboratory frame: since the majority of the electrons in the projectile frame is emitted in the direction of the motion of the incident atom then the main part of the electrons in the laboratory frame has energies which are less than $\varepsilon'_{max} = mc^2\gamma$. The second point states that a considerable part of the emitted electrons in the projectile frame has relatively low kinetic energies and this makes it clearer why the electron spectrum in the laboratory frame has a maximum near $\varepsilon'_{max} = mc^2\gamma$. In addition, the fact that the main part of the emitted electrons has low energies in the rest frame of the projectile shows that the Compton profile of $Pb^{81+}(1s)$ is not a relevant physical quantity for the loss process. The Compton profile of an initial state would be reflected directly in the ionization (loss) spectra only if the electron would be ejected mainly in collisions where the momentum transfer to the electron (in the projectile frame) is large compared to the typical electron momentum in the initial bound state. This would lead to the population of high-energy continuum states of the ion which could be approximated by plane waves. As a result, the Compton profile of the initial state would follow from the corresponding transition matrix elements. However, since, according to Fig. 8.16, the emitted electrons have relatively low energies, the above scenario is certainly not the case here.

As a typical example of the electron loss from relatively light ions, let us now consider the electron loss from $160\,GeV\,u^{-1}\,S^{15+}$ colliding with Al atoms. The electron loss spectrum in the laboratory frame is displayed in Fig. 8.17. One can note two main differences between the spectra displayed in Figs. 8.15a and 8.17. First, the width of the energy distribution of the electrons emitted from the sulphur ions is much smaller than that shown in Fig. 8.15a. Second, the spectrum given in Fig. 8.17 is more symmetric compared to that shown in Fig. 8.15a.

The origin of these differences can be found by inspecting the double differential loss spectra in the rest frame of the projectile which are shown in Fig. 8.18. Similarly to the loss from the Pb^{81+} ions the number of the emitted electrons rapidly decreases with increasing electron kinetic energy. Again the main part of the emitted electrons has kinetic energies smaller or of the order of the initial binding energy of the electron. Since now this energy ($\approx 3.5\,keV$) is much less than that in Pb^{81+} ($\approx 100\,keV$) the spectrum of the electron emitted from $160\,GeV\,u^{-1}\,S^{15+}$ is much narrower in energy than that originating from $160\,GeV\,u^{-1}\,Pb^{81+}$. In Fig. 8.18 one also sees that the angular spectra of the emitted electrons are nearly symmetrical in the rest frame of the projectile with respect to the direction $\theta = \pi/2$ and that this is the case for the whole range of emission energies of importance which gives practically all the contribution to the total loss. This is in contrast to the angular spectra displayed in Fig. 8.16. The reason for this contrast is the following. The energy and minimum momentum which are transferred to the ion in an ultrarelativistic collision are related by $q_{min} = \frac{\varepsilon_k - \varepsilon_0}{c}$, where one has $\varepsilon_k - \varepsilon_0 \sim Z_I^2$ for the majority of electrons emitted from the ions with a charge Z_I. For light ions

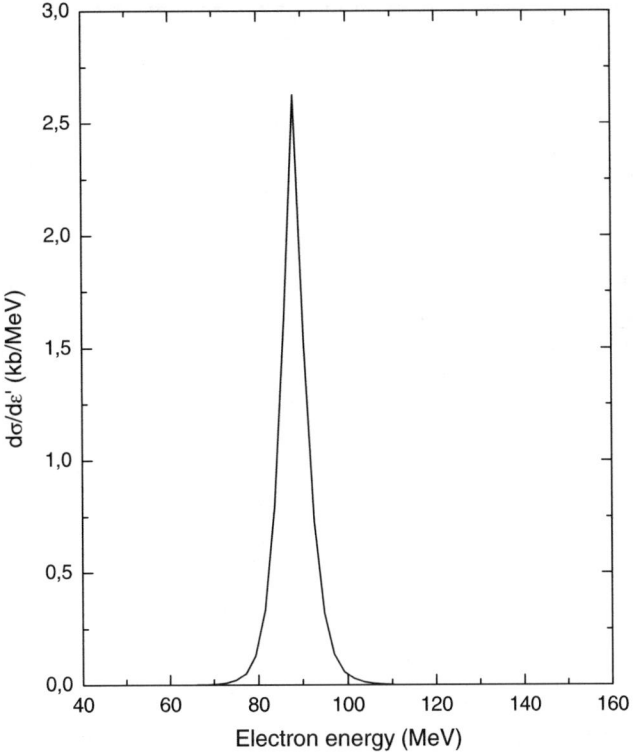

Fig. 8.17. Cross section differential in energy for the electron loss from $160\,\mathrm{GeV\,u^{-1}}$ S^{15+} colliding with Al atoms. The cross section is given in the laboratory frame, where the atoms are at rest. From [210].

one has $q_{min}/\overline{k_z} \sim Z_I/c \ll 1$, where $\overline{k_z} \sim Z_I$ is the typical absolute value for the z-component of the momentum of the emitted electron. A similar relation also holds between the absolute value of the transverse momentum transfer $q_\perp$ and the typical absolute value of the transverse component of the electron momentum. Therefore, in the case of the emission from light ions the electron momentum is balanced mainly by the recoil of the residual ion resulting in dipole-like angular spectra in the rest frame of the ion [211]. For very heavy ions, where $Z_I \sim c$, typical values of the momentum transfers to the ion are already rather close to typical values of the momentum of the emitted electron in the ion frame. Therefore, the emitted electron momentum is no longer balanced by the recoil of the residual nucleus and the angular spectra show considerable shifts to angles less than $\pi/2$.

The nearly symmetrical shape of the loss spectra in the rest frame of the projectile for light projectiles is reflected in the electron loss spectrum in the laboratory frame resulting in a nearly symmetrical distribution of the electron energies with respect to the 'central' energy $\varepsilon'_{max} = mc^2\gamma$ (Fig. 8.17).

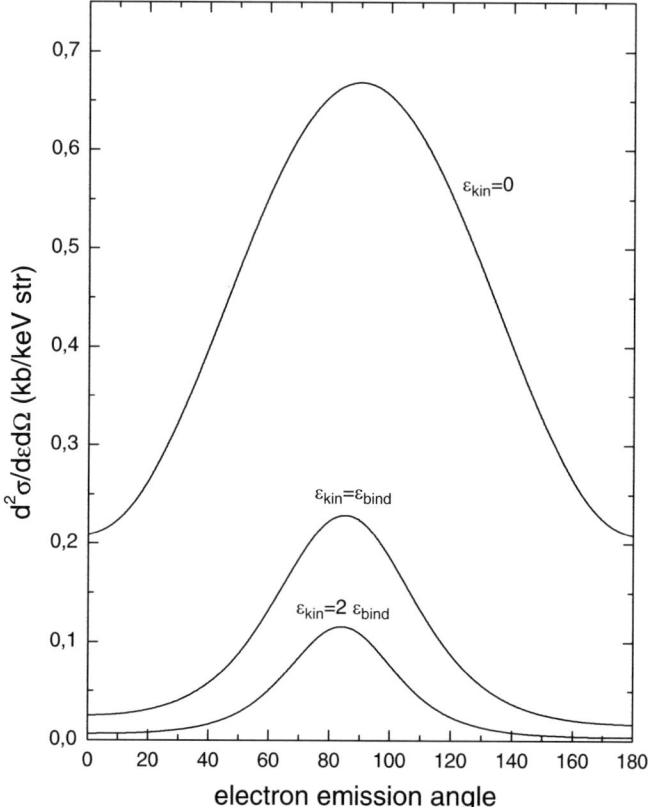

Fig. 8.18. Doubly differential cross section for the electron emission from $160\,\mathrm{GeV\,u^{-1}}$ S^{15+} colliding with Al atoms. The cross section is given in the projectile frame as a function of the emission angle for several kinetic energies of the emitted electron. The zero angle corresponds to the direction of the velocity of the incident atom. From [210].

8.7.2 The Spectrum of Electrons Emitted by 33 TeV Lead Ions Penetrating Thin Foils. The Role of Excited States of the Projectile

As was already mentioned, the first experimental results on the spectra of electrons emitted by ultrarelativistic heavy ions were reported in [209]. In that experiment beams of incident 33 TeV $Pb^{81+}(1s)$ and 33 TeV Pb^{82+} ions were penetrating Al and Au foils, respectively. In both cases it was found that the penetration is accompanied by the emission of ultrarelativistic electrons whose energy distributions have the form of a cusp with a maximum at an energy corresponding to the electrons moving in the laboratory frame with velocities equal to that of the ions.

One of the interesting results reported in [209] was that the measured distribution of the high-energy electrons produced under the bombardment of an aluminum foil by the incident $Pb^{81+}(1s)$ ions was found to be much narrower than it was expected [209]. Indeed, as seen in Fig. 8.15, the electron spectrum calculated assuming the single-collision conditions turns out to be much broader than that observed in the experiment [209].

Another intriguing finding of [209] was that for 33 TeV Pb^{82+} ions incident on a gold foil the shape of the measured energy distribution of high-energy electrons emerging from the foil was very similar to that obtained for the beam of 33 TeV $Pb^{81+}(1s)$ ions incident on the Al foil.

Since it is well known, that the form of the electron emission spectrum depends on the bound state from which the electron leaves the ion, it was speculated in [209] that in the case of the incident 33 TeV Pb^{82+} ions the very narrow shape of the electron cusp might be a signature of the electron capture into excited states. However, for the $Pb^{81+}(1s)$ ions incident on the Al foil the possible influence of excited states of these ions on the electron cusp was not considered seriously.[11]

In this subsection we shall follow the consideration of the electron energy spectra given in [212]. Similarly to the two-step consideration for the charge states of the projectiles, which was described in Sect. 8.6, this consideration assumes that the foil materials are amorphous and is based on the calculations of the total and differential cross sections. Besides, it includes also some analysis of the passage of the emitted high-energy electrons through the foils.

Once the functions $P_j(z)$ entering (8.7) are known, the (preliminary) estimate for the energy spectrum of the electrons emitted from the ion traversing a solid foil of a thickness L is given by [212]

$$\frac{dn_e}{d\varepsilon_p} = n_a \sum_{j=1}^{N_{max}} \frac{d\sigma_j^{loss}}{d\varepsilon_p} \int_0^L dz P_j(z), \qquad (8.12)$$

where ε_p is the total electron energy in the laboratory frame and $\frac{d\sigma_j^{loss}}{d\varepsilon_p}$ is the energy distribution of the electrons emitted from the internal state j.

The third step of the consideration deals with the transport of the emitted electrons through the foil. The detailed analysis of this step represents in general quite a delicate task but in the case under consideration is substantially simplified by the fact that the electrons leaving the ions have in the laboratory frame extremely high values of energy. There are two main effects which can influence the shape of the electron energy distribution when the electrons penetrate the foil.

The first concerns energy losses of the ultrarelativistic electrons traversing the foil. These losses are caused by (a) the excitation of the electrons of the foil and (b) the emission of the radiation by the ultrarelativistic electrons

---

[11] Because of the reasons which were already mentioned in the discussion of the projectile charge states in Sect. 8.6.

because of their acceleration during the interactions with the atoms in the foil. However, for the foil parameters used in the experiment [209] the energy losses can simply be ignored because they are very small ($\lesssim 0.5\%$) compared to the initial energies of these electrons [212].

The second effect which may possibly influence the shape of the measured energy distributions is that collisions in the foil broaden the distribution of the ultrarelativistic electrons along the transverse components (p_x, p_y) of their momenta. For the foil parameters used in [209], the multiple collisions suffered by the ultrarelativistic electrons inside the foil substantially increase the width of their (p_x, p_y)-distribution compared to that which these electrons have when leaving the 33 TeV nuclei.

Nevertheless, even after this increase the transverse components ($\sim 10^2$ a.u.) remain very small compared to the total electron momenta ($\simeq 2 \times 10^4$ a.u.). This means that the broadening of the (p_x, p_y)-distribution may have an impact on the measured electron momentum distribution only if special geometric conditions are employed in an experiment.[12] Since the authors of [212] did not possess all necessary information about the real conditions of the experiment [209], in the calculations for the energy spectra performed in [212] simply all electrons (whichever angle they have after leaving the foil) were taken into account.

Under such conditions the changes in the electron momenta during the electron transport through the foil do not have an impact on the final electron energy distribution. Therefore, the main difference between the calculation for the shape of the electron cusp in the case of 33 $Pb^{81+}(1s)$ incident on Al atoms, which was discussed in Sect. 8.7.1 under the assumption that the electron loss occurs in the single-collision regime, and the model of [212] is that the latter takes into account electron transitions to the continuum not only directly from the ground state of the ions but also via the intermediate excitations to higher bound states occurring when the ions penetrate the foil.

In Fig. 8.19 results are shown for the electron energy spectrum in the case of 33 $Pb^{81+}(1s)$ ions incident on an aluminum foil having a thickness of 2.85×10^{-2} cm. The expectations of [209, 210], that in the case of very heavy ions their excitations are of minor importance for the formation of the electron cusp, seems to be just confirmed when we compare in this figure

---

[12] For instance, if an experiment detects only those high-energy electrons, which finally move inside a very narrow cone centered along v, the expansion of the electron (p_x, p_y)-distribution inside the foil will first of all impact the detected numbers of the electrons emitted from the ground state of the lead ions (both in absolute and relative proportions) because these electrons right after the emission from the ions have larger transverse momenta and, besides, are the first to appear in the continuum. Compared to the emission from excited states the electrons ejected from the ground state have the largest width of the energy distribution. Therefore, a comparatively stronger removal of those electrons, which were emitted from the ground state, occurring when the electron beam traverses the foil will effectively decrease the width of the measured electron energy distribution.

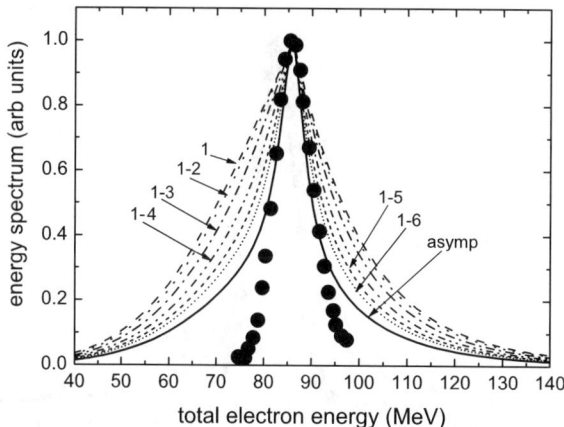

Fig. 8.19. The energy distribution of the electron cusp produced in collisions between an incident beam of $33\,\text{TeV}$ $Pb^{81+}(1s)$ with Al foil with a thickness of $2.85 \times 10^{-2}\,\text{cm}$ (for more explanation see the text). *Circles* show the electron energy distribution measured in [209] for $33\,\text{TeV}$ Pb^{81+} colliding with the Al foil of the same thickness. All the distributions are given in the laboratory frame and are normalized to 1 at the maximum. From [212].

curves labeled '1' and '1–2'. The curve '1' was obtained by ignoring all excited bound states while in the calculation resulting in the curve '1–2' the states with the principal quantum number $n = 2$ were also taken into account. Yet, there is just a tiny difference in the widths of these two curves.

However, when the states with $n = 3$ are added into the analysis (the curve in Fig. 8.19 labeled '1–3') the width-reducing effect becomes quite visible. Adding into the analysis the states with $n = 4$ leads to a further reduction in the calculated width and this reduction is even larger than that observed when the states with $n = 3$ were added. The reduction of the width continues further when the states with $n = 5$ and $n = 6$ are included (see Fig. 8.19). However, it proceeds at a smaller pace compared to that when the states with $n = 3$ and $n = 4$ were added.

It is of course not possible to increase indefinitely the number of bound states in the analysis. Therefore, an extrapolation procedure was applied in [212] in order to get the asymptotic limit for the electron cusp shape which effectively corresponds to taking into account all bound states ($n = 1$–∞). The result of this extrapolation is shown in Fig. 8.19 by the solid curve labeled '*asymp*'.

Comparing the energy distributions in Fig. 8.19 we see that their asymptotic width is about a factor of 3 smaller than the width obtained by assuming that the cusp is produced under the single-collision conditions. This strong effect is caused by the excitation of the ions inside the foil which involves rather highly lying bound states: when the ions move in the foil the electron cloud surrounding the ionic nuclei has enough time to expand tremendously

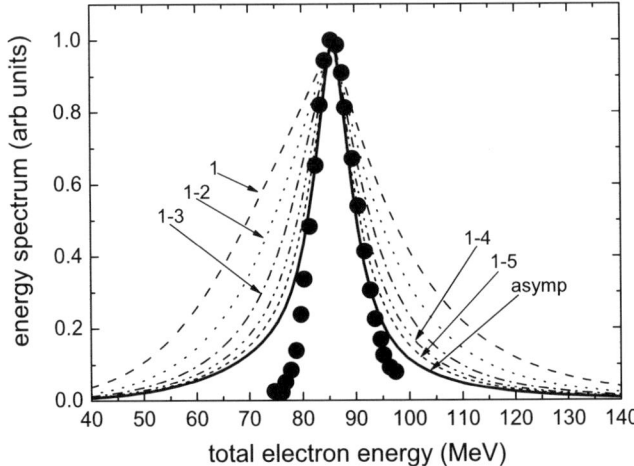

Fig. 8.20. Same as in Fig. 8.15 but for an incident beam of 33 TeV Pb^{82+} penetrating Au foil with a thickness of 8.81×10^{-4} cm corresponding to the conditions of the capture experiment [209]. For more information, *circles* show the electron spectrum measured in [209] for 33 TeV $Pb^{81+}(1s)$ ions incident on the Al foil. From [212].

in size before it will almost completely disappear due to the transitions to the continuum. The key factor making this possible is the relativistic time dilatation. It effectively decreases the spontaneous decay rates of the excited states of the ions by a factor of ≈ 170 making them, thus, much more 'visible' in the process of the electron emission.

Compared to the ground state, the excited states have larger loss cross sections (and, thus, shorter free paths with respect to the loss) and narrower Compton profiles which, as well as the relative decrease in the population of the ground state due to the excitations, lead to the narrowing of the electron energy distribution.

One more point which should be mentioned is that cross sections for the electron capture are relatively very small. As a result, in the formation of the electron cusp in the case of the hydrogen-like ions incident on the Al foil the capture channels do not play any noticeable role.

In Fig. 8.20 results are shown for the energy spectrum calculated for 33 TeV Pb^{82+} incident on a gold foil. Of course, now the electron capture becomes of paramount importance for the very existence of the electron cusp. One should note, however, that the capture cross sections decrease very rapidly when n and j_e increase (j_e is the total angular momentum of the electron in a bound state). Therefore, the most of the excited bound states having a very important impact on the energy spectrum are populated not by capturing the electron directly from the vacuum but via the excitations from few states with the lowest values of n and j_e for which the capture is efficient. This indirect way becomes especially effective because in collisions with Au atoms the excitation cross sections are much larger than in the case with Al.

Comparing the spectra shown in Fig. 8.20 with those displayed in Fig. 8.19 we see that the changes in the form of the calculated spectrum in Fig. 8.20 (occurring when more bound states are taken into account in the analysis) are accumulating at a different pace. Besides, the asymptotic cusp shape in Fig. 8.20 has less pronounced wings. These differences are related to two basic reasons: (a) the excitation/loss cross sections in a gold foil are much larger while the spontaneous decay rates remain exactly the same as in the case of an aluminum foil and (b) the initial step in the cusp formation is now represented by the capture process which also somewhat increases the relative population of the excited states compared to the case when the beam of $Pb^{81+}(1s)$ ions was incident on the Al foil.

Curiously, however, that the asymptotic width in Fig. 8.20 is again about three times smaller than the 'initial' width and the shape of the asymptotic spectra in both cases looks similar (which is also in agreement with the experimental observations of [209]). In general such a similarity will not hold when the foil parameters (for instance, their thicknesses) are changed and, in this sense, is accidental. Yet, in both cases the strong reductions in the widths of the energy distributions are caused by the excitations of the electrons to rather high lying bound ionic states occurring when the ions penetrate the foils.[13]

Thus, the energy spectra of the ultrarelativistic electrons emitted when incident 33 TeV $Pb^{81+}(1s)$ and Pb^{82+} ions penetrate thin foils are much narrower than those which would be produced under the single-collision conditions and the strong width reduction is caused by the excitations of the ions when they penetrate the target foils suffering multiple collisions with the target atoms. In the case under consideration the excitations become so effective because of the relativistic time dilatation which decreases very strongly the spontaneous decay rates of excited states in the ions moving with velocities closely approaching the speed of light.

Although the results discussed above shed some light on the origin of the unexpectedly narrow shape of the electron cusp produced by the ultrarelativistic heavy ions, a more careful analysis taking into account all real conditions of the experiment [209] is necessary in order to make a detailed comparison between the experiment and theory.

8.8 On the Longitudinal and Transverse Contributions to the Total Loss Cross Section

When the potentials of the electromagnetic field are written in the Coulomb gauge, the corresponding expression for the scalar potential does not contain retardation and, independently of how fast the source of the electromagnetic

---

[13] In the case of the incident Pb^{82+} ions the electron capture into excited states (mainly $2s$ and $3s$) also leads to a noticeable contribution to the reduction of the width of the cusp.

field is, the form of the scalar potential remains like if the speed of light were infinity. Therefore, the coulomb gauge can, to certain extend, be viewed as the most 'nonrelativistic' gauge and is especially suited to discuss differences between the relativistic and nonrelativistic considerations.

As was discussed in Sect. 5.8, the total loss cross section calculated in the Coulomb gauge can be presented as an incoherent sum of the longitudinal and transverse contributions which arise in this gauge due to the coupling of the charge and current densities of the electron to the atomic scalar and vector potentials, respectively.

By calculating separately these contributions some additional information of interest can be obtained about the process of the projectile-electron loss in collisions with neutral atoms. For example, for the electron loss from $200\,\mathrm{GeV\,u^{-1}}\,Pb^{81+}$ in collisions with Au^0 the exchange of the transverse photon accounts for more than 60% of the total loss. However, for the electron loss from $200\,\mathrm{GeV\,u^{-1}}\,S^{15+}$ and $200\,\mathrm{GeV\,u^{-1}}\,O^{7+}$ ions the transverse part contributes only about 4% and less than 1%, respectively, to the total loss cross section. In collisions of light hydrogen-like ions with neutral atoms the exchange of the transverse virtual photon always represents the minor mechanism for the total electron loss from the ions. For collisions at low γ ($\gamma \sim 1$) the exchange of the longitudinal photon dominates in the total loss because *in the ion frame* the motion of the electron of the ion is nonrelativistic both in the initial and final states of the ion[14] and γ is small compared to v/v_e where $v_e \sim Z_I$ is a 'typical velocity' of the electron of the ion in the process. In collisions with charged particles at larger values of γ the relative contribution to the loss cross section, caused by the exchange of the transverse photon, becomes much more important and with a further increase of γ will eventually dominate the cross section. However, the exchange of the transverse photon becomes very efficient only at large impact parameters. Therefore, in collisions with neutral atoms the coupling of the electron of a light ion with the atom via the transverse photon is essentially cut off by the screening effect of the atomic electrons.

The latter may have important consequences for 'practical' calculations. Indeed, according to the above discussion, in order to estimate the cross section for the electron loss from light ions (and also from outer shells of heavy multiply charged ions) in collisions with neutral atoms at any collision energy, one can take the interaction with the instantaneous (unretarded) scalar potential of the incident atom as the full interaction acting on the electron of the ion. The form of the corresponding transition amplitude will remain exactly the same as in the nonrelativistic first order theory.[15]

---

[14] Formally the electron in the final state could acquire a relativistic velocity with respect to the nucleus of the ion. Such a situation, however, is rather unlikely and contributes negligibly to the total loss cross section.

[15] It is worth noting that the approximate light-cone amplitude (6.80) also coincides in form with the amplitude obtained in the sudden approximation within the framework of the purely nonrelativistic consideration.

Compared to the relativistic treatment, the nonrelativistic description of atomic collisions is normally much easier to deal with. Therefore, its application can be especially convenient when one has to estimate cross sections for multi-electron losses occurring from outer shells of many-electron (but comparatively low-charged) ions colliding with neutral targets at high and very high impact energies. Reliable estimates for such cross sections are of great importance for predicting the lifetimes of beams of heavy many-electron ions [214–219].

Of course, all this is in sharp contrast with the case of the loss (ionization) occurring in collisions with relativistic charged particles. In the latter collisions the exchange of the transverse virtual photon gives the very important contribution, which is asymptotically dominant at $\gamma \to \infty$ for the loss from both heavy and light ions.

8.9 Loss Cross Sections at Asymptotically High γ: Saturation Effect

In Figs. 8.21 and 8.22 the cross sections for the electron loss from 1–$2,000\,\text{GeV}\,\text{u}^{-1}$ $\text{Au}^{78+}(1s)$ and 1–$100\,\text{GeV}\,\text{u}^{-1}$ $\text{Ne}^{9+}(1s)$ ions in collisions with neutral atoms are shown as functions of the collision energy.

For a comparison we also display the loss cross section in collisions with bare nuclei. In the latter case the cross sections show a continuous logarithmic increase with energy. Such an increase of ionization cross sections in ultrarelativistic collisions is well known and is due to the Lorentz contraction of the electromagnetic field generated by a bare nucleus moving at velocities approaching the speed of light. Because of this contraction the effective time for a collision with a point-like charge is not given by $T(b) \sim b/v$ as in the nonrelativistic case but is estimated according to $T(b) \sim b/(\gamma v)$ (b is the impact parameter) and this time continues to decrease with increase of the collision energy even at $v \approx c$ where the collision velocity cannot be noticeably increased further. The external time-dependent field of the incident atomic nucleus is effective in inducing electron transitions in the ion only provided this field contains high enough frequency components. Therefore, the electron can make a transition with a noticeable probability only if the typical transition time $\tau \sim \omega_{n0}^{-1}$, where $\omega_{n0} = \varepsilon_n - \varepsilon_0$ is the energy transfer to the electron, does not exceed substantially the effective collision time $T(b)$. The latter condition means that the impact parameter range, contributing most to the loss, is given by $b \lesssim \gamma v/\omega_{\text{eff}}$, where ω_{eff} is of the order of the binding energy of the electron in the ion. This range of impact parameters gives rise to the dependence $\sigma_{\text{loss}} \sim \ln \gamma$ (see e.g. [99, 139, 140, 211, 220]).

Compared to the electron removal by collisions with bare nuclei, the distinct feature of the loss process in collisions with neutral atoms is the saturation of the loss cross section at sufficiently high impact energies where

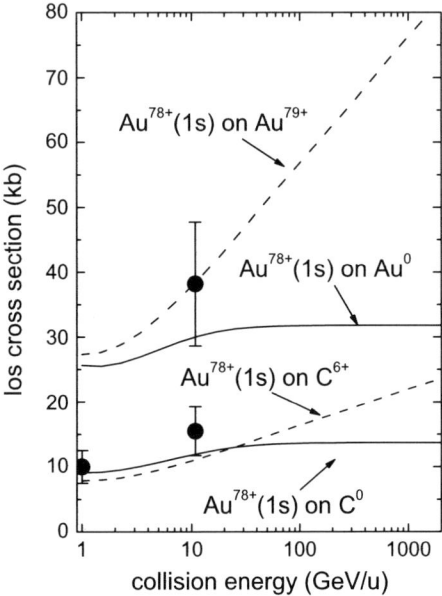

Fig. 8.21. Cross section for the electron loss from Au^{78+} shown as a function of the incident energy for collisions with neutral atoms of carbon and gold. The experimental points for lower and higher energies are from [112, 176], respectively. For a comparison the loss cross sections in collisions with bare gold and carbon nuclei, Au^{79+} and C^{6+}, are also displayed. All the results for collisions with carbon have been multiplied by a factor of 50. The calculated results were obtained by using the Coulomb–Dirac wave functions for the electron states in Au^{78+}.

this cross section becomes independent of the collisional Lorentz factor γ. One can denote this domain of collision energies as the region of asymptotically high γ.

In the elastic target mode, which dominates the total loss cross section in collisions with heavy atoms, the saturation of the loss cross section is caused by the following two factors.

One of them is that the collision velocity has the natural upper limit $v \leq c$. In practical terms this limit is reached already at $\gamma \sim 5$–10 which means that the collision velocity entering expressions for the form-factors and momentum transfers becomes a constant and no longer varies when the impact energy increases further.

The second factor leading to the saturation is that the net charge of a neutral atom is zero. In the elastic target mode the screening effect of the atomic electrons, whose total negative charge counter balances the positive charge of the atomic nucleus, simply 'puts out of play' collisions with impact parameters noticeably larger than the geometrical size of a neutral atom. Therefore, in contrast to collisions with bare atomic nuclei, the range of impact parameters contributing to the elastic part of the loss cross section in collisions

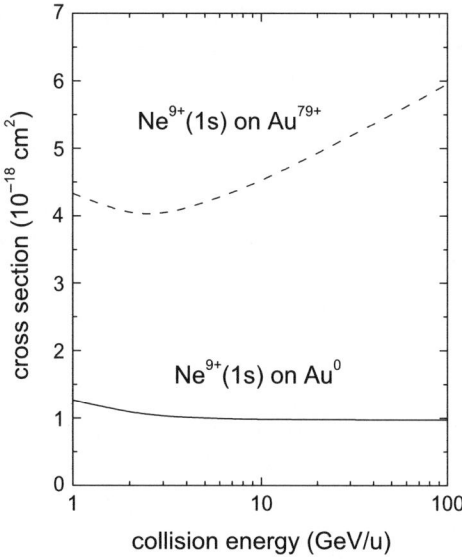

Fig. 8.22. Cross section for the electron loss from $Ne^{9+}(1s)$ shown as a function of the incident energy for collisions with neutral atoms of gold. For a comparison the loss cross section by a bare gold nucleus, Au^{79+}, is also displayed.

with neutral atoms cannot increase indefinitely with the increase in the impact energy. This, as well as the 'stabilization' of the effective strength of the ion–atom interaction in the limit $v \to c$, keeps the cross section constant at very high impact energy.

The screening effect becomes very substantial in such collisions where the impact parameters of importance are not too small and the electron of the ion, when penetrating the atom, 'sees' that a considerable part of the atomic electron cloud is situated between this electron and the atomic nucleus. In the electron loss from different ions, colliding with the same atom and at the same value of γ, relatively smaller impact parameters would contribute to the loss from a heavier ion. As a result, the screening effect in such collisions is smaller for heavier ions. On the other hand, if the electron loss from an ion occurs in collisions with different atoms but at the same collision energy per nucleon, then the screening effect is strongest for collisions with the heaviest atom.

At the asymptotically high γ-s the saturation effect is also present in the inelastic target mode of the ion–atom collisions. Compared to the case with the elastic mode, the origin of the saturation effect in the inelastic mode is somewhat less obvious. Although the existence of the upper limit for the collision velocity plays the same role here, now, according to the first order theory, the inelastic target mode of the collision is governed by the two-center electron-electron interaction. This means that the atomic nucleus 'drops out' of the electron loss process and, hence, one cannot argue that the saturation is caused by the electric neutrality of the atom.[16]

The reasons for the saturation of the loss cross sections at the asymptotically high γ were discussed in detail in Sect. 5.14.1, where the relativistic peculiarities in the two-center dielectronic interaction were considered. Therefore, here we just note that the explanations for the saturation effect, which were obtained there in terms of the momentum transfers, can be also interpreted as showing that the saturation in the inelastic target mode appears because the electrons, which are initially bound in the different colliding atomic particles, have to have collisions with not too large impact parameters in order that they would be able to undergo simultaneous transitions with a noticeable probability.

Let us now make two remarks concerning the behavior of the loss cross sections before they enter the saturation region and the values of the impact energies where the saturation is practically reached.

First, Fig. 8.21 shows that for the collision energies under consideration the cross section for the loss from very heavy ions increases with increase of the collision energy before this cross section enters the saturation region. In contrast, for the loss from much lighter ions like Ne^{9+} the loss cross section decreases before reaching a constant value (see Fig. 8.22).

Second, for different projectile–target pairs the region of the asymptotically high γ-s means in general different energies. For instance, while for collisions between very highly charged ions and light atoms this region approximately begins with impact energies of $\sim 100\,GeV\,u^{-1}$, for the loss from light ions the asymptotically high γ-s can be reached already at 5–$10\,GeV\,u^{-1}$.

We conclude this section by remarking that the saturation effect is obviously also present for the pair production processes occurring in collisions with neutral atoms. Since, compared to the projectile-electron loss, these processes are characterized by much larger momentum transfers, the saturation in the cross sections for the pair production begins at noticeably higher impact energies.

8.10 Excitation and Break-Up of Pionium in Relativistic Collisions with Neutral Atoms

The DIRAC experiment at CERN, aimed at measuring the lifetime of pionium [221], has sparked considerable interest in the study of excitation and break-up of pionium colliding with neutral atoms at relativistic velocities ($\gamma \sim 15$–20). Since the pionium-atom collisions occur predominantly via the electromagnetic interaction, the excitation and break-up of pionium in such collisions are closely related to the ion–atom collisions and we will briefly comment on these processes.

---

[16] In the inelastic target mode the saturation effect exists also for ion–ion collisions.

From the point of view of the common atomic physics, pionium, which is a bound state of π^+ and π^- both having zero spin, represents rather an exotic object. The lifetime of pionium in the ground state is of the order of 10^{-15} s. Compared to a 'normal' hydrogen-like system, consisting of a heavy nucleus and a light electron, pionium has other important differences. The masses of π^+ and π^- are equal, $m_{\pi\pm} \simeq 270m_e$ ($m_e = 1$ is the electron mass), that may bring in considerable features into the dynamics of excitation and break-up of pionium, which would be absent in the case of the excitation or 'ionization' of a hydrogen-like ion. Since the reduced mass of pionium is large, $\mu_\pi \simeq 137m_e = 137$ a.u., the typical dimension of the ground state of the pionium is even smaller than that of the electron orbit in the ground state of U^{91+}. At the same time, the relative velocity of π^+ and π^- in the ground state of pionium is of the order of 1 a.u. and in this sense pionium represents, with an excellent accuracy, a nonrelativistic object.

Excitation and break-up of pionium by collisions with atoms is conveniently described using the pionium frame [222–227]. In this reference frame the motion of the π^+ and π^- is practically always nonrelativistic. Because of this, in collisions with neutral atoms, the exchange of the transverse virtual photon affects pionium transitions much weaker than the exchange of the longitudinal photon. In particular, using the first order perturbation approach, it was shown in [224, 226] that the relative contributions of the exchange of the transverse photon to pionium transitions is substantially less than 1%.

In [224–226] it was argued that, within the first order approximation, one can calculate cross sections for pionium with accuracy better than 1% for both the elastic and inelastic modes. However, substantial deviations from predictions of the first order consideration can occur in collisions with heavy atoms. In such a case it becomes important to take into account the exchange of more than one longitudinal photon between the pionium and the incident atom. Using the Glauber approximation it was shown in [222, 223, 227] that the account of the many-photon exchanges may reduce cross sections by much more than 10% (see Fig. 8.23).[17]

All the calculations of the pionium cross sections briefly discussed above are atomic physics calculations and in general can be directly compared with experiment only provided the pionium interacts with a dilute gas target. In experiment, however, the created pionium atoms propagate in solids. Discussions of some questions concerning the propagation of pionium through the media can be found in [229–231].

---

[17] Concerning the use of the Glauber approximation in relativistic collisions one can also note that (a particular version of) this approximation was incorporated in an approach proposed in [228] to treat the ionization of light atoms by relativistic highly charged ions.

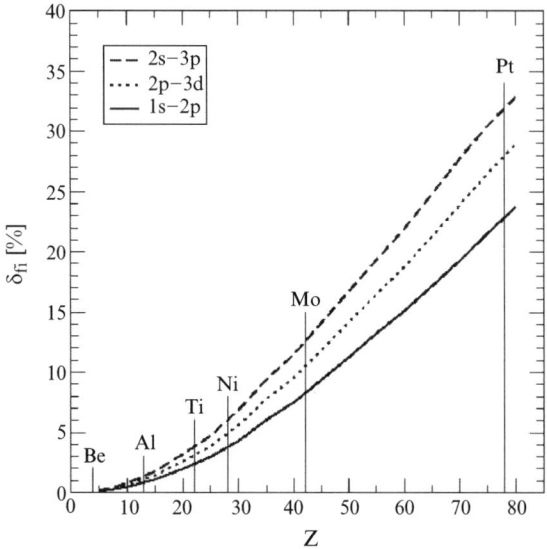

Fig. 8.23. The relative correction δ_{fi} to the excitation cross sections as a function of the atomic number Z of the target. The correction δ_{fi} is the difference between the first-order and Glauber cross sections normalized to the first-order cross section. In the figure this correction is shown for the $1s - 2p$, $2p - 3d$ and $2s - 3p$ transitions by *solid, dot* and *dash curves*, respectively. From [227].

8.11 Higher-Order Effects at Asymptotically High γ

When we discussed the probabilities for the projectile-electron loss in collisions with neutral atoms (see Chap. 7), it was already pointed out (see Sect. 7.4) that the differences between results of the first order and light-cone approximations in general do not disappear when the impact energy increases.

This point is illustrated by results shown in Table 8.1. This table contains the total cross sections for the electron loss from different hydrogen-like ions colliding with neutral atoms of uranium at the infinite impact energy ($\gamma = \infty$). These cross sections were calculated using the first order approximation and the light-cone approach. The contribution to the cross section arising due to the correlations between the electron of the projectile and the electrons of the atom (the antiscreening atomic contribution) was neglected.

It follows from the table that the discrepancy between the first order and light-cone results increases when the atomic number of the projectile decreases reaching about 33% for the $Ne^{9+}(1s)$ projectiles. Thus, one may conclude that even the total loss cross section in collisions between a relatively light ion and a very heavy atom cannot be well described within the first order approach, no matter how high the impact energy is.

As we have seen in the beginning of this chapter, in relativistic collisions with heavy atoms at relatively low energies ($\gamma \sim 1$) the difference between

Table 8.1. The total cross section (in kb) for the electron loss from hydrogen-like projectile-ions in collisions with neutral uranium atoms at the infinite impact energy. The *first column*: the incident projectile. The *second column*: the target. The it third column: results of the first order approximation. The *fourth column*: results of the light-cone approximation

Projectile	First order	Light-cone
$U^{91+}(1s)$	29	27
$Xe^{54+}(1s)$	90.9	80.7
$Kr^{35+}(1s)$	187.4	162.7
$Ar^{18+}(1s)$	629	478
$Ne^{9+}(1s)$	1,450	1,090

the experimental data and first order results for the total cross sections for the electron loss may reach an order of magnitude. At extreme high collision energies such very large differences are not possible. However, even at asymptotically high energies there still may remain a substantial deviation between the first order and exact results for the total loss cross section.

Similar conclusions can be also drawn in the case of the excitation and break-up of pionium. The asymptotically high impact energies are in essence reached when the interaction between the pionium and the atom becomes 'sudden' for the pionium for all impact parameters which contribute to the cross sections, i.e. for all impact parameters which are of order or less than the atomic size.

The condition of the suddenness is given by $T(b) \ll \tau$, where $T(b) \sim b/(\gamma v)$ is the effective collision time and $\tau \sim 1/\omega_{n0}$ is the typical transition time. Typical pionium transition frequencies are of order of $\mu_\pi v_\pi^2 \sim 100$ a.u. Taking all this into account one can conclude that in the case of the excitation and break-up of pionium the asymptotically high impact energies are reached already at $\gamma \simeq 10$. A further increase of γ changes neither the cross sections nor the deviations between predictions of the first order and Glauber approximations.

This, in particular, means that the calculations for the pionium cross sections, reported in [222–227], can actually be viewed as performed for the infinite impact energy, $\gamma \to \infty$. Correspondingly, the results displayed in Fig. 8.23 clearly show the necessity to treat such collisions beyond the first order approximation.

The above discussion for the projectile-electron loss was concerned with projectiles which initially carry only one electron. Highly charged projectiles may initially carry two and more electrons. When such a projectile collides with an atom, more than one projectile electron may undergo transitions. In such a case the range of the asymptotically high impact energies is characterized by much more substantial deviations from the first order predictions (see [174, 232, 233]).

A

Appendix

A.1 Nonrelativistic Atom Approximation for the Screening Mode

For simplicity we shall restrict our discussion to single electron targets. In such a case we have to consider the following three matrix elements:

$$\langle \phi_0 | \alpha_x \exp(i\mathbf{q}_0 \cdot \boldsymbol{\xi}) | \phi_0 \rangle,$$

$$\langle \phi_0 | \alpha_y \exp(i\mathbf{q}_0 \cdot \boldsymbol{\xi}) | \phi_0 \rangle,$$

and

$$\frac{1}{c} \langle \phi_0 | v\alpha_z \exp(i\mathbf{q}_0 \cdot \boldsymbol{\xi}) | \phi_0 \rangle,$$

where $\mathbf{q}_0 = (-\mathbf{q}_\perp, -(\varepsilon_n - \varepsilon_0)/(v\gamma))$.

Let us start with the matrix element containing α_z. First we rewrite the term $v\alpha_z$ as follows:

$$v\alpha_z = \mathbf{v} \cdot \boldsymbol{\alpha} = \left(\mathbf{v} - \frac{\Omega}{q_0^2} \mathbf{q}_0 \right) \cdot \boldsymbol{\alpha} + \frac{\Omega}{q_0^2} \mathbf{q}_0 \cdot \boldsymbol{\alpha}, \tag{A.1}$$

where Ω is a parameter to be determined below. Taking into account (5.56) we immediately see that $\langle \phi_0 | \mathbf{q}_0 \cdot \boldsymbol{\alpha} \exp(i\mathbf{q}_0 \cdot \boldsymbol{\xi}) | \phi_0 \rangle = 0$ and obtain

$$\frac{1}{c} \langle \phi_0 | v\alpha_z \exp(i\mathbf{q}_0 \cdot \boldsymbol{\xi}) | \phi_0 \rangle = \frac{1}{c} \langle \phi_0 | \mathbf{e} \cdot \boldsymbol{\alpha} \exp(i\mathbf{q}_0 \cdot \boldsymbol{\xi}) | \phi_0 \rangle, \tag{A.2}$$

where $\mathbf{e} = \mathbf{v} - \frac{\Omega}{q_0^2} \mathbf{q}_0$. Now we define the parameter Ω by demanding that the vector $\mathbf{e}$ is perpendicular to the momentum transfer $\mathbf{q}_0$ that yields $\Omega = \mathbf{v} \cdot \mathbf{q}_0 = -\frac{\varepsilon_n - \varepsilon_0}{\gamma}$. With such a choice of the 'polarization' vector $\mathbf{e}$ the structure of the matrix element on the right-hand side of (A.2) becomes quite similar to that appearing in the study of the interaction between the atom and a real photon which has linear polarization $\sim\mathbf{e}$ and momentum $\mathbf{q}_0$.

Let us now turn to the consideration of $\langle \phi_0 | \alpha_y \exp(i\mathbf{q}_0 \cdot \boldsymbol{\xi}) | \phi_0 \rangle$. Without any loss of generality we can assume that the total momentum transfer $\mathbf{q}_0$ is in the plane xz. But then, by writing $\alpha_y = \boldsymbol{\lambda} \cdot \boldsymbol{\alpha}$ with $\boldsymbol{\lambda} = (0, 1, 0)$ and, thus, $\boldsymbol{\lambda} \perp \mathbf{q}_0$, we arrive at the same situation as that discussed in the previous paragraph.

The continuity equation (5.56) for the atomic current in the elastic mode reads

$$\langle \phi_0 | q_{0x} \alpha_x \exp(i\mathbf{q}_0 \cdot \boldsymbol{\xi}) | \phi_0 \rangle + \langle \phi_0 | q_{0y} \alpha_y \exp(i\mathbf{q}_0 \cdot \boldsymbol{\xi}) | \phi_0 \rangle$$
$$+ \langle \phi_0 | q_{0z} \alpha_z \exp(i\mathbf{q}_0 \cdot \boldsymbol{\xi}) | \phi_0 \rangle = 0. \quad \text{(A.3)}$$

Taking into account that the ratio q_{0x}/q_{0z} can be arbitrary we conclude that the x-component of the elastic atomic form-factor will not be zero provided that the z component is also not zero.

Thus, for the elastic mode all three space parts of the atomic form-factor of hydrogen-like targets will not be zero only when the known selection rules for the interaction with a real photon permit the 'transition' $\phi_0 \rightarrow \phi_0$. If such a 'transition' is forbidden, then the nonrelativistic atom 'approximation' becomes in fact exact for the elastic mode.

A.2 The Schrödinger–Pauli Equation and Relativistic Collisions

It is known that, in relativistic collisions involving light atomic systems, the electron dynamics in the processes of excitation and ionization/loss (described in appropriate reference frames) remains practically purely nonrelativistic. Therefore, in an attempt to make simplifications in the treatment of relativistic atomic collisions, it was proposed in [103] to use the Schrödinger–Pauli equation for calculating cross sections for ionization and excitation of atoms by relativistically moving nuclei provided the atoms are relatively light.

On this way, however, the authors of [103] had encountered a problem. Namely, they found out that the contribution to the ionization cross section, which arose in their first order perturbation theory due to the term proportional to A^2 (**A** is the vector potential of the field generated by the incident nucleus), possesses wrong asymptotic dependences on the collision energy and the nucleus charge. In order to solve this problem it was argued [103] that a more careful analysis is needed for the transition from the relativistic Dirac to the nonrelativistic Schrödinger–Pauli equation. After analyzing such a transition, the authors of [103] had concluded that the correct wave equation for a nonrelativistic electron should include an additional term which is proportional to $-\Phi^2$, where Φ is the scalar potential of the field generated by the incident nucleus.

Another group of authors [226] studied the problem of the excitation and break-up of pionium in relativistic collisions. Since pionium consists of π^+

and π^-, which are particles having spin zero, the consideration of this problem should in general be based on the Klein–Gordon equation. In a reference frame, where initially the pionium is at rest, the velocities of both pions during and after the collision remain to be much less than the speed of light. Taking this into account, the authors of [226] attempted to describe the excitation and break-up of pionium by using the Schröidnger equation but faced the same problem with the contribution from the term proportional to A^2, which was previously encountered in [103]. They, therefore, decided to reconsider the transition from the Klein–Gordon to the Schrödinger equation. By analyzing this transition the authors of [226] have come to to the same conclusion that the term proportional to $-\varPhi^2$ should be added into the Schrödinger equation.

The form of a wave equation is a matter of principle and any attempt to alter the Schrödinger (or Schrödinger–Pauli) equation for a nonrelativistic electron should be very carefully analyzed. Therefore, following [111] in this section we consider in detail some delicate points concerning the form of a wave equation which enables one to describe the motion of a nonrelativistic electron in collisions between an atomic system, where the electron is initially bound, and a point-like charge moving with a relativistic velocity with respect to the atomic system. Besides, we shall also very briefly comment on the considerations for the transition from the Klein–Gordon equation to the Schrödinger equation given in well known textbooks [74, 125].

The Schrödinger–Pauli equation for an electron, which moves in the field of the nucleus with a charge Z_I and is subjected in the collision to the field of a relativistically moving nucleus with a charge Z_A, reads

$$i\frac{\partial \varPsi_s}{\partial t} = \left(\hat{H}_s^0 + \hat{W}_s(t) \right) \varPsi_s, \qquad (A.4)$$

where the nonrelativistic electronic Hamiltonian $\hat{H}_s^0$ for the undistorted atom is given by (5.105) and the interaction $\hat{W}_s$ between the electron and the projectile reads

$$\hat{W}_s(t) = -\varPhi(\mathbf{r}, t) + \frac{1}{2c} \left(\mathbf{A}(\mathbf{r}, t) \cdot \hat{\mathbf{p}} + \hat{\mathbf{p}} \cdot \mathbf{A}(\mathbf{r}, t) \right) + \frac{\mathbf{A}^2(\mathbf{r}, t)}{2c^2} + \frac{1}{2c} \boldsymbol{\sigma} \cdot \mathbf{H}(\mathbf{r}, t). \qquad (A.5)$$

Here $\boldsymbol{\sigma} = (\sigma_x, \sigma_y, \sigma_z)$ are the Pauli matrices, $\varPhi$ and $\mathbf{A}$ are the scalar and vector potentials of the electromagnetic field generated by the nucleus Z_A and $\mathbf{H} = \boldsymbol{\nabla} \times \mathbf{A}$ is the magnetic part of this field.

Similarly to the applications of the Schrödinger equation to relativistic collisions, reported in the literature, we adopt for the moment that the first order atomic transition amplitude in the impact parameter space is given by

$$a_{fi}(\mathbf{b}) = -i \int_{-\infty}^{+\infty} dt \exp(i(\epsilon_f - \epsilon_i)t) \langle \chi_f \mid \hat{W}_s(t) \mid \chi_i \rangle. \qquad (A.6)$$

where χ_i and χ_f are the initial and final nonrelativistic states, respectively, of the electron in the undistorted atom, which are eigenstates of the nonrelativistic atomic Hamiltonian $\hat{H}_s^0$ with corresponding eigenvalues denoted by ϵ_i and ϵ_f. We shall assume that the spin of the electron in the initial state is quantized along the velocity $\mathbf{v}$.

The momentum–space transition amplitudes for no-spin-flip and spin-flip electron transitions are obtained from (A.5), (A.6) and the relation (3.13) which connects the amplitudes in the $\mathbf{b}$ and $\mathbf{q}_\perp$ spaces. Choosing the potentials of the projectile field in the Lienard–Wiechert form (see (5.86) we obtain that the amplitude for electron transitions without spin-flip is written as

$$S_{fi}^{\text{no-flip}}(\mathbf{q}_\perp) = S_{fi}^{(1)}(\mathbf{q}_\perp) + S_{fi}^{(2)}(\mathbf{q}_\perp), \tag{A.7}$$

where $S_{fi}^{(1)}$ is given by (5.98)) and

$$S_{fi}^{(2)}(\mathbf{q}_\perp) = \frac{-i\pi Z_A^2 \gamma v}{c^4} \frac{1}{\sqrt{\mathbf{q}_I^2 - (\varepsilon_f - \varepsilon_i)^2/c^2}} \langle \varphi_f \mid \exp(i\mathbf{q}_I \cdot \mathbf{r}) \mid \varphi_i \rangle. \tag{A.8}$$

Here φ_i and φ_f are the space parts of the initial and final electron wavefunctions and $\mathbf{q}_I$ is defined by (5.84). Further, the amplitude for electron spin-flip transitions is obtained to be

$$S_{fi}^{\text{flip}}(\mathbf{q}_\perp) = -\frac{iZ_A}{c^2} \frac{q_{I,x} + iq_{I,y}}{\mathbf{q}_I^2 - (\varepsilon_f - \varepsilon_i)^2/c^2} \langle \varphi_f \mid \exp(i\mathbf{q} \cdot \mathbf{r}) \mid \varphi_i \rangle. \tag{A.9}$$

The amplitude for the electron transition without spin flip contains the part $S_{fi}^{(2)}$ which is proportional to Z_A^2 and which arises due to the contribution from the $A^2/2c^2$ term in the interaction (A.5). It was noticed long ago in [103] that at asymptotically high impact energies this term leads to wrong dependences of calculated cross sections on the collision energy and the projectile charge. This does not represent a big obstacle for 'practical' calculations since the wrong term dominates cross sections only at very high values of γ. However, from the basic point of view such a behavior is quite interesting and deserves a detailed discussion.

A.2.1 Wave Equation for a Nonrelativistic Electron and the $-\frac{\Phi^2}{2c^2}\Psi$ Term

As was already mentioned, there have been two attempts to address the problem with the $\mathbf{A}^2/2c^2$-term. The authors of [103] pointed to the fact that in collisions at high γ the scalar potential given in (5.86) may become comparable with or even exceed the rest energy of the electron. They argued that because of this the transformation of the relativistic Dirac equation into the nonrelativistic Schrödinger–Pauli equation must be done more carefully compared to the standard text-book analysis. According to [103], when the potentials entering the Schrödinger equation describe the electromagnetic field generated by a relativistically moving projectile, the canonical form of this equation has to be corrected by introducing the additional term $-\Phi^2/2c^2$ into the Hamiltonian. Then the corrected wave equation for a nonrelativistic electron would read

$$
i\frac{\partial \Psi_{\mathrm{s}}}{\partial t} = \left(\frac{\hat{\mathbf{p}}^2}{2m} - \frac{Z_{\mathrm{I}}}{r} - \Phi + \frac{1}{2c}\left(\mathbf{A}\cdot\hat{\mathbf{p}} + \hat{\mathbf{p}}\cdot\mathbf{A} \right) + \frac{\mathbf{A}^2}{2c^2} - \frac{\Phi^2}{2c^2} + \frac{1}{2c}\boldsymbol{\sigma}\cdot\mathbf{H} \right)\Psi_{\mathrm{s}}.
$$

$$(A.10)$$

The same correction $-\Phi^2/2c^2$ to the nonrelativistic Hamiltonian has been also proposed in [226], where the excitation and break-up of pionium in relativistic collisions with atoms was studied and where the transition from the Klein–Gordon to the Schrödinger equation was analyzed.

Both the authors of [103] and the authors of [226] employed in their considerations the Lienard–Wiechert potentials (5.86). For these potentials one has $A^2/2c^2 - \Phi^2/2c^2 = -\Phi^2/(2c^2\gamma^2)$. Therefore, at $\gamma \gg 1$ the two terms quadratic in the potentials mutually 'neutralize' each other and the term $\sim Z_{\mathrm{A}}^2$ in the transition amplitude calculated by using (A.10) becomes of minor importance also at asymptotically high collision energies. As a result, the problem seems to be solved (see also a brief discussion of this subject in [3], p. 157).

However, a closer look at this problem leads to serious doubts about the validity of the wave equation (A.10). In order to see this, let us consider in some detail the transition from the relativistic to the nonrelativistic description of the electron. Let us start with the Dirac equation and let $\Psi_{\mathrm{u}}(\mathbf{r},t)$ and $\Psi_{\mathrm{d}}(\mathbf{r},t)$ be the upper and lower components, respectively, of the Dirac spinor $\Psi(\mathbf{r},t)$. These components satisfy the following system of coupled equations

$$
\left(i\frac{\partial}{\partial t} + \Phi_t - c^2 \right)\Psi_{\mathrm{u}} = c\boldsymbol{\sigma}\cdot\left(\hat{\mathbf{p}} + \frac{1}{c}\mathbf{A} \right)\Psi_{\mathrm{d}}
$$
$$
\left(i\frac{\partial}{\partial t} + \Phi_t + c^2 \right)\Psi_{\mathrm{d}} = c\boldsymbol{\sigma}\cdot\left(\hat{\mathbf{p}} + \frac{1}{c}\mathbf{A} \right)\Psi_{\mathrm{u}}, \qquad (A.11)
$$

where $\Phi_t = \Phi + Z_{\mathrm{I}}/r$. By acting from the left with the operator $\left(i\frac{\partial}{\partial t} + \Phi_t + c^2 \right)$ on both sides of the first equation in (A.11) we obtain

$$
i\frac{\partial \Psi^0}{\partial t} + i\frac{\Phi_t}{c^2}\frac{\partial \Psi^0}{\partial t} - \frac{1}{2c^2}\frac{\partial^2 \Psi^0}{\partial t^2} = \left(\frac{\hat{\mathbf{p}}^2}{2} - \Phi_t + \frac{\mathbf{A}}{c}\cdot\hat{\mathbf{p}} - \frac{i}{2c}(\boldsymbol{\nabla}\cdot\mathbf{A}) \right)\Psi^0
$$
$$
+ \left(\frac{\mathbf{A}^2}{2c^2} + \frac{1}{2c}\boldsymbol{\sigma}\cdot\mathbf{H} \right)\Psi^0 - \left(\frac{i}{2c^2}\frac{\partial \Phi_t}{\partial t} + \frac{\Phi_t^2}{2c^2} \right)\Psi^0
$$
$$
- i\frac{1}{2c}\boldsymbol{\sigma}\cdot\mathbf{E}\,\Psi_{\mathrm{d}}\exp(ic^2 t), \qquad (A.12)
$$

where $\Psi^0 = \Psi_{\mathrm{u}}\exp(ic^2 t)$ and $\mathbf{E} = -\boldsymbol{\nabla}\Phi_1 - \frac{1}{c}\frac{\partial \mathbf{A}}{\partial t}$ is the strength of the total electric field acting on the electron. No approximations have been done on the way from (A.11) to (A.12) and, hence, the latter is still exact.

Had we started from the Klein–Gordon equation, we would have arrived at a result which can formally be obtained from (A.12) by omitting the spin terms. Since in the present context the spin term in the Schrödinger–Pauli equation does not represent any problem, we may simply ignore those terms in (A.12), which contain the Pauli matrices, obtaining

$$
i\frac{\partial \Psi^0}{\partial t} = \left(\frac{\hat{\mathbf{p}}^2}{2} - \Phi_t + \frac{\mathbf{A}}{c} \cdot \hat{\mathbf{p}} - \frac{i}{2c}(\nabla \cdot \mathbf{A}) + \frac{\mathbf{A}^2}{2c^2} \right) \Psi^0
$$
$$
- \left(\frac{i}{2c^2}\frac{\partial \Phi_t}{\partial t} + \frac{\Phi_t^2}{2c^2} \right) \Psi^0 - i\frac{\Phi_t}{c^2}\frac{\partial \Psi^0}{\partial t} + \frac{1}{2c^2}\frac{\partial^2 \Psi^0}{\partial t^2} \qquad (A.13)
$$

and analyze the transformation of the Dirac and Klein–Gordon equations in a unified manner.

In (A.13) the first line coincides in form with the Schrödinger equation while the second line contains the additional terms which do not appear in the latter. Comparing (A.13) and (A.10), we observe that the modified wave equation proposed in [103, 226] includes only the term $\sim(-\Phi^2/c^2)$ whereas the other additional terms contained in (A.13) are absent in (A.10).

The basic condition underlying our consideration (as well as those of [103, 226]) is that the motion of the quantum particle is *nonrelativistic*. Taking into account the relation between Ψ^0 and Ψ_u, this condition implies that

$$
\frac{1}{c^2} \left| i\frac{\partial \Psi^0}{\partial t} \right| \ll |\Psi^0|, \quad \frac{1}{c^2} \left| \frac{\partial^2 \Psi^0}{\partial t^2} \right| \ll \left| i\frac{\partial \Psi^0}{\partial t} \right|. \qquad (A.14)
$$

In order to get an idea about the character of the difference between (A.13) and the Schrödinger equation in the case of the description of a nonrelativistic electron and, in particular, about the relative importance of the different terms in the second line of (A.13), we first assume that the magnitude of the scalar potential is not large, $|\Phi| \ll c^2$. Then, taking into account (A.14), it is obvious that the terms in the second line of (A.13) represent just small corrections to those in its first line. Therefore, the second line of (A.13) can be estimated by setting in this line $\Psi^0 = \psi_s$, where ψ_s is the solution of the Schrödinger equation

$$
i\frac{\partial \psi_s}{\partial t} = \left(\frac{\left(\hat{\mathbf{p}} + \frac{\mathbf{A}}{c}\right)^2}{2} - \Phi_t \right) \psi_s. \qquad (A.15)
$$

With the help of (A.15), after somewhat lengthy but straightforward calculations, one can show that

$$
\frac{1}{2c^2}\frac{\partial^2 \psi_s}{\partial t^2} - i\frac{\Phi_t}{c^2}\frac{\partial \psi_s}{\partial t} = \left(\frac{i}{2c^2}\frac{\partial \Phi_t}{\partial t} + \frac{\Phi_t^2}{2c^2} \right) \psi_s
$$
$$
- \frac{1}{8c^2}\left(\hat{\mathbf{p}} + \frac{\mathbf{A}}{c}\right)^4 \psi_s + \frac{i}{2c^2}\mathbf{E} \cdot \left(\hat{\mathbf{p}} + \frac{\mathbf{A}}{c}\right) \psi_s
$$
$$
+ \frac{\nabla \cdot \mathbf{E}}{4c^2} \psi_s. \qquad (A.16)
$$

For a nonrelativistic electron, taking into account that its kinetic momentum $\hat{\mathbf{k}} = \hat{\mathbf{p}} + \frac{\mathbf{A}}{c}$ is much smaller than c, the last three terms on the right-hand side of (A.16) represent corrections $\sim 1/c^2$ and should not be taken into account in the description of a nonrelativistic electron. Concerning the first two terms with the scalar potential on the right-hand side of (A.16) we see that these terms being inserted into the second line of (A.13) cancel the first two terms in this line. This cancellation clearly shows that for $\mid \Phi \mid /c^2 \ll 1$ the difference between the Schrödinger equation and (A.13) does not contain the term $\sim \Phi^2/c^2$ (and even does not explicitly depend on the scalar potential).

Further, as long as the basic condition of the nonrelativistic character of the electron motion remains fulfilled, the above consideration and its conclusion are in fact fully valid also in the case when the absolute magnitude of the scalar potential Φ approaches (or exceeds) the electron rest energy. Indeed, this magnitude *per se* does not determine the real strength of the field of the incident nucleus and cannot serve as a measure of the difference between the Schrödinger equation and (A.13) because (i) (A.13) as well as the Schrödinger equation are gauge invariant and (ii) one can always find a gauge in which the absolute magnitude of the scalar potential can be made as small as desired for the whole space and time (see e.g. [8]). Therefore, for any Φ one can always resort to an appropriate gauge transformation obtaining in a new gauge $\mid \Phi' \mid \ll c^2$. Then one can repeat the above analysis and arrive at the same conclusion that in the case of the description of a *nonrelativistic* electron the difference between the Schrödinger equation and (A.13) essentially vanishes. Since both these wave equations are gauge invariant, the sum of the terms in the second line of (A.13) is gauge invariant as well and it is obvious that for any magnitude of the scalar potential the description of the motion of a *nonrelativistic* electron may not contain the term $\sim \Phi^2$.

In addition to the above discussion one should also note that the basic properties of the nonrelativistic Schrödinger equation (for instance, its compatibility with the minimal coupling of electrodynamics and gauge invariance) are violated in the wave equation (A.10). Taking all this into account it becomes clear that from the fundamental point of view the attempts to 'amend' the Schrödinger equation by incorporating the interaction term $-\Phi^2/2c^2$ into the Hamiltonian cannot be regarded as satisfactory and by no means represent a solution of the problem with the unphysical contribution of the $\mathbf{A}^2/2c^2$-term into the transition amplitudes (A.6)–(A.7).

Moreover, the wave equation (A.10) cannot be viewed even as a 'technical' solution of this problem. Indeed, at high collision energies the near cancellation between A^2 and $-\Phi^2$ terms occurs only provided the potentials of the projectile field are taken in the Lienard–Wiechert form. The latter, however, merely represents a particular choice within the Lorentz family of gauges. For instance, it is not difficult to convince oneself that in the Coulomb gauge given by (5.96) the addition of the term $-\Phi^2/2c^2$ does not lead to the near cancellation. Calculations performed in this gauge using the amplitude (A.6) again lead to wrong cross section dependences on the energy and charge.[1] What is more, when using the gauge described by (5.93) one encounters an even more severe problem with the term $A^2/2c^2$ whose contribution into the amplitude (A.7) in this gauge can be shown to be infinite for any projectile charge and collision energy! In this gauge $\Phi = 0$ and the incorporation of the term $-\Phi^2/2c^2$ would have no impact on the result. This example also clearly shows that the problem with the $A^2/2c^2$ term in the amplitude (A.7) is not specific for fields generated by relativistically moving projectiles but has a more general character.

A.2.2 'First Order' Amplitude and Nonconserved Electron Current

In order to get insight into the true root of the problem with the term $\mathbf{A}^2/2c^2$ let us consider the coupling between an electron and an electromagentic field generated by an external source. The variation in the interaction between the electron and the field is given by (see e.g. [9])

$$\delta w = \rho(\mathbf{r}, t)\, \delta\Phi(\mathbf{r}, t) - \frac{1}{c}\mathbf{j}(\mathbf{r}, t)\, \delta\mathbf{A}(\mathbf{r}, t), \tag{A.17}$$

where ρ and $\mathbf{j}$ are the charge density and the current density, respectively, and $\delta\Phi$ and $\delta\mathbf{A}$ are infinitesimally small variations in the scalar and vector potentials of the field. For a nonrelativistic electron the quantities ρ and $\mathbf{j}$ are evaluated by using the Schrödinger–Pauli equation. Expression (A.17) depends explicitly on the field potentials and, as is well known, leads to gauge independent results only provided the electron charge and current densities satisfy the continuity equation

$$\frac{\partial\rho(\mathbf{r}, t)}{\partial t} + \boldsymbol{\nabla} \cdot \mathbf{j}(\mathbf{r}, t) = 0, \tag{A.18}$$

i.e. if the electron charge is conserved.

Equation (A.18) will certainly be fulfilled if the charge and current densities are constructed by using exact solutions of the Schrödinger–Pauli equation. However, in first order considerations the exact solutions are replaced by states of the undistorted atom. If, in an attempt to get the transition charge density and current which could be associated with the first order consideration, we simply replace the exact solutions by the undistorted atomic states, the corresponding transition charge density and current will be given by

$$\rho_{fi}^{(1)}(\mathbf{r}, t) = \exp(\mathrm{i}(\varepsilon_\mathrm{f} - \varepsilon_\mathrm{i})t)\, \chi_\mathrm{f}^\dagger(\mathbf{r})\chi_\mathrm{i}(\mathbf{r})$$

$$\mathbf{j}_{fi}^{(1)}(\mathbf{r}, t) = \exp(\mathrm{i}(\varepsilon_\mathrm{f} - \varepsilon_\mathrm{i})t)\left(\frac{1}{2\mathrm{i}}\left(\chi_\mathrm{f}^\dagger(\mathbf{r})\boldsymbol{\nabla}\chi_\mathrm{i}(\mathbf{r}) - \left(\boldsymbol{\nabla}\chi_\mathrm{f}^\dagger(\mathbf{r})\right)\chi_\mathrm{i}(\mathbf{r})\right)\right.$$

$$\left. - \frac{1}{c}\mathbf{A}(\mathbf{r}, t)\chi_\mathrm{f}^\dagger(\mathbf{r})\chi(\mathbf{r}) + \frac{1}{2c}\boldsymbol{\nabla} \times \left(\chi_\mathrm{f}^\dagger(\mathbf{r})\boldsymbol{\sigma}\chi_\mathrm{i}(\mathbf{r})\right)\right). \tag{A.19}$$

---

[1] Although the result in the Coulomb gauge does not coincide with that obtained by using the Lienard–Wiechert potentials.

According to the form of the interaction (A.17) it is natural to expect that, within the first order coupling between the atomic electron and the field of the projectile, the electron transition charge and current densities should be independent of the field of the projectile. It indeed would be true for the charge and current densities appearing in the description based on the Dirac equation. It is also true for the nonrelativistic charge density in (A.19) but not for the nonrelativistic transition current which in (A.19) explicitly depends on the vector potential. Since this is the only dependence on the field which this current possesses, it cannot be compensated by any other terms in the current. As a result, because of the presence of the term proportional to $\mathbf{A}$, the current (A.19) is gauge dependent and, moreover, the condition (A.18) of the charge conservation is violated.

However, it is the part of the electron current (A.19) proportional to $\mathbf{A}$, which is directly associated with the term $\sim\mathbf{A}^2$ in the Schrödinger equation. Therefore, the root of the problem with the unphysical contribution of the $A^2/2c^2$ term into the amplitude (A.6) can be identified as the nonconservation of the electron charge and the concomitant gauge dependence.

A.2.3 Correct Form of the First Order Amplitude

The obvious way to cure this problem is to notice that the current in the form given by (A.19) is simply not compatible with the rigorous first order consideration in which only a one-photon exchange between the projectile and the electron may be treated in a consistent way. Therefore, within the first order projectile-electron coupling, the $\mathbf{A}$-dependent term in the electron transition current (A.19) has to be omitted. This step, provided exact states of the undistorted atom are employed, restores the charge conservation and yields gauge independent results for the transition amplitudes and cross sections. In particular, the latter ones possess the correct asymptotic dependences on the projectile charge and energy and at $Z_I/c \ll 1$ coincide with the cross sections calculated by using the relativistic transition amplitude (5.83) (or (5.88) and (5.90)).

In other words, within the self-consistent first order treatment the term of the Schrödinger equation quadratic in the vector potential should not be taken into account just because it is alien to such a treatment. In our case this simply means that the correct first order amplitude for the electron transitions without spin flip is given merely by the first term in (A.7).

In order to avoid any possible misinterpretation of the above discussion, one should add that the term $\sim\mathbf{A}^2/2c^2$ must of course be kept in considerations of relativistic atomic collisions which go beyond the first order in the projectile-electron coupling. Within such considerations this term is not expected to lead to any unphysical results. For instance, it is known that no problem with this term arises when the projectile-electron interaction is treated within the symmetric eikonal approximation [234, 235].

As soon as the **A**-dependent term in the current (A.19) is omitted and exact states of the undistorted atom are employed, the charge conservation is restored and for electron transitions without spin-flip the following continuity equation holds

$$(\epsilon_f - \epsilon_i)\,\langle \varphi_f \mid e^{i\mathbf{q_I}\cdot\mathbf{r}} \mid \varphi_i\rangle = \frac{1}{2}\langle \varphi_f \mid \left(\mathbf{q_I}\cdot\hat{\mathbf{p}}\,e^{i\mathbf{q_I}\cdot\mathbf{r}} + e^{i\mathbf{q_I}\cdot\mathbf{r}}\,\mathbf{q_I}\cdot\hat{\mathbf{p}}\right) \mid \varphi_i\rangle. \quad (A.20)$$

Equation (A.20) expresses the conservation of charge in the case of a non-relativistic electron and enables to manipulate easily with the form of the nonrelativistic first order transition amplitude. In particular, by using (A.20) and (5.98) one can obtain the amplitudes (5.99) and (5.100).

A.2.4 Few Remarks on the Treatment of the Transformation from the Klein–Gordon Equation to the Schrödinger Equation given in Some Textbooks

The relativistic (Dirac and Klein–Gordon) and nonrelativistic (Schrödinger) wave equations have been discovered very long ago. In particular, the reductions of the relativistic wave equations into the nonrelativistic one in the case, when the velocity of a quantum particle is much less than the speed of light, seem to be well established procedures which enter most textbooks on quantum mechanics. However, sometimes this question is improperly treated not just in a specialized literature but even in well known (and otherwise very good) textbooks.

An example of the latter situation can be found in [74] where on pages 50–51 there is an attempt to argue that the Schrödinger equation for a zero-spin particle in the presence of an electromagnetic field should read

$$i\hbar\frac{\partial\varphi}{\partial t} = \left[\frac{1}{2m}\left(i\hbar\boldsymbol{\nabla} + \frac{e}{c}\mathbf{A}\right)^2 + e\Phi + \frac{i\hbar e}{2mc^2}\frac{\partial\Phi}{\partial t}\right]\varphi \quad (A.21)$$

(see (1.140) on p. 51 of [74]). It is further claimed in [74] that in the Coulomb gauge ($\boldsymbol{\nabla}\cdot\mathbf{A} = 0$) the term proportional to $\frac{\partial\Phi}{\partial t}$ must be kept while in the Lorentz family of gauges $\left(\boldsymbol{\nabla}\cdot\mathbf{A} + \frac{\partial\Phi}{\partial t} = 0\right)$, the corresponding Schrödinger equation must not contain the term proportional to $\boldsymbol{\nabla}\cdot\mathbf{A} = 0$ because in (A.21) it is canceled by the term proportional to $\frac{\partial\Phi}{\partial t} = 0$.

Two remarks concerning (A.21) should be made. First of all, the obvious defects of the 'Schrödinger equation' (A.21) is that (i) it is not compatible with the minimum coupling of Electrodynamics and (ii) it is not gauge invariant. Therefore, (A.21) cannot represent the correct form of a wave equation for a nonrelativistic electron.

The second remark concerns the more technical question about where there was a mistake in the transformation of the Klein–Gordon equation into the Schrödinger equation, performed in [74], which had lead to the incorrect (A.21). This point can be answered by comparing the transition from the Klein–Gordon to the Schrödinger equation discussed in Sect. A.2.1 (see (A.13)–(A.16)) with the corresponding discussion given in [74]. The latter simply does not take into account the delicate cancellation which occurs between the terms in the second line of (A.13) under the assumption that the motion of a quantum particle is nonrelativistic.

Finally let us note that a mistake quite similar to that made in [74] had also been done in [125] (see (58.7) and (58.9) on p. 210 of that book).

A.3 On the Existence of the 'Overlap' Region

A.3.1 Collisions with a Point-Like Charge

Let us first consider collisions with a point-like charge Z_A. In this case the light-cone transition amplitude is given by (6.61), the first order amplitude is given by (6.72) where one should set $M_j = 0$. Due to the presence of the ground state in the transition matrix element, the electron coordinates are effectively restricted to $r \overset{<}{\sim} \frac{1}{Z_I}$. Therefore, for collisions with impact parameters $b \gg \frac{1}{Z_I}$ one can approximately write

$$K_0 \left(B_0 \mid \mathbf{r}_\perp - \mathbf{b} \mid \right) \approx K_0(B_0 b) + \frac{B_0 K_1(B_0 b)}{b} \mathbf{b} \cdot \mathbf{r}_\perp, \tag{A.22}$$

where $B_0 = \frac{\omega_{n0}}{\gamma v}$ and K_1 is a modified Bessel function. Further, one can also expand

$$\ln \frac{\mid \mathbf{b} - \mathbf{r}_\perp \mid}{b} \approx -\frac{\mathbf{b} \cdot \mathbf{r}_\perp}{b^2}. \tag{A.23}$$

Using (A.22) and the condition $\sum_j A_j = 1$ one obtains for the first order transition amplitude (6.72)

$$a_{0n}^p(\mathbf{b}) \approx \frac{2iZ_A}{v} K_0(B_0 b) < \psi_n \mid \left(1 - \frac{v}{c}\alpha_z \right) \exp\left(i\frac{\omega_{n0} z}{v} \right) \mid \psi_0 >$$
$$+ \frac{2iZ_A}{vb} B_0 K_1(B_0 b) < \psi_n \mid \left(1 - \frac{v}{c}\alpha_z \right) \exp\left(i\frac{\omega_{n0} z}{v} \right) (\mathbf{r}_\perp \cdot \mathbf{b}) \mid \psi_0 > . \tag{A.24}$$

Applying the identity $< \psi_n \mid \alpha_z \exp\left(i\frac{\omega_{n0} z}{v} \right) \mid \psi_0 > \equiv \frac{v}{c} < \psi_n \mid \exp\left(i\frac{\omega_{n0} z}{v} \right) \mid \psi_0 >$ one sees that the first term in (A.24) is proportional to $\frac{1}{\gamma^2}$. We will neglect this term and choose b to satisfy not only the relation $b \gg \frac{1}{Z_I}$ but also $b \ll \frac{\gamma v}{\omega_{n0}}$. Estimating $\omega_{n0} \sim Z_I^2$ one can see that it is always possible to find the range $\frac{1}{Z_I} \ll b \ll \frac{\gamma v}{\omega_{n0}}$ for ultrarelativistic collisions when one has $\gamma c \gg Z_I$ for any Z_I. Since $B_0 = \frac{\omega_{n0}}{\gamma v}$, it is easy to see that in this range of impact parameters $B_0 b \ll 1$. Correspondingly, one can approximate $K_1(B_0 b) \approx \frac{1}{B_0 b}$ [108] and the first order transition amplitude reads

$$a_{0n}^{p}(\mathbf{b}) \approx \frac{2iZ_A}{cb^2} < \psi_n \mid (1 - \alpha_z) \exp\left(i\frac{\omega_{n0}z}{c}\right)(\mathbf{r}_\perp \cdot \mathbf{b}) \mid \psi_0 >, \quad (A.25)$$

where we set $v \approx c$.

On the other hand, taking into account (A.23), the light-cone transition amplitude (6.61) becomes

$$a_{0n}^{Coul}(\mathbf{b}) \approx < \psi_n \mid (1 - \alpha_z) \exp\left(i\frac{\omega_{n0}z}{c}\right) \exp\left(\frac{2iZ_A}{c}\frac{\mathbf{b} \cdot \mathbf{r}_\perp}{b^2}\right) \mid \psi_0 > . \quad (A.26)$$

Since $r_\perp \sim \frac{1}{Z_I}$, then, for $b \gg \frac{Z_A}{Z_I c}$, one can expand the exponential function in (A.26) and the light-cone transition amplitude (A.26) recovers the first order transition amplitude (A.25). Thus, one can conclude that, for collisions with a point-like charge

1. The first order perturbation theory can be used for $b \gg \frac{Z_A}{Z_I c}$ and
2. The light-cone and first order transition amplitudes are approximately equal at $\frac{1}{Z_I} \ll b \ll \frac{\gamma v}{\omega_{n0}}$.

A.3.2 Collisions with a Neutral Atom

Let us now discuss briefly the electron excitation and loss in ultrarelativistic collisions with neutral atoms. Since for collisions with a neutral atom having atomic number Z_A the screened atomic field for any impact parameter is not stronger than the field of a point-like charge Z_A then the conclusion (1) is applicable for collisions with neutral atoms as well. In the light-cone amplitude

$$a_{0n}^{eik}(\mathbf{b}) = \left\langle \psi_n \mid (1 - \alpha_z) \exp\left(i\frac{\omega_{n0}z}{c}\right) \right.$$

$$\left. \times \exp\left(\frac{2iZ_A}{c}\sum_j A_j K_0\left(M_j \mid \mathbf{r}_\perp - \mathbf{b} \mid\right)\right) \mid \psi_0 \right\rangle. \quad (A.27)$$

we expand the functions $K_0(M_j \mid \mathbf{r}_\perp - \mathbf{b} \mid)$ for $b \gg \frac{1}{Z_I}$ similarly to (A.22). Since $b \gg \frac{1}{Z_I} > \frac{Z_A}{Z_I c}$ one can further expand the exponential function in (A.27) and obtain

$$a_{0n}^{eik}(\mathbf{b}) \approx \frac{2iZ_A}{cb}\sum_j A_j M_j K_1(M_j b) < \psi_n \mid (1 - \alpha_z) \exp\left(i\frac{\omega_{n0}z}{c}\right)\mathbf{b} \cdot \mathbf{r}_\perp \mid \psi_0 > .$$
$$(A.28)$$

For the same region of impact parameters $b \gg \frac{1}{Z_I}$ the first order transition amplitude is approximately given by

$$a_{0n}^{p}(\mathbf{b}) \approx \frac{2iZ_A}{cb}\sum_j A_j B_j K_1(B_j b) < \psi_n \mid (1 - \alpha_z) \exp\left(i\frac{\omega_{n0}z}{c}\right)\mathbf{b} \cdot \mathbf{r}_\perp \mid \psi_0 > .$$

$$(A.29)$$

As it follows from (A.28) and (A.29) the light-cone and first order amplitudes are approximately equal for $b \gg \frac{1}{Z_I}$ if $B_j \simeq M_j$. If the latter condition is not fulfilled the amplitudes (A.28) and (A.29) can still be approximately equal if there exists an overlap between $b \gg \frac{1}{Z_I}$ and $b \ll \frac{1}{M_j}$, and $b \gg \frac{1}{Z_I}$ and $b \ll \frac{1}{B_j}$. In the ranges $b \ll \frac{1}{M_j}$ and $b \ll \frac{1}{B_j}$ the amplitudes (A.28) and (A.29) can be further simplified using for small arguments $K_1(x) \approx \frac{1}{x}$. This yields

$$a_{0n}^{eik}(\mathbf{b}) \approx a_{0n}^{p}(\mathbf{b}) \approx \frac{2\mathrm{i}Z_A}{cb^2} < \psi_n \mid (1 - \alpha_z) \exp\left(\mathrm{i}\frac{\omega_{n0}z}{c}\right) \mathbf{b} \cdot \mathbf{r}_\perp \mid \psi_0 >. \quad (A.30)$$

The inspection of the screening constants given in [102] shows that the strict conditions $\frac{1}{Z_I} \ll b \ll \frac{1}{M_j}$ and $\frac{1}{Z_I} \ll b \ll \frac{1}{B_j}$ are in general not fulfilled. However, the less restrictive conditions for the overlap $\frac{1}{Z_I} < b < \frac{1}{M_j}$ and $\frac{1}{Z_I} < b < \frac{1}{B_j}$ are fulfilled for very heavy projectile-ions where Z_I is considerably larger than $max\{M_j\}$.

In general the cross section (6.73) can be calculated according to the following simple rule. At any impact parameter the transition amplitude should be represented by the value obtained either from the light-cone or the first order transition amplitudes whichever gives the smallest transition probability.

A.4 Radiative Atomic Processes and Galilean and Gauge Transformations

A relativistically covariant quantum-electrodynamic description of processes, in which atomic particles possessing both internal and translational degrees of freedom interact with the electromagnetic field, is in general quite a nontrivial task. On the other hand, particles with nonzero rest masses can often be well described by the nonrelativistic equations, which are in general much simpler to deal with compared to their relativistic counterparts. Therefore, atomic processes, in which charged particles moving with nonrelativistic velocities interact with photons, are normally considered by using the nonrelativistic Schrödinger equation.

However, within the scope of such an approach one encounters the principal difficulty connected with the problem of covariance of the calculated results. Indeed, the nonrelativistic equations of motion are covariant under a Galilean transformation whereas the Maxwell equations describing electromagnetic fields are Lorentz covariant. Since any calculation for the radiative processes is to be performed in a certain reference frame, it is in general not clear how reliable results of the calculation are.

One example of the problem of covariance arises in the theoretical studies of the radiative electron capture in nonrelativistic ion–atom collisions. The common statement, which can be found in the literature (see e.g. [4, 77, 236]), is that the application of the Schrödinger equation to treat this process yields values for the total cross section which in general strongly depend on a particular choice of a Galilean reference frame and that the rest frame of the ion is much more appropriate to calculate the cross section than other frames.

Amongst other radiative processes, for which the same problem of covariance arises, are the radiative electron–ion recombination in crossing nonrelativistic beams of electrons and ions and the spontaneous radiative decay of a moving excited atom. All the radiative processes mentioned above have much in common. In particular, the radiative recombination and capture are rather closely related and can be viewed as different types of one-photon bremsstrahlung.

In this section, following the discussion given in [237, 238], we consider the radiative electron capture in ion–atom collisions, the electron–ion(atom) recombination and the spontaneous radiative decay of an excited atom(ion). Based on the Schrödinger equation we shall explore the behavior of the total cross sections (of the total decay rate in the case of the spontaneous decay) under a Galilean transformation. We shall also consider in detail the important and delicate interrelation between Galilean and gauge transformations. In particular, we will show that this interrelation can easily cause the confusion of the problem of gauge dependence with the problem of Galilean covariance if a sufficient care is not taken in the analysis.

A.4.1 One Radiating Atomic System and Two Reference Frames: Galilean Invariance

The relativistic effects related to the reference frame transformations (like e.g. time dilation and Lorentz contraction) begin with terms $\sim (v/c)^2$, where v is the velocity of one frame with respect to the other and c is the speed of light. The relativistic corrections to the energy and momentum of a particle with a nonzero rest mass, which moves with a velocity v_0 with respect to the origin of a given reference frame, start with terms $\sim (v_0/c)^2$. Therefore, in our consideration all terms of the order of $(v/c)^2$, $(v_0/c)^2$, vv_0/c^2 (and higher) will be neglected. The speed of light will be taken as reference frame-independent. One should remark that for the analysis given below the latter assumption is in fact not necessary and is made here just for the sake of simplicity. One could regard the speed of light as frame dependent, however, as long as terms $\sim (v/c)^2$, $(v_0/c)^2$ and vv_0/c^2 are neglected, this would have no impact on the final result.

We begin with the detailed consideration of the transition amplitudes for the radiative processes using the semi-classical approach. Within this approach the heavy subsystem (consisting for the processes in question of one or two nuclei) is regarded as classical and assumed to move along a given trajectory generating the Coulomb potential for the electron. Because of comparatively very large nuclear mass, the coupling of the heavy subsystem to the radiation field is much weaker than that of the electron and will be neglected.

Within the semi-classical approach the transition amplitude, obtained in the first order approximation in the interaction between the electron and the electromagnetic field, is given by

$$a_{fi} = -i \int_{-\infty}^{+\infty} dt \langle \Psi_f(t) | \hat{W} | \Psi_i(t) \rangle. \qquad (A.31)$$

Here

$$\hat{W} = -\frac{e}{2mc} (\hat{\mathbf{p}} \cdot \mathbf{A} + \mathbf{A} \cdot \hat{\mathbf{p}}) + e\varphi \qquad (A.32)$$

is the interaction between the electron and the electromagnetic field, where $\hat{\mathbf{p}}$ is the operator of the electron momentum, $\mathbf{A}$ and φ are the vector and scalar potentials of the electromagnetic field, e is the electron charge and m its mass. The term $e^2\mathbf{A}^2/2mc^2$ does not contribute to one-photon processes and, therefore, is omitted in (A.32).[2] Further, $\Psi_i(t)$ and $\Psi_f(t)$ are the initial and final states of the electron which are solutions of the Schrödinger equation

$$i\frac{\partial}{\partial t}\Psi = \left(\frac{\hat{\mathbf{p}}^2}{2m} + V_c(t) \right) \Psi, \qquad (A.33)$$

where $V_c(t)$ is the interaction between the electron and the heavy subsystem.

Spontaneous Radiative Recombination and Decay

We start with the consideration of the spontaneous radiative transition occurring in the atomic system consisting of an electron and a nucleus. This can be either the transition between continuum and bound states of the system (the radiative recombination) or the transition between two bound states (the spontaneous radiative decay).

At first we consider the radiative transition using a reference frame K' where the nucleus, for the sake of definiteness, is supposed to be at rest (this restriction is not essential and can easily be removed). We take the position of the nucleus as the origin of this frame and denote the electron coordinates with respect to the origin by $\mathbf{s}$. The initial and final states describing the electron, which interacts with the nucleus, read

$$\Psi_i'(t) = e^{-i\varepsilon_i t}\chi_i(\mathbf{s})$$
$$\Psi_f'(t) = e^{-i\varepsilon_f t}\chi_f(\mathbf{s}), \qquad (A.34)$$

where ε_i and ε_f are the initial and final energies of the electron.

In order to describe the potentials of the electromagnetic field in the frame K' we employ the radiation gauge. In this gauge the operators for the vector and scalar potentials of the field are given by (see e.g. [94])

---

[2] In our analysis we also disregard the spin degrees of freedom of the electron which are known to be not important for the processes considered here.

$$\hat{\mathbf{A}}_{K'}(\mathbf{s}, t) = \sum_{\boldsymbol{\kappa}\rho} \boldsymbol{\alpha}_{\boldsymbol{\kappa}\rho} \left(c^+_{\boldsymbol{\kappa}\rho} e^{i(\omega_\kappa t - \boldsymbol{\kappa}\cdot\mathbf{s})} + C.C. \right)$$

$$\hat{\varphi}_{K'}(\mathbf{s}, t) = 0, \tag{A.35}$$

where $\boldsymbol{\alpha}_{\boldsymbol{\kappa}\rho} = \sqrt{\frac{2\pi c^2}{V\omega_\kappa}} \mathbf{e}_{\boldsymbol{\kappa}\rho}$, $c^+_{\boldsymbol{\kappa}\rho}$ is the photon creation operator, $\mathbf{e}_{\boldsymbol{\kappa}\rho}$ is the polarization vector $(\mathbf{e}_{\boldsymbol{\kappa}\rho_1} \cdot \mathbf{e}_{\boldsymbol{\kappa}\rho_2} = \delta_{\rho_1,\rho_2}$, $\mathbf{e}_{\boldsymbol{\kappa}\rho} \cdot \boldsymbol{\kappa} = 0)$ and V is the normalization volume for the field. The sum in (A.35) runs over all photon modes.

The field potentials $\mathbf{A}$ and φ, which enter expressions (A.31)–(A.32), are 'generated' on the transition between the photon vacuum state $| \, 0 \rangle$ and a state $| \, \mathbf{k}', \lambda \rangle$ describing one photon which has in the frame K' a momentum $\mathbf{k}'$, frequency $\omega' = ck'$ and polarization $\mathbf{e}_{\mathbf{k}'\lambda}$. These potentials in K' are obtained according to

$$\mathbf{A}_{K'}(\mathbf{s}, t) = \langle \mathbf{k}', \lambda \, | \, \hat{\mathbf{A}}_{K'}(\mathbf{s}, t) \, | \, 0 \rangle = \boldsymbol{\alpha}_{\mathbf{k}'\lambda} \exp(i(\omega' t - \mathbf{k}' \cdot \mathbf{s}))$$

$$\varphi_{K'}(\mathbf{s}, t) = \langle \mathbf{k}', \lambda' \, | \, \hat{\varphi}_{K'}(\mathbf{s}, t) \, | \, 0 \rangle = 0. \tag{A.36}$$

Taking into account (A.31)–(A.32) and (A.36), for the transition amplitude in the frame K' we obtain

$$a_{K'} = -i \int_{-\infty}^{+\infty} dt \langle \Psi_f(t) | \hat{W}_{K'} | \Psi_i(t) \rangle$$

$$= 2\pi i \frac{e}{mc} \, \delta(\omega' + \varepsilon_f - \varepsilon_i) \int d^3\mathbf{s} \, \chi_f^*(\mathbf{s}) e^{-i\mathbf{k}'\cdot\mathbf{s}} \boldsymbol{\alpha}_{\mathbf{k}'\lambda} \cdot \hat{\mathbf{p}}\chi_i(\mathbf{s}), \tag{A.37}$$

where the delta-function expresses the energy conservation for the emission process observed in the frame K'.

Now we assume that in another reference frame K the motion of the nucleus is represented by a classical straight-line trajectory $\mathbf{R}(t) = \mathbf{b} + \mathbf{v}t$, where $\mathbf{v}$ is the velocity of the nucleus and $\mathbf{b}$ is its impact parameter with respect to the origin of the frame K. We denote by $\mathbf{r}$ the electron coordinates with respect to the origin. The electron coordinates with respect to the nucleus are as before given by $\mathbf{s}$. Note that $\mathbf{r} = \mathbf{s} + \mathbf{R}$.

In the frame K the initial and final states of the electron are given by

$$\Psi_i(t) = \chi_i(\mathbf{r} - \mathbf{R}) e^{im\mathbf{v}\cdot\mathbf{r}} e^{-iE_i t}$$

$$\Psi_f(t) = \chi_f(\mathbf{r} - \mathbf{R}) e^{im\mathbf{v}\cdot\mathbf{r}} e^{-iE_f t}, \tag{A.38}$$

where $E_i = \varepsilon_i + mv^2/2$, and $E_f = \varepsilon_f + mv^2/2$.

For the evaluation of the transition amplitude in the frame K one should take into account that the transformation from K' to K alters not only the electron wave functions but also the potentials of the electromagnetic field. Indeed, with accuracy of up to the order of v/c the transformation of the potentials (A.35) to the frame K are given by $\mathbf{A}_K = \mathbf{A}_{K'}$ and $\varphi_K = \frac{1}{c}\mathbf{v}\cdot\mathbf{A}_{K'}$ (see e.g. [7]), where $\mathbf{A}_K$ and φ_K are the vector and scalar potentials in the frame K. This transformation yields

$$\mathbf{A}_K(\mathbf{r}, t) = \boldsymbol{\alpha}_{\mathbf{k}\lambda} \exp(\mathrm{i}\mathbf{k} \cdot \mathbf{b}) \exp(\mathrm{i}(\omega t - \mathbf{k} \cdot \mathbf{r}))$$

$$\varphi_K(\mathbf{r}, t) = \frac{1}{c}\mathbf{v} \cdot \mathbf{A}_K(\mathbf{r}, t), \tag{A.39}$$

where $\omega = \omega' + \mathbf{k} \cdot \mathbf{v}$. The latter relation describes the nonrelativistic Doppler shift. Note also that the transformation (A.39) does not affect the photon momentum: $\mathbf{k} = \mathbf{k}'$.

The interaction between the electron and the nucleus should also be transformed to the new reference frame. As a result, in the frame K this interaction, in addition to the term V_c, will include the current–current term (the so-called Darwin or Breit term, see e.g. [7]). The latter one, however, is a correction to V_c of the order of vv_0/c^2 and has to be omitted.

Taking into account (A.31)–(A.32) and (A.38)–(A.39), for the transition amplitude in the frame K we obtain

$$a_K = -\mathrm{i}\int_{-\infty}^{+\infty} \mathrm{d}t\langle\Phi_\mathrm{f}(t)|\hat{W}_K|\Phi_\mathrm{i}(t)\rangle$$

$$= \mathrm{i}\frac{e}{mc}\mathrm{e}^{\mathrm{i}\mathbf{k}\cdot\mathbf{b}}\int_{-\infty}^{+\infty} \mathrm{d}t\,\mathrm{e}^{\mathrm{i}(\omega+\varepsilon_\mathrm{f}-\varepsilon_\mathrm{i})t}$$

$$\times \left\{ \int \mathrm{d}^3\mathbf{r}\,\chi_\mathrm{f}^*(\mathbf{r} - \mathbf{R})\mathrm{e}^{-\mathrm{i}m\mathbf{v}\cdot\mathbf{r}}\mathrm{e}^{-\mathrm{i}\mathbf{k}\cdot\mathbf{r}}\,\overline{\boldsymbol{\alpha}}_{\mathbf{k}\lambda}\cdot\hat{\mathbf{p}}\left(\mathrm{e}^{\mathrm{i}m\mathbf{v}\cdot\mathbf{r}}\chi_\mathrm{i}(\mathbf{r} - \mathbf{R})\right) \right.$$

$$\left. -\overline{\boldsymbol{\alpha}}_{\mathbf{k}\lambda}\cdot\mathbf{v}\int \mathrm{d}^3\mathbf{r}\,\chi_\mathrm{f}^*(\mathbf{r} - \mathbf{R})\mathrm{e}^{-\mathrm{i}\mathbf{k}\cdot\mathbf{r}}\chi_\mathrm{i}(\mathbf{r} - \mathbf{R}) \right\}$$

$$= 2\pi\mathrm{i}\frac{e}{mc}\delta(\omega - \mathbf{k}\cdot\mathbf{v} + \varepsilon_\mathrm{f} - \varepsilon_\mathrm{i})\int \mathrm{d}^3\mathbf{s}\,\chi_\mathrm{f}^*(\mathbf{s})\mathrm{e}^{-\mathrm{i}\mathbf{k}\cdot\mathbf{s}}\,\overline{\boldsymbol{\alpha}}_{\mathbf{k}\lambda}\cdot\hat{\mathbf{p}}\chi_\mathrm{i}(\mathbf{s}), \tag{A.40}$$

where $\overline{\boldsymbol{\alpha}}_{\mathbf{k}\lambda} = \sqrt{\frac{2\pi c^2}{V(\omega - \mathbf{v}\cdot\mathbf{k})}}\mathbf{e}_{\mathbf{k}\lambda}$. The delta-function in (A.40) expresses the energy conservation for the observer in the frame K.

It is seen in (A.40) that in the consideration given in the frame K an important cancellation occurs between the term arising from the action of the electron momentum operator on the electron translational factor $\mathrm{e}^{\mathrm{i}m\mathbf{v}\cdot\mathbf{r}}$ and the term due to the scalar potential.

Radiative Electron Capture

Now we turn to the radiative capture in ion–atom collisions. For simplicity we assume that the colliding atomic system consists just of one electron and two nuclei. Let an inertial frame K' have the origin in the point O'. Let the initial and final electronic wave functions be given in this frame by $\psi_\mathrm{i}(\mathbf{s}, t)$ and $\psi_\mathrm{f}(\mathbf{s}, t)$, respectively, where $\mathbf{s}$ is the electron coordinate with respect to the point O'. $\psi_{\mathrm{i,f}}$ are solutions of the Schrödinger equation for the electron moving in the combined time-dependent potential $V_c(t)$ of the nuclei of the atom and ion. In the frame K' the initial and final states, which describe the electron moving in the potential of the nuclei, read

$$\Psi'_i(t) = \psi_i(\mathbf{s}, t)$$
$$\Psi'_f(t) = \psi_f(\mathbf{s}, t). \qquad (A.41)$$

Using (A.31)–(A.32) and (A.41) and assuming that the field potentials in K' are given by (A.36), for the capture transition amplitude in this frame we obtain

$$a_{K'} = i\frac{e}{mc} \int_{-\infty}^{+\infty} dt\, e^{i\omega' t} \int d^3s\, \psi_f^*(\mathbf{s}, t) e^{-i\mathbf{k}'\cdot\mathbf{s}} \boldsymbol{\alpha}_{\mathbf{k}'\lambda} \cdot \hat{\mathbf{p}}\psi_i(\mathbf{s}, t). \qquad (A.42)$$

Now we take another inertial frame K, with respect to which the frame K' moves with a velocity $\mathbf{v}$, and let the point O be the origin of the new frame. The coordinates of the origin O' are given in K by $\mathbf{R}(t) = \mathbf{b} + \mathbf{v}t$ and we denote by $\mathbf{r}$ the electron coordinates with respect to the origin O. In the frame K the initial and final electron states are given by

$$\Psi_i(t) = \psi_i(\mathbf{r} - \mathbf{R}, t) e^{im\mathbf{v}\cdot\mathbf{r}} e^{-imv^2 t/2}$$
$$\Psi_f(t) = \psi_f(\mathbf{r} - \mathbf{R}, t) e^{im\mathbf{v}\cdot\mathbf{r}} e^{-imv^2 t/2}. \qquad (A.43)$$

The potentials of the radiation field in the frame K are related to those in K' by the transformation (A.39).

Using (A.31)–(A.32), (A.39) and (A.43) we obtain that the capture transition amplitude in the frame K is given by

$$a_K = i\frac{e}{mc} e^{i\mathbf{k}\cdot\mathbf{b}} \int_{-\infty}^{+\infty} dt\, e^{i\omega t}$$

$$\times \left\{ \int d^3r\, \psi_f^*(\mathbf{r} - \mathbf{R}, t) e^{-im\mathbf{v}\cdot\mathbf{r}} e^{-i\mathbf{k}\cdot\mathbf{r}}\, \overline{\boldsymbol{\alpha}}_{\mathbf{k}\lambda} \cdot \hat{\mathbf{p}} \left(e^{im\mathbf{v}\cdot\mathbf{r}}\psi_i(\mathbf{r} - \mathbf{R}, t) \right) \right.$$

$$\left. - \overline{\boldsymbol{\alpha}}_{\mathbf{k}\lambda} \cdot \mathbf{v} \int d^3r\, \psi_f^*(\mathbf{r} - \mathbf{R}, t) e^{-i\mathbf{k}\cdot\mathbf{r}}\psi_i(\mathbf{r} - \mathbf{R}, t) \right\}$$

$$= i\frac{e}{mc} \int_{-\infty}^{+\infty} dt\, e^{i(\omega - \mathbf{k}\cdot\mathbf{v})t} \int d^3s\, \psi_f^*(\mathbf{s}, t) e^{-i\mathbf{k}\cdot\mathbf{s}}\, \overline{\boldsymbol{\alpha}}_{\mathbf{k}\lambda} \cdot \hat{\mathbf{p}}\psi_i(\mathbf{s}, t), \qquad (A.44)$$

where $\overline{\boldsymbol{\alpha}}_{\mathbf{k}\lambda} = \sqrt{\frac{2\pi c^2}{V(\omega - \mathbf{v}\cdot\mathbf{k})}}\mathbf{e}_{\mathbf{k}\lambda}$. The relation $\omega - \mathbf{k}\cdot\mathbf{v} = \omega'$ between the photon frequencies in K and K', which follows from the comparison of the last line of (A.44) with (A.42), describes the Doppler shift.

Analogously to the case of the radiative recombination and decay, we observe in (A.44) that in the consideration of the radiative capture given in the frame K the term, which arises from the action of the electron momentum operator on the electron translational factor $e^{im\mathbf{v}\cdot\mathbf{r}}$, is canceled by the term, which appears due to the scalar potential. Comparing (A.44) and (A.42) we see that the form of the capture transition amplitude in both frames is similar.

One can show that the form of the transition amplitudes for the radiative processes, described in different Galilean reference frames K' and K, remains similar also in the case when these amplitudes are obtained in the full quantum considerations in which all the particles are described quantum mechanically. Besides, the same result also follows if the coupling of the heavy nuclues (nuclei) to the electromagnetic field is taken into account.

The similarity in the form of the transition amplitudes in different reference frames originates from the mutual cancellation between the terms, which arise due to the change in the wave functions of the charged particles, and the terms, which appear because of the change in the potentials of the electromagnetic field. This cancellation (particular cases of which was explicitly demonstrated in (A.40) and (A.44)) is of course not fortuitous. The reason for this cancellation is that, with accuracy of up to v/c, a Galilean transformation does not affect the form of the first-order interaction between the charged particles and the radiation field

$$\hat{W} = \hat{\rho}\hat{\varphi} - \hat{\mathbf{j}} \cdot \hat{\mathbf{A}}/c, \tag{A.45}$$

where $\hat{\rho}$ and $\hat{\mathbf{j}}$ are (the operators for) the total charge and current densities of the particles and $\hat{\varphi}$ and $\hat{\mathbf{A}}$ are the scalar and vector potentials of the field. Indeed, under a Galilean transformation from K' to K one has $\hat{\rho} = \hat{\rho}'$ and $\hat{\mathbf{j}} = \hat{\mathbf{j}}' + \hat{\rho}\mathbf{v}$, where $\mathbf{v}$ is the velocity of K' with respect to K. Taking also into account the corresponding transformation of the potentials we obtain

$$\hat{\rho}\hat{\varphi} - \frac{1}{c}\hat{\mathbf{j}} \cdot \hat{\mathbf{A}} = \rho \left(\hat{\varphi}' + \frac{\mathbf{v}}{c} \cdot \hat{\mathbf{A}}' \right) - \frac{1}{c} \left(\hat{\mathbf{j}}' + \hat{\rho}'\mathbf{v} \right) \cdot \left(\hat{\mathbf{A}}' + \frac{\mathbf{v}}{c}\hat{\varphi}' \right)$$
$$= \hat{\rho}'\hat{\varphi}' - \hat{\mathbf{j}}' \cdot \hat{\mathbf{A}}'/c. \tag{A.46}$$

Note that in the last line of (A.46) we omitted the terms $\sim(v/c)^2$ and $\sim v v_0/c^2$, where $v_0 \ll c$ are characteristic velocities of the charged particles in K'.

The similarity of the radiative transition amplitudes established above of course does not mean that these amplitudes are covariant under a Galilean transformation. The reason is obvious: it is the presence of the photon or, more precisely, the momentum of the photon. However, taking into account that $\omega - \mathbf{v} \cdot \mathbf{k} = \omega'$ and keeping in mind that the transformation (A.39) does not change the photon momentum and thus one has $\omega' = ck$, it becomes clear that in both reference frames the corresponding amplitudes yield identical total cross sections and decay rates. We, therefore, may conclude that the total cross sections for the radiative electron recombination and capture and the total decay rate for the spontaneous radiative decay are invariant under a Galilean transformation.

One remarkable point in the establishing of the Galilean invariance for the radiative processes, which should be especially mentioned, is that we have arrived at this result without imposing any conditions on the initial and final wave functions describing the charged particles (except the obvious one that these wave functions are nonrelativistic). Therefore, this result is valid, no matter whether the wave functions are exact or approximate and, in the latter case, which particular approximations for these functions are employed.

A.4.2 Two Radiating Systems and One Reference Frame:
The Problem of Gauge Dependence

In the previous subsection, by describing a single radiating system using two different reference frames, we explored the behavior of the radiative transition amplitudes and cross sections under a Galilean transformation. Now we consider the complementary situation where one reference frame is used to describe two radiating systems which have different translational velocities but otherwise are identical.

For the sake of simplicity in this section we shall give the detailed consideration only for the spontaneous radiative decay and at first treat this process for an atom which rests at the origin of a given reference frame K_0. The atom is initially in its excited internal state χ_i with an energy ε_i and by emitting a photon makes a transition into an internal state χ_f with an energy ε_f. The amplitude for the spontaneous transition with emission of a photon having momentum $\mathbf{k}$ and polarization vector $\mathbf{e}_{\mathbf{k}\lambda}$ reads

$$a_{RA} = 2\pi i \frac{e}{mc} \sqrt{\frac{2\pi c^2}{V\omega}} \, \delta(\omega - \omega_{if}) \, \langle \chi_f(\mathbf{r}) \mid e^{-i\mathbf{k}\cdot\mathbf{r}} \, \mathbf{e}_{\mathbf{k}\lambda} \cdot \hat{\mathbf{p}} \mid \chi_i(\mathbf{r}) \rangle, \quad (A.47)$$

where $\omega = c \mid \mathbf{k} \mid$ and $\omega_{if} = \varepsilon_i - \varepsilon_f$ is the atomic transition frequency. The total decay rate for the atom at rest is then given by

$$
\begin{aligned}
W_{RA} &= \lim_{T \to \infty} \frac{V}{8\pi^3} \sum_{\lambda=1}^{2} \int d^3 k \frac{\mid a_{RA} \mid^2}{T} \\
&= \frac{e^2 \omega_{if}}{2\pi m^2 c^3} \sum_{\lambda=1}^{2} \int d\Omega_{\mathbf{k}} \mid \langle \chi_f(\mathbf{r}) \mid e^{-i\mathbf{k}\cdot\mathbf{r}} \, \mathbf{e}_{\mathbf{k}\lambda} \cdot \hat{\mathbf{p}} \mid \chi_i(\mathbf{r}) \rangle \mid^2 . \quad (A.48)
\end{aligned}
$$

The integration and the sum run over the solid angle $d\Omega_{\mathbf{k}}$ and polarization, respectively, of the emitted photon. Note that in the second line of (A.48) $\mid \mathbf{k} \mid = \omega_{fi}/c$.

Let us now consider the spontaneous decay of an atom which moves in the frame K_0 with a constant velocity $\mathbf{u}$ along a classical trajectory $\mathbf{S}(t) = \mathbf{u}t$. With respect to its inner degrees of freedom the moving atom is fully identical to that considered above and initially and finally is in the same internal states χ_i and χ_f.

In the frame K_0 the initial and final states of the moving atom read

$$
\begin{aligned}
\psi_i(\mathbf{r}, t) &= \chi_i(\mathbf{r} - \mathbf{S}) e^{-i\varepsilon_i t} e^{im\mathbf{u}\cdot\mathbf{r}} e^{-imu^2 t/2} \\
\psi_f(\mathbf{r}, t) &= \chi_f(\mathbf{r} - \mathbf{S}) e^{-i\varepsilon_f t} e^{im\mathbf{u}\cdot\mathbf{r}} e^{-imu^2 t/2} . \quad (A.49)
\end{aligned}
$$

Using (A.49) the amplitude for the spontaneous transition of the moving atom with emission of a photon having momentum $\mathbf{k}'$ and polarization vector $\mathbf{e}_{\mathbf{k}'\lambda}$ is obtained to be

$$a_{\mathrm{MA}} = 2\pi i \frac{e}{mc} \sqrt{\frac{2\pi c^2}{V\omega'}} \, \delta(\omega' - \omega_{if} - \mathbf{k}' \cdot \mathbf{u})$$
$$\times \left(\langle \chi_{\mathrm{f}}(\mathbf{r}) \mid e^{-i\mathbf{k}' \cdot \mathbf{r}} \, \mathbf{e}_{\mathbf{k}'\lambda} \cdot \hat{\mathbf{p}} \mid \chi_{\mathrm{i}}(\mathbf{r}) \rangle + m \, \mathbf{e}_{\mathbf{k}'\lambda} \cdot \mathbf{u} \, \langle \chi_{\mathrm{f}}(\mathbf{r}) \mid e^{-i\mathbf{k}' \cdot \mathbf{r}} \mid \chi_{\mathrm{i}}(\mathbf{r}) \rangle \right).$$
$$(A.50)$$

Compared to the amplitude (A.47) the transition amplitude for the moving atom contains the additional term proportional to the product of the transition density and the atomic velocity which describes the contribution to the electron current caused by the motion of the atom. Taking into account the consideration of the previous section, this term can also be viewed from a different perspective as reflecting the presence of the nonzero scalar potential of the radiation field in the rest frame of the atom. The amplitude (A.50) can be cast into the form pretty similar to that of (A.47) by using the condition of charge conservation. According to the latter the atomic transition charge density and current have to obey the continuity equation which, being written in the momentum space, in our case yields

$$\left(\omega_{if} - \frac{q^2}{2m} \right) \langle \chi_{\mathrm{f}}(\mathbf{r}) \mid e^{-i\mathbf{q} \cdot \mathbf{r}} \mid \chi_{\mathrm{i}}(\mathbf{r}) \rangle = \frac{1}{m} \langle \chi_{\mathrm{f}}(\mathbf{r}) \mid e^{-i\mathbf{q} \cdot \mathbf{r}} \, \mathbf{q} \cdot \hat{\mathbf{p}} \mid \chi_{\mathrm{i}}(\mathbf{r}) \rangle. \quad (A.51)$$

With the help of (A.51) (in which the term $q^2/2m \approx \omega_{if}^2/mc^2$ has to be neglected compared to ω_{if}) the transition amplitude (A.50) can be rewritten as

$$a_{\mathrm{MA}} = 2\pi i \frac{e}{mc} \sqrt{\frac{2\pi c^2}{V\omega'}} \, \delta(\omega' - \omega_{if} - \mathbf{k}' \cdot \mathbf{u}) \, \langle \chi_{\mathrm{f}}(\mathbf{r}) \mid e^{-i\mathbf{k}' \cdot \mathbf{r}} \, \mathbf{E}_{\mathbf{k}'\lambda} \cdot \hat{\mathbf{p}} \mid \chi_{\mathrm{i}}(\mathbf{r}) \rangle,$$
$$(A.52)$$

where

$$\mathbf{E}_{\mathbf{k}'\lambda} = \mathbf{e}_{\mathbf{k}'\lambda} + \frac{\mathbf{u} \cdot \mathbf{e}_{\mathbf{k}'\lambda}}{\omega_{if}} \mathbf{k}'. \quad (A.53)$$

The term in (A.50), which is proportional to the translational velocity $\mathbf{u}$ of the atom, arises because the moving atom actually does not 'see' the same field potentials as the atom at rest and, as was already mentioned, reflects the appearance of the nonzero scalar potential of the radiation field in the rest frame of the moving atom. After the relation (A.51) had been employed, the term related to the scalar potential has disappeared in (A.52) which implies that the field was in fact subjected to a certain gauge transformation. Within the accuracy of the nonrelativistic consideration, in which terms $\sim 1/c^2$ are neglected, the vector $\mathbf{E}_{\mathbf{k}'\lambda}$ has the properties (i) $\mathbf{E}_{\mathbf{k}'\lambda}^2 = \mathbf{e}_{\mathbf{k}'\lambda}^2 = 1$, (ii) $\mathbf{E}_{\mathbf{k}'\lambda} \cdot \mathbf{k}' = 0$ and represents the polarization vector for the gauge transformed field. Using (A.52) the total decay rate for the moving atom is obtained to be

$$W_{\mathrm{MA}} = \frac{e^2 \omega_{if}}{2\pi m^2 c^3} \sum_{\lambda=1}^{2} \int d\Omega'_{\mathbf{k}} \frac{\mid \langle \chi_{\mathrm{f}}(\mathbf{r}) \mid e^{-i\mathbf{k}' \cdot \mathbf{r}} \, \mathbf{E}_{\mathbf{k}'\lambda} \cdot \hat{\mathbf{p}} \mid \chi_{\mathrm{i}}(\mathbf{r}) \rangle \mid^2}{(1 - u \cos\theta_{\mathbf{k}'}/c)^2}, \quad (A.54)$$

where $\theta_{\mathbf{k}'}$ is the angle between the momentum of the emitted photon and the atomic velocity $\mathbf{u}$. Within the accuracy of the nonrelativistic consideration the absolute value of the photon momentum in (A.54) is given by $\mid \mathbf{k}' \mid = \omega_{fi}/c$, i.e. is the same as that in (A.48).

In order to show that, within the accuracy of the nonrelativistic treatment, the decay rates (A.48) and (A.54) are equal one should take the following points into account. First, for a nonrelativistic atom one has $\mathbf{k}' \cdot \mathbf{r} \sim \mathbf{k} \cdot \mathbf{r} \sim v_0/c$, where v_0 is the characteristic electron velocity inside the atom. Therefore, in order to be consistent with the nonrelativistic consideration, one has to expand the transition matrix elements in (A.48) and (A.54) and to keep there only the absolute square of the dipole transition element and the correction of the order of v_0/c to the dipole term. Second, within the nonrelativistic consideration one also has to expand the term $(1 - u\cos\theta_{\mathbf{k}'}/c)^{-2}$ in (A.54) and to neglect in this expansion all terms of the order of $(u/c)^2$ and higher. Then, taking into account that the dipole transition matrix element cannot depend on the direction of the photon momentum, it is a straightforward task to show that in the nonrelativistic consideration the total decay rate is independent of the velocity of the atom.

One should emphasize that we would not have arrived at this natural conclusion without using the relationship (A.51). However, the latter, being a particular case of the continuity equation for the conservation of charge, generally holds only if χ_i and χ_f are exact atomic eigenstates. Therefore, if a numerical calculation employs approximate atomic states one could come to the spurious result that the total decay rate in the nonrelativistic theory depends on the atomic velocity[3]. As we have seen above, the replacement of the charge density via the current density in the expression for the transition amplitude and the corresponding change in the form of the transition amplitude for the moving atom (compare (A.50) and (A.52)) amounts to a gauge transformation. Therefore, the spurious velocity dependence of the total decay rate, which is obtained when approximate wave functions are used, is the direct consequence of the problem of gauge dependence.

Considerations, which are basically similar to that presented above for the spontaneous decay, can be also given for the radiative recombination and capture. They lead to the same result: if these processes are treated within one reference frame the total cross section will in general be independent of the center-of-mass velocity only provided exact wave functions are used to describe the initial and final states of the charged particles. This velocity-independence is directly related to the gauge independence with the general

---

[3] Note that this spurious velocity dependence would by no means contradict to the invariance of the decay rate under a Galilean transformation from one reference frame to another. It would also be fully acceptable from the purely logic point of view (implying, for instance, that there is a special Galilean frame ('ether') in which the *absolute* translational motion of atoms is 'marked' by 'setting' their decay rates according to their velocities with respect to this frame and that all other frames simply 'copy' these rates).

condition for the latter to be fulfilled given by the conservation of the total electric charge in the process and expressed as the continuity equation

$$\frac{\partial \rho_{fi}}{\partial t} + \operatorname{div} \mathbf{j}_{fi} = 0 \tag{A.55}$$

for the transition charge density ρ_{fi} and current $\mathbf{j}_{fi}$.

If approximate wave functions are used, the condition (A.55) will in general be violated, the charge will not be conserved and the results will be gauge dependent. This will not violate the invariance of the calculated total cross sections and decay rates under a Galilean transformation. However, if the consideration is restricted to one reference frame, it may lead to the wrong conclusion that the nonrelativistic cross sections and decay rates depend on the velocity of the center-of-mass motion of the atomic system in this frame.

Since exact solutions of the three-body problem are not known, the problem of gauge dependence is especially critical for calculations of the radiative electron capture.

A.4.3 Example: Radiative Electron Capture

As it follows from the above analysis, the Galilean and gauge transformations are very intimately interrelated. This interrelation, if overlooked, can easily lead to the confusion of the problem of covariance of calculated results under a Galilean transformation with the problem of gauge dependence of these results. The latter seems to have happened in the studies of the radiative capture in ion–atom collisions and below we shall discuss this point in some detail.

In the literature on the ion–atom collisions there had been a long-standing and widely spread opinion (see, for instance, [4, 236]) that the Galilean invariance of the total cross sections for the radiative electron capture is possible only if the initial and final wave functions $\psi_{i,f}$ obey some special conditions. This opinion had arisen from the first calculations of the total capture cross section performed in [239] and has gained a strong support due to the analysis of this problem undertaken in [240].

The authors of [239] calculated the capture cross section in the rest frames of the ion-projectile and the target-atom and found that the results may strongly differ. The obvious reason for this difference is that, while they took into account the Galilean transformation of the wave functions, they regarded the electromagnetic field and its potentials as frame-independent (and used the radiation gauge in both frames). As a result, in their calculation the terms $\sim v/c$ arising from the Galilean transformation of the wave functions to a new frame were included but the corrections of the same order v/c, appearing due to the change in the field potentials, were ignored.

The authors of [240] approached the problem of Galilean invariance by considering the radiative recombination and capture just in *one* Galilean frame and focusing on the dependence of the amplitudes and the cross sections on the translational velocity of the atomic system. They found that, unless the wave functions are exact (or satisfy some special conditions), the total cross section will be velocity-dependent but did not relate this result to the problem of gauge dependence which was not even mentioned. Instead, this result was interpreted as the problem of Galilean noninvariance.

The confusion of the problems related to the frame and gauge transformations also arises when two different Galilean frames are considered but the field potentials are regarded as frame-independent (as, for instance, it was done in [239]). Indeed, a Galilean transformation alters the potentials of the radiation field. Therefore, in order to keep these potentials unchanged after the transformation of a reference frame, the potentials in the new frame have to be subjected to an appropriate gauge transformation.

The radiation gauge is extremely popular in atomic physics calculations. In particular, this gauge was used in all treatments of the radiative capture which we are aware of. However, taking into account what has been said above, it becomes clear that if in a theoretical study the radiation gauge is imposed on the field potentials *simultaneously* in different reference frames, then such a study does not address the problem of covariance under a Galilean transformation but represents a certain check of gauge independence. The latter is also useful since it can serve as a test for theoretical models of the radiative electron capture.

A.4.4 A Gauge Test for the 1B and CDW Models of the Radiative Electron Capture

In order to illustrate the latter point we show in Fig. A.1 results of our calculation for the total cross sections for the radiative electron capture into the K shell occurring in fast collisions between a highly charged ion-projectile and a molecular hydrogen target. In the calculation molecular hydrogen was regarded as two independent hydrogen atoms which is a very good approximation to evaluate the total cross section for the radiative capture in high-velocity collisions.

The calculations were performed as follows. The transition amplitude for the capture was evaluated in the rest frame of the projectile-ion and the rest frame of the target using the same radiation gauge for *each reference frame*. Two models were considered to approximate the electron states: the so called first Born (1B) and the Continuum Distorted Wave (CDW) approximations.

Within the 1B model the initial and final electron states are approximated by undistorted eigenstates of the target and projectile, respectively. For instance, in the projectile-ion reference frame with the origin taken at the position of the projectile nucleus the first Born states are given by

$$\psi_i^{(+)}(t) = \exp\left(-i(\epsilon_i + v_a^2/2)t\right)\exp(i\mathbf{v}_a \cdot \mathbf{s})\phi_i(\mathbf{r})$$
$$\psi_f^{(-)}(t) = \exp(-i\varepsilon_f t)\chi_f(\mathbf{s}). \tag{A.56}$$

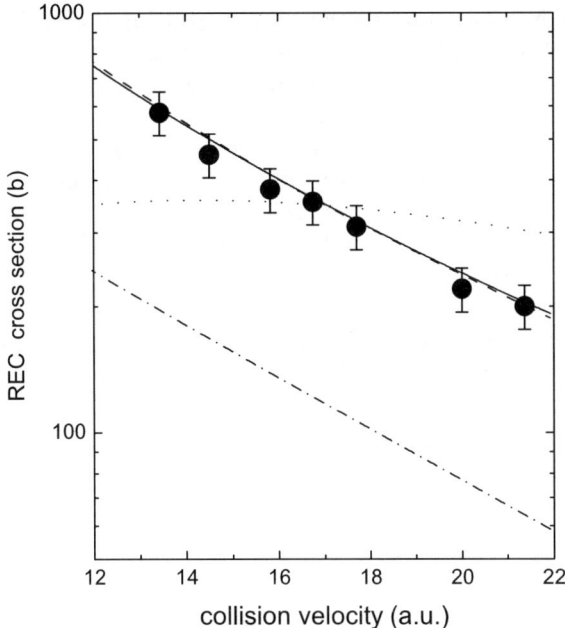

Fig. A.1. The total cross section for the radiative electron capture into the K-shell of the projectile in collisions between $Ge^{31+}(1s)$ and molecular hydrogen. The cross section is given as a function of the collision velocity. *Solid* and *dashed curves* (which almost coincide) are results of the CDW model calculated in the projectile and target frames, respectively (both results are scaled by a factor of 0.75). *Dot* and *dash-dot curves* are results of the 1B model calculated in the projectile and target frames, respectively (note that in the figure the latter results are multiplied by a factor of 100). *Circles* are experimental data reported in [241]. From [238].

Here $\mathbf{r}$ is the electron coordinate with respect to the nucleus of the atom, which moves in the projectile frame with a velocity $\mathbf{v}_a$, and $\mathbf{s}$ is the electron coordinate with respect to the nucleus of the ion. Further, $\phi_i(\mathbf{r})$ is the initial bound state of the electron in the atom with an energy ϵ_i and $\chi_f(\mathbf{s})$ is the final bound state of the electron in the ion with an energy ε_f.

Within the CDW model the initial and final electron wave functions are approximated by two-center states. For instance, the initial and final states in the rest frame of the projectile are given by

$$\psi_i^{(+)}(t) = \phi_i(\mathbf{r})\chi_{\mathbf{v}_a}^{(+)}(\mathbf{s}) \exp\left(-i(\epsilon_i + v_a^2/2)t\right),$$
$$\psi_f^{(-)}(t) = \chi_f(\mathbf{s}) \exp(i\mathbf{v}_a \cdot \mathbf{r})\phi_{-\mathbf{v}_a}^{(-)}(\mathbf{r}) \exp(-i\varepsilon_f t), \qquad (A.57)$$

where $\chi_{\mathbf{v}_a}^{(+)}(\mathbf{s})$ $(\phi_{-\mathbf{v}_a}^{(-)}(\mathbf{r}))$ is the Coulomb continuum state of the electron moving in the field of the ionic (atomic) nucleus with an asymptotic momentum $m\mathbf{v}_a$ $(-m\mathbf{v}_a)$ which satisfies the corresponding 'in' ('out') boundary condition (see e.g. [9]). For more details on the CDW approximation and its applications to the Coulomb (nonradiative) capture in nonrelativistic ion–atom collisions we refer to [38, 39], a detailed discussion of distorted-wave models in the case of relativistic collisions can be found in [4].

It is seen in Fig. A.1 that the 1B model strongly failed to pass the test yielding cross section values which differ by orders of magnitude and have quite different dependencies on the collision velocity. The enormous difference between results for the total capture cross section obtained in the first Born approximation by performing calculations in the rest frames of the projectile-ion and the target (a particular case of which is seen in Fig. A.1) has been for very long time attributed to the Galilean noninvariance of this approximation [4, 77, 236]. However, since these calculations always employed the same

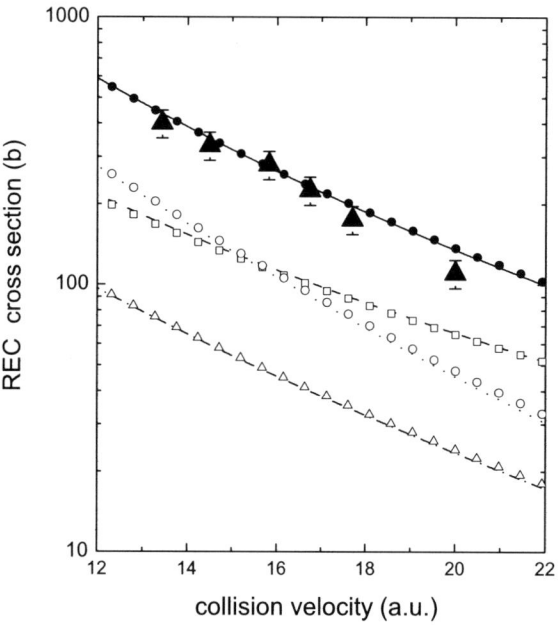

Fig. A.2. The cross sections for the radiative electron capture into the L-shell of the projectile in collisions between $Ge^{31+}(1s)$ and molecular hydrogen. The cross sections are given as a function of the collision velocity. *Solid triangles* with error bars are experimental data from [241]. Results of the CDW calculations in the target frame are shown by *solid curve* for the total capture into the L-shell), by *dash curve* for the capture into the $2s$-state), by *dot curve* for the capture into the $2p_0$-state) and by *dash-dot curve* for the sum of the capture into the $2p_{+1}$ and $2p_{-1}$ states. (Note that the states are quantized along the collision velocity.) The corresponding results of the CDW calculations in the projectile frame are displayed by *solid circles*, *open squares*, *open circles* and *open triangles*. *Solid triangles* with error bars are experimental data from [241] Note that all the theoretical results are scaled by a factor of 0.75. From [238].

radiation gauge for both reference frames, the difference is simply a signature of the very strong gauge dependence of this approximation.

In contrast to the first Born approximation, the application of the CDW model was quite successful leading to very close values of the capture cross section which, in addition, are in a reasonable agreement with the experimental data (see Fig. A.1).

In Fig. A.2 the cross sections for the radiative capture into the L-shell are plotted as a function of the collision velocity for the same colliding system as in Fig. A.1. It is seen in the figure that the CDW model has again passed the gauge test quite successfully yielding close results not only for the total capture into the L-shell but also for the 'partial' capture cross sections into states with definite values of the angular momentum and magnetic quantum number. In contrast, results of the 1B model (now shown in the figure) are very strongly gauge-dependent also for the capture to the L-shell.

Thus, in the case of a particular gauge transformation considered above the CDW approach turned out to be almost gauge independent and has clearly demonstrated its great advantage over the simple first Born approximation.

References

1. N. Stolterfoht, R.D. DuBois, R.D.Rivarola, *Electron Emission in Heavy Ion-Atom Collisions* (Springer, Berlin Heidelberg New York, 1997)
2. J.H. McGuire, *Electron Correlation Dynamics in Atomic Collisions* (Cambridge University Press, Cambridge, 1997)
3. J. Eichler, W. Meyerhof, *Relativistic Atomic Collisions* (Academic, San Diego, 1995)
4. D.S.F. Crothers, *Relativistic Heavy-Particle Collision Theory* (Kluwer, London, 2000)
5. J. Eichler, *Lectures on Ion-Atom Collisions* (Elsevier, Amsterdam, 2005)
6. B.H. Bransden, M.R.C. McDowell, *Charge Exchange and the Theory of Ion-Atom Collisions* (Clarendon Press, Oxford, 1992)
7. J.D. Jackson, *Classical Electrodynamics* (Wiley, New York, 1975)
8. L.D. Landau, E.M. Lifshitz, *Classical Field Theory* (Pergamon, Oxford, 1971)
9. L.D. Landau, E.M. Lifshitz, *Quantum Mechanics* (Pergamon, Oxford, 1977)
10. H.A. Bethe, E.E. Salpeter, *Quantum Mechanics of One- and Two-Electron Atoms* (Academic, New York, 1957)
11. J.J. Sakurai, *Advanced Quantum Mechanics* (Addison-Wesley, Reading, MA, 1967)
12. M.E. Rose, *Relativistic Electron Theory* (Wiley, New York, 1961)
13. J.D. Bjorken, S.D. Drell, *Relativistic Quantum Mechanics* (McGraw-Hill, New York, 1964)
14. M.L. Goldberger, K.M. Watson, *Collision Theory* (Wiley, New York, 1964)
15. B.H. Bransden, *Atomic Collision Theory* (W.A. Benjamin, New York, 1970)
16. M.R.C. McDowell, J.P. Coleman, *Introduction to the Theory of Ion-Atom Collisions* (North-Holland, Amsterdam, 1970)
17. C.J. Joachain, *Quantum Collision Theory* (North-Holland, Amsterdam, 1975)
18. D.R. Bates, G. Griffing, Proc. Phys. Soc. London Sect. A **66**, 961 (1953)
19. D.R. Bates, G. Griffing, Proc. Phys. Soc. London Sect. A **67**, 663 (1954)
20. D.R. Bates, G. Griffing, Proc. Phys. Soc. London Sect. A **68**, 90 (1955)
21. J.H. McGuire, Adv. At. Mol. Opt. Phys. **29**, 217 (1991)
22. E.C. Montenegro, W.E. Meyerhof, J.H. McGuire, Adv. At. Mol. Opt. Phys. **34**, 249 (1994)
23. N. Bohr, Kgl. Danske Videnskab Selskab, Mat. Fys. Medd. **18**(8), (1948)
24. E.C. Montenegro, T.J.M. Zouros, Phys. Rev. **A 50**, 3186 (1994)

25. E.C. Montenegro, W.E. Meyerhof, Phys. Rev. A **46**, 5506 (1992)
26. D.H. Jakubassa-Amundsen, J. Phys. B **23**, 3335 (1990)
27. D.H. Jakubassa-Amundsen, Z. Phys. D **22**, 701 (1992)
28. D.H. Jakubassa-Amundsen, J. Phys. B **26**, 2853 (1993)
29. H.M. Hartley, H.R. Walters, J. Phys. B **20**, 3811 (1987)
30. I.M. Cheshire, Proc. Phys. Soc. **84**, 89 (1964)
31. D. Belkic, J. Phys. B **11**, 3529 (1978)
32. D. Belkic, R. Gayet, A. Salin, Phys. Rep. **56**, 279 (1979)
33. D.S.F. Crothers, J.F.J McCann, J. Phys. B **16**, 3229 (1983)
34. J.M. Maidagan, R.D. Rivarola, J. Phys. B **17**, 2477 (1984)
35. G.R. Deco, P.D. Fainstein, R.D. Rivarola, J. Phys. B **19**, 213 (1986)
36. P.D. Fainstein, R.D. Rivarola, J. Phys. B **20**, 1285 (1987)
37. P.D. Fainstein, V.H. Ponce, R.D. Rivarola, J. Phys. B **24**, 3091 (1991)
38. D.P. Dewangan, J. Eichler, Phys. Rep. **247**, 59 (1994)
39. D. Belkic, J. Comp. Meth. Sci. Eng. **1**, 1 (2001)
40. A.B. Voitkiv, B. Najjari, J. Phys. B **38**, 3587 (2005)
41. A.B. Voitkiv, Phys. Rev. A **72**, 062705 (2005)
42. M.M. Sant'Anna, W.S. Melo, A.C.F. Santos, G.M. Sigaud, E.C. Montenegro, Nucl. Instr. Meth. B **99**, 46 (1995)
43. P.L. Grande, G. Schiwietz, G.M. Sigaud, E.C. Montenegro, Phys. Rev. A **54**, 2983 (1996)
44. T. Kirchner, M. Horbatch, H.J. Lüdde, J. Phys. B **37**, 2379 (2004)
45. T. Kirchner, Nucl. Instr. Meth. B **233**, 151 (2005)
46. T. Kirchner, A.C.F. Luna, H. Luna, M.M. Santanna, W.S. Melo, G.M. Sigaud, E.C. Montenegro, Phys. Rev. A **72**, 012707 (2005)
47. A.B. Voitkiv, N. Grün, W. Scheid, J. Phys. B **33**, 3431 (2000)
48. W. Magnus, Commun. Pure Appl. Math. **7**, 649 (1954)
49. A.M. Dykhne, G.L. Yudin, Usp. Fiz. Nauk **125**, 377 (1978)
50. A. Messiah, *Quantum Mechanics*, vol. 2 (Wiley, New York, 1962)
51. E. Merzbacher, *Quantum Mechanics*, 2nd edn. (Wiley, New York, 1970), p. 486
52. A.B. Voitkiv, G.M. Sigaud, E.C. Montenegro, Phys. Rev. A **59**, 2794 (1999)
53. R.J. Glauber, in *Lectures in Theoretical Physics*, vol. 1, ed. by W.E. Brittin et al. (Unterscience, New York, 1959)
54. E. Gerjuoy, B.K. Thomas, Rep. Progr. Phys. **37**, 1345 (1974)
55. R. Abrines, I.C. Percival, Proc. Phys. Soc. **88**, 861, 873 (1966)
56. R.E. Olson, in *Atomic, Molecular & Optical Physics Handbook*, ed. by G.W.F. Drake (AIP Press, Woodbury, New York, 1996), p. 664
57. C.O. Reinhold, J. Burgdörfer, J. Phys. B **26**, 3101 (1993)
58. S. Hofstetter, N. Grün, W. Scheid, Z. Phys. D **37**, 1 (1996)
59. C.J. Wood, R.E. Olson, W. Schmidt, R. Moshammer, J. Ullrich, Phys. Rev. A **56**, 3746 (1997)
60. J. Fiol, R.E. Olson, A.C.F. Santos, G.M. Sigaud, E.C. Montenegro, J. Phys. B **34**, L503 (2001)
61. R. Dörner, V. Mergel, O. Jagutzki, L. Spielberger, J. Ullrich, R. Moshammer, H. Schmidt-Böcking, Phys. Rep. **330**, 95 (2000)
62. J. Ullrich, R. Moshammer, A. Dorn, R. Dörner, L. Schmidt, H. Schmidt-Böcking, Rep. Prog. Phys. **66**, 1463 (2003)
63. W. Wu, K.L. Wong, R. Ali, C.Y. Chen, C.L. Cocke, V. Frohne, J.P. Giese, M. Raphaelian, B. Walch, R. Dörner, V. Mergel, H. Schmidt-Bocköking and W.E. Meyerhof, Phys. Rev. Lett. **72**, 3170 (1994)

64. W. Wu, K.L. Wong, E.C. Montenegro, R. Ali, C.Y. Chen, C.L. Cocke, R. Dörner, V. Frohne, J.P. Giese, V. Mergel, W.E. Meyerhof, M. Raphaelian, H. Schmidt-Bocköking, B. Walch, Phys. Rev. A **55**, 2771 (1997)
65. H. Kollmus, R. Moshammer, R.E. Olson, S. Hagmann, M. Schulz, J. Ullrich, Phys. Rev. Lett. **88**, 103202 (2002)
66. H.S.W. Massey, *Negative Ions*, 3rd edn. (Cambridge University Press, London, 1976)
67. F. Melchert, W. Debus, M. Liehr, R.E. Olson, E. Salzborn, Europhys. Lett. **9**, 433 (1989)
68. F. Melchert, M. Benner, S. Krüdener, R. Schulze, S. Meuser, K. Huber, E. Salzborn, D.B. Uskov, L.P. Presnyakov, Phys. Rev. Lett. **74**, 888 (1995)
69. A.B. Voitkiv, N. Grün, W. Scheid, J. Phys. B **32**, 101 (1999)
70. A.B. Voitkiv, N. Grün, W. Scheid, J. Phys. B **33**, 1533 (2000)
71. T. Ferger, D. Fischer, M. Schulz, R. Moshammer, A.B. Voitkiv, B. Najjari and J. Ullrich, Phys. Rev. A **72**, 062709 (2005)
72. K.G. Dedrick, Rev. Mod. Phys. **34**, 429 (1962)
73. V.B. Berestetskii, E.M. Lifshitz, L.P. Pitaevskii, *Quantum Electrodynamics* (Pergamon, Oxford, 1982)
74. W. Greiner, *Relativistic Quantum Mechanics*, 3rd edn. (Springer, Berlin Heidelberg New York, 2000)
75. C.A. Bertulani, G. Baur, Phys. Rep. **163**, 299 (1988)
76. R. Anholt, H. Gould, Adv. At. Mol. Phys. **22**, 315 (1986)
77. J. Eichler, Phys. Rep. **193**, 165 (1990)
78. A.B. Voitkiv, Phys. Rep. **392**, 191 (2004)
79. A.B. Voitkiv, Int. J. Mod. Phys. B **20**(1), 1 (2006)
80. R. Anholt, Phys. Rev. A **31**, 3579 (1985)
81. R. Anholt, W.E. Meyerhof, H. Gould, C. Munder, J. Alonso, P. Thienberger, H.E. Wegner, Phys. Rev. A **32**, 3302 (1985)
82. D.M. Davidovic, B.L. Moiseiwitsch, P.H. Norrington, J. Phys. B **11**, 847 (1978)
83. R. Anholt, Phys. Rev. A **19**, 1004 (1979)
84. R. Anholt, U. Becker, Phys. Rev. A **36**, 4628 (1987)
85. H.F. Krause, C.R. Vane, S. Datz, P. Grafström, H. Knudsen, S. Scheidenberger, R.H. Schuch, Phys. Rev. Lett. **80**, 1190 (1998)
86. A.H. Sørensen, Phys. Rev. A **58**, 2895 (1998)
87. A. Belkacem, A.H. Sørensen, Rad. Phys. Chem. A **75**, 656 (2006)
88. C.F. von Weizsäcker, Z. Physik **58**, 1934
89. E.J. Williams, Phys. Rev. **45**, 729 (1934)
90. A.B. Voitkiv, N. Grün, W. Scheid, Phys. Lett. A **260**, 240 (1999)
91. A.B. Voitkiv, N. Grün, W. Scheid, Phys. Rev. A **61**, 052704 (2000)
92. A.B. Voitkiv, M. Gail, N. Grün, J. Phys. B **33**, 1299 (2000)
93. C. Itzykson, J.-B. Zuber, *Quantum Field Theory* (McGraw-Hill, New York, 1980)
94. F. Mandl, G. Shaw, *Quantum Field Theory* (Wiley, London, 1984)
95. T. Stöhlker, D.C. Ionescu, P. Rymuza, F. Bosch, H. Geissel, C. Kozhuharov, T. Ludziejwski, P.H. Mokler, C. Scheidenberger, Z. Stachura, A. Warczak, R.W. Dunford, Phys. Lett. A **238**, 43 (1998)
96. T. Stöhlker, D.C. Ionescu, P. Rymuza, F. Bosch, H. Geissel, C. Kozhuharov, T. Ludziejwski, P.H. Mokler, C. Scheidenberger, Z. Stachura, A. Warczak, R.W. Dunford, Phys. Rev. A **57**, 845 (1998)

97. T. Ludziejewsky, T. Stöhlker, C.D. Ionesku, P. Rymuza, H. Beyer, F. Bosch, C. Kozhuharov, A. Krämer, D. Liesen, P.H. Mokler, Z. Stachura, P. Swiat, A. Warczak, R.W. Dunford, Phys. Rev. A **61**, 052706 (2000)
98. D. Ionesku, T. Stölker, Phys. Rev. A **67**, 022705 (2002)
99. U. Fano, Phys. Rev. **102**, 385 (1956)
100. H. Bethe, E. Fermi, Z. Physik **77**, 296 (1932)
101. G. Moliere, Naturforsch. **2A**, 133 (1947)
102. F. Salvat, J.D. Martinez, R. Mayol, J. Parellada, Phys. Rev. A **36**, 467 (1987)
103. P.A. Amundsen, K. Aashamar, J. Phys. B **14**, 4047 (1981)
104. B. Najjari, A.B. Voitkiv, J. Ullrich, J. Phys. B **35**, 533 (2002)
105. C.G. Darwin, Proc. Roy. Soc. A **118**, 654 (1928)
106. W.H. Furry, Phys. Rev. A **46**, 391 (1934)
107. A. Sommerfeld, A.W. Maue, Ann. Physik **22**, 629 (1935)
108. M. Abramowitz, I. Stegun, *Handbook of Mathematical Functions* (Dover Publications, New York, 1965)
109. I.S. Gradshteyn, I.M. Ryzhik, *Table of Integrals, Series, and Products*, 5th edn. (Academic, New York, 1994)
110. H.A. Bethe, L. Maximon, Phys. Rev. **93**, 768 (1954)
111. A.B. Voitkiv, J. Phys. B **40**, 2885 (2007)
112. T. Stöhlker, C.D. Ionesku, P. Rymuza, T. Ludziejewski, P.H. Mokler, C. Scheidenberger, F. Bosch, B. Franzke, H. Geissel, O. Klepper, C. Kozhuharov, R. Moshammer, F. Nickel, H. Reich, Z. Stachura, A. Warczak, Nucl. Instr. Meth. B **124**, 160 (1997)
113. H. Bräuning et al., Phys. Scripta **T92**, 43 (2001)
114. A.L. Nikishov, N.V. Pichkurov, Sov. J. Nucl. Phys. **35**, 561 (1982)
115. F. Decker, Phys. Rev. A **44**, 2883 (1991)
116. D. Ionesku, J. Eichler, Phys. Rev. A **48**, 1176 (1993)
117. A.B. Voitkiv, J. Ullrich, J. Phys. B **34**, 4513 (2001)
118. A.B. Voitkiv, B. Najjari, J. Ullrich, Phys. Rev. Lett. **92**, 213202 (2004)
119. A.B. Voitkiv, B. Najjari, J. Phys. B **37**, 3339 (2004)
120. A.B. Voitkiv, B. Najjari, J. Ullrich, Phys. Rev. Lett. **94**, 163203 (2005)
121. A.B. Voitkiv, J. Phys. B **38**, 1773 (2005)
122. A.B. Voitkiv, Phys. Rev. A **74**, 012728 (2006)
123. E. Fermi, Phys. Rev. **57**, 485 (1940)
124. L.D. Landau, E.M. Lifshitz, *Electrodynamics of Continuous Media* (Pergamon, Oxford, 1971)
125. A.S. Davydov, *Quantum Mechanics* (Pergamon, Oxford, 1965)
126. T. Stöhlker, J. Eichler, Phys. Rep. **439**, 1 (2007)
127. A.B. Voitkiv, B. Najjari, J. Ullrich, Phys. Rev. Lett. **99**, 193210 (2007)
128. B. Segev, J.C. Wells, Phys. Rev. A **57** 1849 (1998)
129. A.J. Baltz, L. McLerran, Phys. Rev. C **58** 1679 (1998)
130. U. Eichmann, J. Reinhardt, S. Schramm, W. Greiner, Phys. Rev. A **59** 1223 (1999)
131. A.J. Baltz, Phys. Rev. Lett. **78**, 1231 (1997)
132. A.B. Voitkiv, C. Müller, N. Grün, Phys. Rev. A **62**, 062701 (2000)
133. D. Amati, N. Ciafaloni, G. Veneziano, Phys. Lett. B **197**, 81 (1987)
134. G. t'Hooft, Phys. Lett. B **198**, 61 (1987)
135. R. Jackiw, D. Kabat, M. Ortiz, Phys. Lett. B **277**, 148 (1992)
136. O. Busic, PhD Thesis, The University of Giessen, Giessen (2000)

137. C.A. Bertulani, Phys. Rev. A **63**, 062706 (2001)
138. A.B. Voitkiv, A.V. Koval', Tech. Phys. **39**, 335 (1994)
139. A.B. Voitkiv, A.V. Koval', J. Phys. B **31**, 499 (1998)
140. A.J. Baltz, Phys. Rev. A **61**, 042701 (2000)
141. M. Bär, B. Jakob, P.-G. Reinhard, G. Toepffer, Phys. Rev. A **73**, 022719 (2006)
142. R. Moshammer et al., Phys. Rev. Lett. **79**, 3621 (1997)
143. A.B. Voitkiv, N. Grün, W. Scheid, J. Phys. B **32**, 3923 (1999)
144. D. Fischer, A.B. Voitkiv, R. Moshammer, J. Ullrich, Phys. Rev. A **68**, 032709 (2003)
145. A.B. Voitkiv, B. Najjari, J. Ullrich, Phys. Rev. A **75**, 062716 (2007)
146. A.B. Voitkiv, B. Najjari, J. Ullrich, Phys. Rev. A **76**, 022709 (2007)
147. A.B. Voitkiv, B. Najjari, J. Phys. B **40**, 3295 (2007)
148. J.F. McCann, J. Phys. B **18**, L569 (1985)
149. G.R. Deco, R. Rivarola, J. Phys. B **21**, L299 (1988)
150. G.R. Deco, R. Rivarola, J. Phys. B **22**, 1043 (1989)
151. G.R. Deco, N. Grün, J. Phys. B **22**, 1357 (1989)
152. G.R. Deco, K. Momberger, N. Grün, J. Phys. B **23**, 1990 (1989)
153. G.R. Deco, N. Grün, J. Phys. B **22**, 3709 (1989)
154. A.J. Baltz, M.J. Rhoades-Brown, J. Weneser, Phys. Rev. A **44** 5569 (1991)
155. A.J. Baltz, Phys. Rev. A **52** 4970 (1995)
156. E. Teubner, G. Terlecki, N. Grün, W. Scheid, J. Phys. B **13**, 523 (1980)
157. U. Becker, PhD Thesis. The University of Giessen, Giessen, (1986)
158. U. Becker, in *Physics of Strong Fields*, ed. by W. Greiner, NATO Advanced Study Institute series B: Physics, **153** (Plenum, New York, 1987), p. 609
159. K. Momberger, N. Grün, W. Scheid, U. Becker, J. Phys. B **23**, 2293S (1990)
160. N. Toshima, J. Eichler, Phys. Rev. A **41**, 5221 (1990)
161. N. Toshima, J. Eichler, Phys. Rev. A **42**, 3896 (1990)
162. M. Gail, N. Grün, W. Scheid, J. Phys. B **36**, 1397 (2003)
163. U. Becker, N. Grün, W. Scheid, J. Phys. B **16**, 1967 (1983)
164. U. Becker, N. Grün, W. Scheid, G. Soff, Phys. Rev. Lett. **56**, 2016 (1986)
165. J. Thiel, A. Bunker, K. Momberger, N. Grün, W. Scheid, Phys. Rev. A **46**, 2607 (1992)
166. K. Momberger, A. Belkacem, A.H. Sörensen, Phys. Rev. A **53**, 1605 (1996)
167. O. Busic, N. Grün, W. Scheid, Phys. Rev. A **70**, 062707 (2004)
168. R. Anholt, W.E. Meyerhof, X.-Y. Xu, H. Gould, B. Feinberg, R.J. McDonald, H.E. Wigner, P. Thieberger, Phys. Rev. A **36**, 1586 (1987)
169. W.E. Meyerhof, R. Anholt, X.-Y. Xu, H. Gould, B. Feinberg, R.J. McDonald, H.E. Wegner, P. Thieberger, NIM A **262**, 10 (1987)
170. H.-P. Hülskötter, W.E. Meyerhof, E. Dillard, N. Guardala, Phys. Rev. Lett. **63**, 1938 (1989)
171. H.-P. Hülskötter, B. Feinberg, W.E. Meyerhof, A. Belkacem, J.R. Alonso, L. Blumenfeld, E.A. Dillard, H. Gould, N. Guardala, G.F. Krebs, M.A. McMahan, M.E. Rhoades-Brown, B.S. Rude, J. Schweppe, D.W. Spooner, K. Street, P. Thieberger, H.E. Wegner, Phys. Rev. A **44**, 1712 (1991)
172. C. Scheidenberger, H. Geissel, Nucl. Instr. Meth. B **135**, 25 (1998)
173. C. Scheidenberger, T. Stölker, W.E. Meyerhof, H. Geissel, P.H. Mokler, B. Blank, Nucl. Instr. Meth. B **142**, 441 (1998)
174. C. Müller, A.B. Voitkiv, N. Grün, Phys. Rev. A **66** 012716 (2002)
175. B. Najjari, A.B. Voitkiv, J. Phys. B **41** 115202 (2008)

176. N. Claytor, A. Belkacem, T. Dinneen, B. Feinberg, H. Gould, Phys. Rev. A **55**, R842 (1997)

177. J.H. Hubbell, W.J. Veigele, E.A. Briggs, R.T. Brown, D.T. Cromer, R.J. Howerton, J. Phys. Chem. Ref. Data **4**, 471 (1975)

178. J.H. Hubbell, I. Øverbø, J. Phys. Chem. Ref. Data **8**, 69 (1979)

179. U. Fano, Annu. Rev. Nucl. Sci. **13**, 1 (1963)

180. H.F. Krause, C.R. Vane, S. Datz, P. Grafström, H. Knudsen, U. Mikkelsen, S. Scheidenberger, R.H. Schuch, Z. Vilakazi, Phys. Rev. A **63**, 032711 (2001)

181. C.R. Vane, H.F. Krause, Nucl. Instr. Meth. B **261**, 244 (2007)

182. G. Baur, K. Hencken, D. Trautmann, J. Phys. G **24**, 1657 (1998)

183. G. Baur, K. Hencken, D. Trautmann, S. Sadovsky, Y. Kharlov, Phys. Rep. **364**, 359 (2002)

184. G. Baur, K. Hencken, D.Trautman, Phys. Rep. **453**, 1 (2007)

185. U. Becker, N. Grün, W. Scheid, J. Phys. B **20**, 2075 (1987)

186. U. Becker, J. Phys. B **20**, 6563 (1987)

187. A. Belkacem, H. Gould, B. Feinberg, R. Bossingham, W.E. Meyerhof, Phys. Rev. Lett. **71**, 1514 (1993)

188. A. Belkacem, H. Gould, B. Feinberg, R. Bossingham, W.E. Meyerhof, Phys. Rev. Lett. **73**, 2432 (1994)

189. A. Belkacem, H. Gould, B. Feinberg, R. Bossingham, W.E. Meyerhof, Phys. Rev. A **56**, 2806 (1997)

190. A. Belkacem, N. Claytor, T. Dinneen, B. Feinberg, H. Gould, Phys. Rev. A **58**, 1253 (1998)

191. B. Najjari, A.B. Voitkiv, submitted to JPB

192. A.B. Voitkiv, N. Grün, W. Scheid, Phys. Lett. A **269**, 325 (2000)

193. J. Wu, J.H. Derrickson, T.A. Parnell, M.R. Strayer, Phys. Rev. A **60**, 3722 (1999)

194. H. Ogawa et al., Phys. Rev. A **75**, 020703 (2007)

195. N. Bohr, J. Lindhart, Kgl. Danske Videnskab Selskab, Mat. Fys. Medd. **28**, 7 (1954)

196. H.D. Betz, Rev. Mod. Phys. **44**, 465 (1972)

197. H. Rothard, D.H. Jakubassa-Amundsen, A. Billebaud, J. Phys. B **31**, 1563 (1998)

198. H. Geissel, H. Weick, C. Scheidenberger, R. Bimbot, D. Gardes, Nucl. Instr. Meth. B **195**, 3 (2002)

199. Y. Takabayshi, T. Ito, T. Azuma, K. Komaki, Y. Yamazaki, H. Tawara, E. Takada, M. Seliger, K. Tökesi, C.O. Reinhold, J. Burgdörfer, Phys. Rev. A **68**, 042703 (2003)

200. V.P. Shevelko, O. Rosmej, H. Tawara, I.Y. Tolstikhina, J. Phys. B **37**, 201 (2004)

201. V.P. Shevelko, H. Tawara, O.V. Ivanov, T. Miyoshi, K. Noda, Y. Sato, A.V. Subbotin, I.Y. Tolstikhina, J. Phys. B **38**, 2675 (2005)

202. T. Miyoshi, K. Noda, Y. Sato, H. Tawara, I.Y. Tolstikhina, V.P. Shevelko, Nucl. Instr. Meth. B **251**, 89 (2006)

203. T. Miyoshi, K. Noda, H. Tawara, I.Y. Tolstikhina, V.P. Shevelko, Nucl. Instr. Meth. B **258**, 329 (2007)

204. C. Kondo, S. Masugi, Y. Nakano, A. Hatakeyama, T. Azuma, K. Komaki, Y. Yamazaki et al., Phys. Rev. Lett. **97**, 135503 (2006)

205. T. Azuma, Y. Takabayshi, C. Kondo, T. Muranaka, K. Komaki, Y. Yamazaki et al., Phys. Rev. Lett. **97**, 145502 (2006)

206. Y. Nakano, S. Masugi, T. Muranaka, T. Azuma, C. Kondo, A. Hatakeyama, K. Komaki et al., J. Phys. Conf. Ser. **58**, 359 (2007)
207. K.Y. Bahmina, V.V. Balashov, A.A. Sokolik, A.V. Stysin, J. Phys. Conf. Ser. **58**, 327 (2007)
208. A.B. Voitkiv, B. Najjari and A. Surzhykov, J. Phys. B **41** 111011 (2008)
209. C.R. Vane, U. Mikkelsen, H.F. Krause, S. Datz, P. Grafström, H. Knudsen, S. Møller, E. Uggerhøj, C. Scheidenberger, R.H. Schuch and Z. Vilikazi, in *Proceedings of XXI ICPEAC*, 2000, p. 709
210. A.B. Voitkiv, N. Grün, J. Phys. B **34**, 267 (2001)
211. A.B. Voitkiv, N. Grün, W. Scheid, J. Phys. B **32**, 3923 (1999)
212. B. Najjari, A. Surzhykov, A.B. Voitkiv, Phys. Rev. A **77** 042714 (2008)
213. S. Datz, G.R. Beene, P. Graftström, H. Knudsen, H.F. Krauze, R.H. Schuch, C.R. Vane, PRL **79**, 3355 (1997)
214. R.L. Watson, Y. Peng, V. Horvat, G.J. Kim, R.E. Olson, Phys. Rev. A **67**, 022706 (2003)
215. R.E. Olson, R.L. Watson, V. Horvat, A.N. Perumal, Y. Peng, T. Stöhlker, J. Phys. B **37**, 4539 (2004)
216. R.D. DuBois, A.C.F. Santos, T. Stöhlker, F. Bosch, A. Bräuning-Demian, A. Gumberidze, S. Hagman, C. Kozhuharov, R. Mann, A. Orsic-Muthig, U. Spillmann, S. Tachenov, W. Bart, L. Dahl, B. Franzke, J. Glatz, L. Gröning, S. Richter, D. Wilms, K. Ullmann, O. Jagutzki, Phys. Rev. A **70**, 032712 (2004)
217. V.I. Matveev, D. Matrasulov, S.V. Ryabchenko, JETP Lett. **82**, 404 (2006)
218. V.I. Matveev, D. Matrasulov, S.V. Ryabchenko, JETP **102**, 1 (2006)
219. R.D. DuBois, O. de Lucio, M. Thomason, G. Weber, T. Stöhlker, K. Beckert, P. Beller, F. Bosch, C. Brandau, A. Gumberidze, S. Hagman, C. Kozhuharov, F. Nolden, R. Reuschl, J. Rzadkjewicz, P. Spiller, U. Spillmann, M. Steck, S. Trotsenko, Nucl. Inst. Meth. B **261**, 230 (2007)
220. A.J. Baltz, Phys. Rev. A **64**, 022718 (2001)
221. L.I. Nemenov, Yad. Fiz. **41**, 980 (1985)
222. A.V. Tarasov, I.U. Christova, JINR P2-91-10, Dubna (1991)
223. L. Afanasyev, A. Tarasov, O. Voskresenskaya, J. Phys. G **25**, B7 (1999)
224. Z. Halabuka, T.A. Heim, K. Hencken, D. Trautmann, R.D. Viollier, Nucl. Phys. B **554**, 86 (1999)
225. T.A. Heim, K. Hencken, D. Trautmann, G. Baur, J. Phys. B **33**, 3583 (2000)
226. T.A. Heim, K. Hencken, D. Trautmann, G. Baur, J. Phys. B **34**, 3763 (2001)
227. M. Schumann, T.A. Heim, K. Hencken, D. Trautmann, G. Baur, J. Phys. B **35**, 2683 (2002)
228. A.B. Voitkiv, B. Najjari, J. Ullrich, J. Phys. B **36**, 2325 (2003)
229. S. Santamarina, M. Schumann, L.G. Afanasyev, T. Heim, J. Phys. B **36**, 4273 (2003)
230. O. Voskresenskaya, J. Phys. B **36**, 3293 (2003)
231. L. Afanasyev, C. Santamarina, A. Tarasov, O. Voskresenskaya, J. Phys. B **37**, 4749 (2004)
232. A.B. Voitkiv, N. Grün, W. Scheid, Phys. Lett. A **265**, 111 (2000)
233. A.B. Voitkiv, C. Müller, N. Grün, Nucl. Instr. Meth. B **205**, 504 (2003)
234. A.V. Selin, A.M. Ermolaev, C.J. Joachain, J. Phys. B **36**, L303 (2003)
235. A.B. Voitkiv, B. Najjari, J. Phys. B **37**, 4831 (2004)
236. R. Shakeshaft, L. Spruch, Rev. Mod. Phys. **51**, 369 (1979)
237. A.B. Voitkiv, Phys. Rev. A **73**, 052714 (2006)

238. A.B. Voitkiv, J. Phys. A **39**, 4275 (2006)
239. J.S. Briggs, K.Dettmann, Phys. Rev. Lett. **33**, 1123 (1974)
240. R. Shakeshaft, L. Spruch, Phys. Rev. Lett. **38**, 175 (1977)
241. T. Stöhlker, C. Kozhuharov, A.E. Livingston, P.H. Mokler, Z. Stachura, A. Warczak, Z. Phys. D **23**, 121 (1992)

Springer Series on
ATOMIC, OPTICAL, AND PLASMA PHYSICS

Springer Series on
ATOMIC, OPTICAL, AND PLASMA PHYSICS